Lecture Notes in Artificial Intelligence 9580

Subseries of Lecture Notes in Computer Science

More information about this series at http://www.springer.com/series/1244

Stefan Michaelis · Nico Piatkowski
Marco Stolpe (Eds.)

Solving Large Scale
Learning Tasks

Challenges and Algorithms

Essays Dedicated to Katharina Morik
on the Occasion of Her 60th Birthday

Springer

Editors
Stefan Michaelis
TU Dortmund
Dortmund
Germany

Nico Piatkowski
TU Dortmund
Dortmund
Germany

Marco Stolpe
TU Dortmund
Dortmund
Germany

Cover illustration: The illustration appearing on the cover belongs to Katharina Morik. Used with permission.

Photograph on p. V: The photograph of the honoree was taken by Jürgen Huhn. Used with permission.

ISSN 0302-9743   ISSN 1611-3349   (electronic)
Lecture Notes in Artificial Intelligence
ISBN 978-3-319-41705-9   ISBN 978-3-319-41706-6   (eBook)
DOI 10.1007/978-3-319-41706-6

Library of Congress Control Number: 2016942885

LNCS Sublibrary: SL7 – Artificial Intelligence

Printed on acid-free paper

This Springer imprint is published by Springer Nature
The registered company is Springer International Publishing AG Switzerland

Foreword

The German word "Festschrift" has made it into international dictionaries because it very succinctly denotes a volume of writings by different researchers that originate from an event and constitutes a tribute to a scholar of extraordinary reputation. As such, a Festschrift offers a unique approach toward a field of science, since at its center, instead of an a priori-defined topical focus, there are the works and scientific vision of an outstanding individual as reflected in the works of collaborators and contributors to the volume.

While this nature of a Festschrift makes it an interesting approach independent of what the field of science is, in the field of machine learning this way of accessing science is of particular interest, since in a certain sense it reflects the very nature of the field itself. In the present volume, which originated at the scientific symposium in honor of Katharina Morik's 60th birthday, you will see that the individual contributions of her colleagues offer an implicit view of her strategic vision of what machine learning should be and how research in machine learning should be conducted, as reflected in her choice of collaborators. You are thus invited to do what any good machine-learning algorithm would do when presented with examples: use the research presented in this book to induce for yourselves the implicit vision that lies at their heart.

In this foreword, I certainly do not want to take away from the pleasure of drawing these conclusions yourselves, so let me just say that in my view, the papers clearly reflect Katharina Morik's commitment and conviction that machine learning should be firmly rooted in fundamental research with all its rigor, while at the same time being turned into software and engineering results and demonstrating its usefulness by applications in various disciplines. As you will see, this vision is clearly shared by the excellent researchers who have contributed to this volume.

Enjoy the book!

December 2015   Stefan Wrobel

Preface

In celebration of Prof. Morik's 60th birthday, this Festschrift covers research areas that Prof. Morik worked in and presents various researchers with whom she collaborated. Articles in this Festschrift volume provide challenges and solutions from theoreticians and practitioners on data preprocessing, modeling, learning, and evaluation. Topics include data-mining and machine-learning algorithms, feature selection, optimization as well as efficiency of energy and communication.

March 2016

Stefan Michaelis
Nico Piatkowski
Marco Stolpe

Biographical Details

Katharina Morik was born in 1954. She earned her PhD (1981) at the University of Hamburg and her habilitation (1988) at the TU Berlin. In 1991, Katharina became a full professor of computer science at the TU Dortmund University (former Universität Dortmund), Germany. Starting with natural language processing, her interest moved to machine learning ranging from inductive logic programming to statistical learning, then to the analysis of very large data collections, high-dimensional data, and resource awareness. She is a member of the National Academy of Science and Engineering and the North Rhine-Westphalia Academy of Science and Art. She is the author of more than 200 papers in acknowledged conferences and journals. Her latest results include spatio-temporal random fields and integer Markov random fields, both allowing for complex probabilistic graphical models under resource constraints.

Throughout her career, Katharina has been passionate about teaching. She has often taught more courses than required, and inspired students with her passion for artificial intelligence and computer science in general.

Her aim to share scientific results strongly supports open source developments. For instance, the first efficient implementation of the support vector machine, SVM_{light}, was developed at her lab by Thorsten Joachims. The leading data-mining platform RapidMiner also started out at her lab, which continues to contribute to it. Currently, the Java streams framework is being developed, which abstracts processes on distributed data streams.

Since 2011, she has been leading the collaborative research center SFB876 on resource-aware data analysis, an interdisciplinary center comprising 14 projects, 20 professors, and about 50 PhD students or postdocs.

Katharina was and is strongly engaged in the data mining and machine learning community. She was one of the founders of the IEEE International Conference on Data Mining together with Xindong Wu, and she chaired the program of this conference in 2004. She was the program chair of the European Conference on Machine Learning (ECML) in 1989 and one of the program chairs of ECML PKDD 2008. Katharina is on the editorial boards of the international journals *Knowledge and Information Systems* and *Data Mining and Knowledge Discovery*.

Contents

Online Social Networks Event Detection: A Survey

Mário Cordeiro[1]([✉]) and João Gama[2]

[1] University of Porto, Porto, Portugal
pro11001@fe.up.pt
[2] INESC TEC Laboratory of Artificial Intelligence and Decision Support,
University of Porto, Porto, Portugal
jgama@fep.up.pt

Abstract. Today online social network services are challenging state-of-the-art social media mining algorithms and techniques due to its real-time nature, scale and amount of unstructured data generated. The continuous interactions between online social network participants generate streams of unbounded text content and evolutionary network structures within the social streams that make classical text mining and network analysis techniques obsolete and not suitable to deal with such new challenges. Performing event detection on online social networks is no exception, state-of-the-art algorithms rely on text mining techniques applied to pre-known datasets that are being processed with no restrictions on the computational complexity and required execution time per document analysis. Moreover, network analysis algorithms used to extract knowledge from users relations and interactions were not designed to handle evolutionary networks of such order of magnitude in terms of the number of nodes and edges. This specific problem of event detection becomes even more serious due to the real-time nature of online social networks. New or unforeseen events need to be identified and tracked on a real-time basis providing accurate results as quick as possible. It makes no sense to have an algorithm that provides detected event results a few hours after being announced by traditional newswire.

Keywords: Event detection · Social networks

1 Introduction

Today, online social networking services like Twitter [102], Facebook [99], Google+ [100], LinkedIn [101], among others, play an important role in the dissemination of information on a real-time basis [91].

Recent observation proves that some events and news emerge and spread first using those media channels rather than other traditional media like the online news sites, blogs or even television and radio breaking news [50,88]. Natural disasters, celebrity news, products announcements, or mainstream event coverage show that people increasingly make use of those tools to be informed, discuss and

© Springer International Publishing Switzerland 2016
S. Michaelis et al. (Eds.): Morik Festschrift, LNAI 9580, pp. 1–41, 2016.
DOI: 10.1007/978-3-319-41706-6_1

exchange information [38]. Empirical studies [50, 88] show that the online social networking service Twitter is often the first medium to break important natural events such as earthquakes often in a matter of seconds after they occur. Being Twitter the "what's-happening-right-now" tool [91] and given the nature of it's data — an real-time flow of text messages (tweets) coming from very different sources covering varied kinds of subjects in distinct languages and locations — makes the Twitter public stream an example of an interesting source of data for "real time" event detection based on text mining techniques. Note that "real time" means that events need to be discovered as early as possible after they start unraveling in the online social networking service stream. Such information about emerging events can be immensely valuable if it is discovered timely and made available.

When some broad major event happens, three factors are the main contributors to the rapidly spread of information materialized in exchanged messages between users of an online social network service. (i) the ubiquity nature of today's social network services, that are available nowadays by any internet connected device like a personal computer or a smartphone; (ii) the ease of use and agility of entering or forward information is also a key factor that lead some messages to be spread very fast on the network and go viral [40]; and (iii) the lifespan of the messages is also an interesting feature of those online social network services. Posted messages tend to be exchanged, forwarded or commented following a time decay pattern, meaning that the information they contain has the importance peak when it is posted in the following hours or days [51]. This statement is coherent with the Barabasi [11] conclusion that the timing of many human activities, ranging from communication to entertainment and work patterns, follow non-Poisson statistics, characterized by bursts of rapidly occurring events separated by long periods of inactivity.

With the purpose of correlating the occurrence of events in the real world and the resulting activity in online social networks, Zhao et al. [108] and Sakaki et al. [88] introduced the concept of "social sensors" where the social text streams are seen as sensors of the real world. The assumption made is that each online social user (i.e.: a Twitter, Facebook, Google+ user) is regarded as a sensor and each message (i.e.: tweet, post, etc.) as sensory information. Zhao et al. [108] pointed two major substantial differences of the social text stream data over general text stream data: (i) social text stream data contains rich social connections (between the information senders/authors and recipients/reviewers) and temporal attributes of each text piece; and (ii) the content of text piece in the social text stream data is more context sensitive. Sakaki et al. [88] went beyond in its concept of "social sensors" by introducing an analogy to a physical sensor network. Some common characteristics of those "virtual" social sensors in comparison with real physical sensors are: (i) some sensors are very active, others are not — the activity of each user is different as some users post more messages than others; (ii) a sensor could be inoperable or malfunctioning sometimes — this means that a user can be offline at a given time i.e.: sleeping, on vacation; or even offline (without internet connection); and (iii) very noisy compared to ordinal physical sensors — the output of the sensor is not normalized, there are many users that are posting messages that can be considered as spammers.

1.1 Event Detection Overview

The event detection problem is not a new research topic, Yang et al. [105] in 1998, investigated the use and extension of text retrieval and clustering techniques for event detection. The main task was to detect novel events from a temporally-ordered stream of news stories automatically. In an evaluation of the system using manually labeled events were obtained values for the F-score[1] of 82 % in retrospective detection and 42 % in on-line detection. Despite the fact of the size of the corpus with 15,836 documents used in the evaluation, the system performed quite well and showed that basic techniques such as document clustering can be highly effective to perform event detection.

Two years later Allan et al. [6] evaluated the UMASS reference system [4] in three of the five Topic Detection and Tracking (TDT) tasks: (i) detection; (ii) first story detection; and (iii) and story link detection [3]. The core of this system used a vector model for representing stories, each story as a vector in term-space, and terms (or features) of each vector were single words, reduced to their root form by a dictionary-based stemmer. The study concluded that the results were acceptable for the three evaluated tasks but not as high quality as authors expected. Allan et al. [5] showed that performing first story detection based upon tracking technology has poor performance and to achieve high-quality first story detection the tracking effectiveness should be improved to a level that experiments showed not to be possible. Therefore Allan et al. [5] concluded that first story detection is either impossible or requires substantially different approaches.

Despite the fact that in the following 10 years, the period between years 2000 and 2010, the event detection problem was a relatively active research topic, it was in the latest 8 years, coinciding with the advent and massification of the online social networks phenomena and big data era that the problem gained more interest from the research community. Just targeting event detection specifically in the online social network service Twitter, Petrovic [73] pointed out and compared major scientific contribution from Hu et al. [39], Jurgens and Stevens [43], Sankaranarayanan et al. [89], Sakaki et al. [88], Popescu and Pennacchiotti [79], Cataldi et al. [20], Mathioudakis and Koudas [62], Phuvipadawat and Murata [76], Becker et al. [14], Weng et al. [98], Cordeiro [23], Li et al. [56], Li et al. [55], Agarwal et al. [1] and Ozdikis et al. [68]. This fact by itself is explanatory on the interest and relevance of the research topic. None of this listed publications managed to solve the problem of event detection in online social networks completely. Some of them assumed to solve the problem partially by defining constraints or limiting the scope of the problem. One year later, Atefeh and Khreich [9] published a survey that classifies the major techniques for Twitter event detection according to the event type (specified or unspecified events), detection method

---

[1] http://en.wikipedia.org/wiki/F1_score.

(supervised or unsupervised learning), and detection task (new event detection or retrospective event detection). Due the fact that the research conducted by Petrovic [73] work was primarily focused on solving online new event detection of unspecified events using unsupervised methods, it did not compare his work with other references to advancements in other specific areas of the event detection. Atefeh and Khreich [9] survey considers work described by Petrovic [14,23,73,74,76,79,88,89,98] and additional advancements like Long et al. [58], Popescu et al. [80], Benson et al. [15], Lee and Sumiya [52], Becker et al. [12], Massoudi et al. [61], Metzler et al. [63] and Gu et al. [35]. This survey also discusses the common used features used in event detection tasks for each one of the listed methods. Imran et al. [41] in a survey under the subject of communication channels during mass convergence and emergency events, gave an overview of the challenges and existing computational methods to process social media messages that may lead to an effective response in mass emergency scenarios. This survey, not being specifically devoted to event detection, includes a full chapter where Retrospective and Online New Event Detection types are addressed.

Most of the techniques described by Petrovic [73], Atefeh and Khreich [9] and Imran et al. [41] lack evaluation or are evaluated empirically. Measuring the accuracy and performance of an event detection methods is hampered by the lack of standard corpora and results leading some authors to create and make publicly available their own datasets with events being annotated manually [73]. In other cases evaluation is made with some automation by comparing directly to a reference system as a baseline [98] by generating a list of detected event that serves as ground truth. The need for public benchmarks to evaluate the performance of different detection approaches and various features was also highlighted by Atefeh and Khreich [9].

1.2 Problem Statement

Most of the described approaches to solving the event detection in text streams are not real-time and use batch algorithms. The good results obtained by reference systems used in the evaluation of the TDT task were obtained from reduced corpus datasets. Latter studies proved that they do not scale to larger amounts of data [72], in fact they were not even designed do deal with text streams. The performance, effectiveness and robustness of those reference systems was acceptable under the specified evaluation conditions at that time. Due the characteristics of today's online social network services data, unbounded massive unstructured text streams, these systems are nowadays considered as being obsolete. Apart from not being designed to handle big amounts of data, the data in online social network services is also dynamic, messages are arriving at high data rates, requiring the adaption of the computing models to process documents as they arrive. Finally today computation time is an issue, in most cases when using this kind of systems it is preferable to have an immediate and approximated solution rather than waiting too much for an exact solution [10].

Online social network text streams seem to be the ideal source to perform real-time event detection applying data mining techniques [74,88,98]. The main benefits of using those sources of data are their real-time or near real-time data availability, contextualization of the messages with additional information (temporal, geospatial, entity, etc. referenced in messages), possible data segmentation at several levels (by communities of user, by regions, by topic, etc.), access to static relations between users (friends, followers, user groups), possibility to build dynamic user relations built from exchanged messages flows, among others.

To perform data mining, every previously mentioned advantage of the online social network text data source reveals, in fact, to have a significant drawback and shortcoming. Performing real-time event detection using online social network services requires dealing and mining massive unstructured text data streams with messages arriving at high data rates. Given this, the approach to deal with this specific problem involves providing solutions that are able to mine continuous, high-volume, open-ended data streams as they arrive [17,82]. Because text data source is not disjointed from the online social network topological properties, it is expected that information retrieved using metrics of networks analysis (nodes, connections and relations, distributions, clusters, and communities) could improve the quality of the solution of the algorithm. In Table 3, for each one of the techniques, is included the collection, corpus size, and temporal scope of the dataset used in the evaluation.

1.3 Scope and Organization

It makes no sense to talk about an event detection system without first specifying an defining exactly what is an event. Section 2 introduces the concepts of story, event and topic. Section 3 defines, under topic detection and tracking task, introduces the origins of event detection as Information Retrieval problem. New Event Detection (NED) and Retrospective Event Detection (RED) tasks are described in Sect. 3.1. Section 3.2 describes the differences of systems designed to detected specified and unspecified events. Pivot techniques are presented in Sect. 3.3. Section 4 presents a taxonomy of event detection systems. The taxonomy was made taking into account the type of event that the system tries to detect (specified or unspecified event), and the type of the detection (unsupervised, supervised or hybrid detection). An overview of the common detection methods is presented in Sect. 4.3. Section 5 includes a list of the datasets, their respective size, and temporal scope used to evaluate each one of the event detection techniques. Finally Sect. 6 presents the conclusions, future and trends of event detection systems.

2 Event Definition

Fiscus and Doddington [30] in the scope of the Topic Detection and Tracking project gave the following definitions of story, event and topic:

story is *"a topically cohesive segment of news that includes two or more declarative independent clauses about a single event"*;

event is *"something that happens at some specific time and place along with all necessary preconditions and unavoidable consequences"*;

topic is *"a seminal event or activity, along with all directly related events and activities"*.

Sakaki et al. [88] defines an event as an arbitrary classification of a space/time region that might have actively participating agents, passive factors, products, and a location in space/time like is being defined in the event ontology by Raimond and Abdallah [81]. The target events in this work are broad events that are visible through messages, posts, or status updates of active users in Twitter online social network service. These events have several properties: (i) they are of large scale because many users experience the event, (ii) they particularly influence people's daily life, being that the main reason why users are induced to mention it, and (iii) they have both spatial and temporal regions, topically the importance of an event is correlated with the distance users have between themselves and the event and with the spent time since the occurrence.

The Linguistic Data Consortium [57] defines the broad topic types denoting the category where an event falls into. As defined by the TDT5 [93] there are the following broad topic type categories: (i) Elections; (ii) Scandals/Hearings; (iii) Legal/Criminal Cases; (iv) Natural Disasters; (v) Accidents; (vi) Acts of Violence or War; (vii) Science and Discovery News; (viii) Financial News; (ix) New Laws; (x) Sports News; (xi) Political and Diplomatic Meetings; (xii) Celebrity and Human Interest News; and (xiii) Miscellaneous News.

3   Earlier Event Detection and Discovery

The Topic Detection and Tracking project was started with the objective to improve technologies related to event-based information organization in 1998, see [3]. The project consisted of five distinct tasks: (i) segmentation; (ii) tracking; (iii) detection; (iv) first story detection; and (v) linking. From the previous list of tasks the tracking, detection, and first story detection are the ones that are relevant for event detection.

tracking: the tracking task detect stories that discuss a previously known target topic. This task is very closely linked to the first story detection. A tracking system can be used to solve a first story detection by finding other on-topic stories in the rest of the corpus. A nearest-neighbour based first story detection system could be used to solve tracking;

detection: the detection task is concerned with the detection of new, previously unseen topics. This task is often also called on-line clustering, every newly received story is assigned to an existing cluster or to a new cluster depending if there is a new story of not;

first story detection: the first story detection is considered the most difficult of the five topic detection and tracking tasks [5]. The aim of the task is to detect the very first story to discuss a previously unknown event. The first story detection can be considered a special case of detection by deciding when to start a new cluster in the on-line clustering problem.

Event detection on social media streams requires significantly different approaches than the ones used for traditional media. Social media data arrives at larger volumes and speed than traditional media. Moreover, most social media data is composed of short, noisy and unstructured content requiring significantly different techniques to solve similar machine learning or information retrieval problems [5]. Taking into account these considerations, Sect. 3.3 presents an overview of Document-Pivot and Feature-Pivot event detection techniques applied to traditional medial [98]. Document-pivot methods detect events by clustering documents based on the semantics distance between documents [105], feature-pivot methods studies the distributions of words and discovers events by grouping words together [46].

3.1 Detection Task

The task of discovering the "first story on a topic of interest" by continuously monitoring document streams is known in the literature as new event detection, first-story detection or novelty detection. Makkonen et al. [59] described first-story detection or novelty detection as an example of "query-less information retrieval" where events can be detected with no prior information available on a topic of interest. Events are evaluated using a binary decision on whether a document reports a new topic that has not been reported previously, or if should be merged with an existent event [103]. Depending on how data is processed, two categories of Event Detection systems were identified [7, 105].

Online New Event Detection (NED). Online New Event Detection refers to the task of identifying events from live streams of documents in real-time. Most new and retrospective event detection techniques rely on the use of well know clustering-based algorithms [2, 16]. Typically new event detection involves the continuous monitoring of Media feeds for discovering events in near real time, hereupon scenarios where the detection of real-world events like breaking news, natural disasters or other of general interest. Events are unknown apriori and in most cases use unspecified event detection. When monitoring specific NED like natural disasters or celebrities related, where specific apriori known information about the event can be used. In these cases, NED is performed using specified event detection.

Retrospective Event Detection (RED). Retrospective Event Detection refers to the process of identifying previously unidentified events from accumulated historical collections or documents that have arrived in the past. In Retrospective Event Detection, most methods are based on the retrieval of event relevant documents by performing queries over a collection of documents or by performing TF-IDF analysis on the document corpus. Both techniques assume that event relevant documents contain the query terms. A variation of the previous approach is the use of query expansion techniques, meaning that some messages relevant to a specific event do not contain explicit event related information, but with the use of enhanced queries messages related to the event can be retrieved.

3.2 Type of Event

Event detection can be classified into specified or unspecified event detection techniques [9, 29]. By using specific pre-known information and features about an event, traditional information retrieval and extraction techniques can be adapted to perform specified event detection (i.e.: filtering, query generation and expansion, clustering, and information aggregation). When no prior information or features are available about the event or even if we don't know a clue about the kind of event we want to detect, most traditional information retrieval and extraction techniques are useless. Unspecified event detection techniques address this issue on the basis that temporal signals constructed via document analysis can detect real work events. Monitoring bursts or trends in document streams, grouping features with identical trends, and classifying events into different categories are among some of the used tasks to perform unspecified event detection.

3.3 Pivot Techniques

Both Document-Pivot and Feature-Pivot techniques are being used in event detection applied to traditional media. The following sections describe how each of them works and how it is being used.

Document-Pivot Techniques. Document-pivot techniques try to detect events by clustering documents using their textual similarity, these techniques consider all documents to be relevant and assume that each of them contain events of interest [5]. The noisy characteristics of social networks, where relevant events are buried by in large amount of noisy data [95], allied with scale and speed processing restrictions [5] make document-pivot techniques not suitable to perform event detection in social media data. Nevertheless, because they were the primordial steps to modern event detection systems, they will be briefly presented here.

The main goal of the TDT research initiative was to provide core technology and tools that by monitoring multiple sources of traditional media are able to keep users updated about news and developments. A particular event detection goal was to discover new or previously unidentified events were each event refers to a specific thing that happens at a specific time and place [7]. Yang et al. [105,106] described the three traditional event detection major phases as data prepossessing, data representation, and data organisation or clustering. Filtering out stop-words and applying words stemming and tokenization techniques are some of the steps done in the data prepossessing phase. Term vectors of bag of words are common use traditional data representations techniques used in event detection. Entries are non-zero if the corresponding term appear in the document

Table 1. Type, technique, detection method and detection task for each one of the references. Column Application refers to the target application of the work: (a) Detecting General Interest Events; (b) Identification of Novel Topics in Blogs; (c) Detecting Controversial Events from Twitter; (d) Calendar of Significant Events; (e) Geo-Social Event Detection system; (f) Detection of Natural Disaster Events; (g) Query-based Event Retrieval; (h) Query-based Structured Event Retrieval; (i) Crime and Disaster related Events; (j) Detection of Breaking News; (k) Emergent topics; (l) Trend Detection; (m) Crisis-related Sub-event Detection; (n) Event Photo Identification; (o) Creating event-specific queries for Twitter

	Type of event		Pivot technique		Detection method		Detection task		Application
	Specified	Unspecified	Document	Feature	Supervised	Unsupervised	NED	RED	
Hu et al. [39]	x		x			x	x		(a)
Jurgens and Stevens [43]	x		x			x	x		(b)
Popescu and Pennacchiotti [79]	x			x	x		x		(c)
Popescu et al. [80]	x			x	x		x		(c)
Benson et al. [15]	x			x	x			x	(d)
Lee and Sumiya [52]	x		x			x	x		(e)
Sakaki et al. [88]	x			x	x		x		(f)
Becker et al. [12]	x		x		x			x	(g)
Becker et al. [13]	x		x					x	(g)
Massoudi et al. [61]	x		x		x			x	(g)
Metzler et al. [63]	x		x			x	x		(h)
Gu et al. [35]	x		x			x	x		(h)
Li et al. [56]	x			x	x		x		(i)
Ozdikis et al. [68]	x		x			x	x		(a)
Sankaranarayanan et al. [89]		x	x	x	x	x	x		(j)
Cataldi et al. [20]		x	x		x	x	x		(k)
Mathioudakis and Koudas [62]		x	x			x	x		(l)
Phuvipadawat and Murata [76]		x	x			x	x		(j)
Petrovic et al. [74]		x	x			x	x		(a)
Becker et al. [14]		x	x		x	x	x		(a)
Long et al. [58]		x	x			x	x		(a)
Weng et al. [98]		x	x			x	x		(a)
Cordeiro [23]		x	x			x	x		(a)
Li et al. [55]		x	x			x	x		(a)
Agarwal et al. [1]		x	x			x	x		(a)
Sayyadi et al. [90]		x	x			x	x		(a)
Zhao et al. [108]		x		x		x	x		(a)
Pohl et al. [78]		x	x			x	x		(m)
Chen and Roy [22]		x		x		x		x	(n)
Ritter et al. [85]		x		x	x	x	x		(d)
Robinson et al. [86]	x			x		x	x		(f)
Corley et al. [24]		x	x			x	x		(a)
Tanev et al. [94]	x		x			x	x		(o)
Dou et al. [25]		x	x			x	x		(a)

Table 2. Event detection approach

	Approach	Event types	Scalable	Real-time	Query type	Statio-temporal	Sub-events
Hu et al. [39]	Online clustering of query profiles	Open domain	Yes	No	Open	No	No
Jurgens and Stevens [43]	Temporal Random Indexing	Open domain	Yes	No	Keywords	No	No
Popescu and Pennacchiotti [79]	Regression machine learning models based on Gradient Boosted Decision Trees	Controversial Events	No	No	Open	No	No
Popescu et al. [80]	Regression machine learning models based on Gradient Boosted Decision Trees	Controversial Events	No	No	Open	No	No
Benson et al. [15]	Factor Graph Model and Conditional Random Field	Concerts in New York City	No	No	Keywords	Yes	No
Lee and Sumiya [52]	K-means clustering method for detecting ROI, measuring statistical variations of a set of geo-tags	Local events such as local festivals	No	No	Open	Yes	No
Sakaki et al. [88]	Support vector machine (SVM)	Natural disaster event	Yes	Yes	Keywords	Yes	No
Becker et al. [12]	Rule-based classifier	Planned Event	No	No	Keywords	No	No
Becker et al. [13]	Precision/Recall Oriented Strategies	Planned Event	No	No	Keywords	No	No
Massoudi et al. [61]	Query Expansion using the top k terms	Topic of interest	No	No	Keywords	No	No
Metzler et al. [63]	Temporal Query Expansion based on temporal co-occurrence of terms	Topic of interest	No	No	Keywords	No	No
Gu et al. [35]	Hierarchical clustering	Topic of interest	No	No	Keywords	No	Yes?
Li et al. [56]	Classification	Crime and Disaster related Events	No	No	Spatial/temporal/keywords	Yes	No

(continued)

Table 2. (*continued*)

	Approach	Event types	Scalable	Real-time	Query type Spatial/keywords/users	Statio-temporal	Sub-events
Ozdikis et al. [68]	Semantic Expansion of Hashtags via agglomerative clustering	Open domain	No	No	Spatial/keywords/users	No	No
Sankaranarayanan et al. [89]	Tweet Naïve Bayes classifier and weighted term vector based online clustering	Breaking-News	Yes	No	Keywords	Yes?	No
Cataldi et al. [20]	Keyword-based topic graph	Breaking-News	No	Yes	Keywords	No	No
Mathioudakis and Koudas [62]	Context extraction algorithms (PCA, SVD) and Keyword Co-Occurrence Grouping	Breaking-News/Topic of interest	Yes	Yes	Open	No	No
Phuvipadawat and Murata [76]	Similarity based grouping via TF-IDF	Breaking-News	No	No	Keywords	No	No
Petrovic et al. [74]	Detection of Events via Locally Sensitive Hashing	Open domain	Yes	Yes	Open	No	No
Becker et al. [14]	Incremental, online clustering/classification via support vector machine (SVM)	Open domain	Yes	No	Open	No	No
Long et al. [58]	Top-down hierarchical divisive clustering on a co-occurrence graph	Open domain	No	No	Open	No	No
Weng et al. [98]	Clustering of Wavelet-based Signals via graph partitioning	Open domain	No	No	Keywords	No	No
Cordeiro [23]	Wavelet-based Signals and Latent Dirichlet Allocation	Open domain	Yes	Yes	Open	No	No

(continued)

Table 2. (*continued*)

	Approach	Event types	Scalable	Real-time	Query type	Statio-temporal	Sub-events
Li et al. [55]	Symmetric Conditional Probability (SCP) for n-grams, bursty detection using binomial distribution, Clustering by k-Nearest Neighbor Graph	Open domain	Yes	No	Open	No	No
Agarwal et al. [1]	Clustering in a Correlated Keyword Graph	Open domain	Yes	Yes	Open	No	No
Sayyadi et al. [90]	Community Detection on a Keyword Graph	Open domain	No	No	Open	No	No
Zhao et al. [108]	Content-Based Clustering where word in the text piece is quantified as the TF.IDF, adaptive time series, information flomodeling	Open domain	No	No	Open	No	No
Pohl et al. [78]	Two-phase clustering: 1. calculation of term-based centroids using geo-referenced data; 2. Assignment of best fitting data points using cosine distance measure	Crisis-related sub-event	No	No	Geo-referenced Data	Yes	Yes

(*continued*)

Table 2. (continued)

	Approach	Event types	Scalable	Real-time	Query type	Static-temporal	Sub-events
Chen and Roy [22]	Discrete Wavelet Transform (DWT), density-based clustering (DBSCAN)	Periodic events/aperiodic events	Yes	No	Keywords	Yes	No
Ritter et al. [85]	Named Entity Segmentation, Conditional Random Fields for learning and inference events, latent variable models to categorize events (LinkLDA)	Open domain	No	No	Keywords	No	No
Robinson et al. [86]	Burst detector using binomial model	Natural disaster event	No	Yes	Keywords	Yes	No
Corley et al. [24]	Detection of Signal Consistency from Social Sensors, Topic Clustering via Pearson correlation coefficient, Autoregressive Integrated Moving Average	Open domain	No	No	Keywords	No	No
Tanev et al. [94]	Query expansion methods	Open domain	No	No	Keywords	No	No
Dou et al. [25]	Topical themes using Latent Dirichlet Allocation (LDA), early event detection using cumulative sum control chart (CUSUM)	Open domain	No	No	Keywords	Yes	No

and zero otherwise. Classical term frequency-inverse document frequency (tf-idf) is used to evaluate how important a word is in a corpus and also to retrieve the list of documents where the word is mentioned. This rudimentary event detection approach does not solve the problem, the term vector model size can grow indefinitely depending on the size of the corpus. Temporal order, the semantics and syntactic features of the of words are discarded. Although this model can find similarities of documents it may not capture the similarity or dissimilarity of related or unrelated events. Exploring other data representation techniques such as semantical and contextual features was also done by Allan et al. [5,6] where they presented an upper bound for full-text similarity. Alternative data representations such as the named entity vector [49] attempt to extract information answering the question who, what, when, and where [64]. Mixed models using term and named entity vectors were also proposed [49,104]. Probabilistic representations including language models were applied by Lee et al. [53] and Li et al. [77] proposed a probabilistic framework McRank that incorporates advanced probabilistic learning models. Traditional metrics like the Euclidean distance, Pearson's correlation coefficient, and cosine similarity were also used to measure the similarity between events. Other similarity measures like the Hellinger distance [19] and the clustering index [42] were also used.

Feature-Pivot Techniques. Modeling an event in text streams as a bursty activity, with certain features rising sharply in frequency as the event emerges is the common approach for Feature-Pivot techniques. Kleinberg [46] show that events may be represented by a number of keywords showing bursts in appearance counts. Moreover, in his work, he developed a formal approach for modeling bursts in a way that they can be robustly and efficiently identified, provide an organizational framework for analyzing the underlying content. Several systems to detect emerging trends in textual data (Emerging Trend Detection systems) were described by Kontostathis et al. [48]. The main goal of a trend detection task over textual data is to identify topic areas that were previously unseen or rapidly growing in importance within the corpus. Kontostathis et al. [48] described, for each system, the components (including linguistic and statistical features), learning algorithms, training and test set generation, visualization, and evaluation. Bursty event detection has been also an active topic in recent years with contributions from Fung et al. [32], He et al. [37], He et al. [36], Wang et al. [97] and Goorha and Ungar [34].

Kleinberg [46] approach is based on modeling the stream using an infinite-state automaton, in which bursts appear naturally as state transitions. The output of the algorithm yields a nested representation of the set of bursts that imposes a hierarchical structure on the overall stream computed in a highly efficient way. Fung et al. [32] proposed a parameter free probabilistic approach, called feature-pivot clustering, that detect a set of bursty features for a burst event detected in a sequence of chronologically ordered documents. The feature-pivot clustering modeled word appearance as a binomial distribution, identified the bursty words according to a heuristic-based threshold, and grouped bursty features to find bursty events. Spectral analysis techniques using the discrete Fourier transformation (DFT) were used by He et al. [37] to categorize features for different event characteristics, i.e.: important or not important, and periodic or aperiodic events. Passing from the time domain to the frequency domain, using the DFT, allows the identification of bursts in signals by monitoring the corresponding spike in the frequency domain. Aware that the DFT cannot identify the period of a burst event, He et al. [36] improved their previous works with Gaussian mixture models to identify the feature bursts and their associated periods. Important work in the domain of multiple coordinated text streams was done by Wang et al. [97]. They proposed a general probabilistic algorithm which can effectively discover correlated bursty patterns and their bursty periods across text streams even if the streams have completely different vocabularies (e.g., English vs. Chinese). An online approach for detecting events in news streams was presented by Snowsill et al. [92], this technique is based on statistical significant tests of n-gram word frequency within a time frame. The online detection was achieved, by reducing time and space constraints, when an incremental suffix tree data structure was applied. Social and mainstream media system monitoring tools are also available in Goorha and Ungar [34]. These tools are focused on the user discovery, query and visualisation process for lists of emerging trends previously collected by using some of the algorithms described in this section.

Like the document-pivot techniques, feature-pivot techniques do not deal well with noise resulting in poor event detection performance. Moreover, not all bursts are relevant events of interest, other ones may be missed due the fact that they happen without explicit burst occurrences.

4 Event Detection Taxonomy

The event detection taxonomy is presented in Table 1. A description of each one of the techniques is included in the present section. A division of each one of the techniques was made by taking into account the type of events they were designed (i.e.: Specified or Unspecified Event Detection) and the respective detection method type (i.e.: supervised, unsupervised or hybrid in case it is a combination of both). A resume of the Approaches used in each one of the techniques is presented in Table 2.

4.1 Specified Event Detection

Specified event detection systems using either unsupervised, supervised and hybrid detection techniques are being described in this section.

Unsupervised Detection. Hu et al. [39] proposed event detection of common user interests from huge volume of user-generated content by assuming that the degree of interest from common users in events is evidenced by a significant surge of event-related queries issued to search for documents (e.g., news articles, blog posts). Defining query profile as a set of documents matching a query at a given time and single streams of query profiles as the integration of a query profile and respective documents, events are therefore detected by applying incremental clustering to the stream of query profiles. A temporal query profile is a set of published documents at a given time matching the queries formulated by users at the same time. Based on the observations regarding the number of documents retrieved, authors were able to associate a query profile to the occurrence of a specific event, correlate different query profiles in the context of the same event and establish their duration and evolution. Event detection uses a simple online clustering algorithm consisting of modules: event-related query identification, event assignment, and event archive.

Lee and Sumiya [52] developed a geo-social event detection system, which attempts to find out the occurrence of local events such as local festivals, by monitoring crowd behaviours indirectly via Twitter. To detect such unusual geo-social events, the proposed method depend on geographical regularities deduced from the usual behaviour patterns of crowds with geo-tagged microblogs. The decision whether or not there are any unusual events happening in the monitored geographical area is done by comparing these regularities with the estimated ones. The method performs event detection in the following steps: collecting geo-tagged tweets; configuration of region-of-interests (RoIs) is done using a clustering-based space partition method based on the geographical coordinates. The K-partitioned regions over a map, obtained via K-means clustering, are then regarded as RoIs; geographical regularity of each RoI crowd behaviours is estimated during a certain time period using following properties of a RoI: number of tweets, number of users, and moving users. Features are accumulated over historical data using 6-h time intervals. Unusual events in the monitored geographical area are detected by comparing statistics from new tweets with the estimated behaviour.

Gu et al. [35] proposed ETree, an effective and efficient event modelling solution for social media network sites. ETree used three key components: an n-gram based content analysis technique for identifying and group large numbers of short messages into semantically coherent information blocks; an incremental and hierarchical modelling technique for identifying and constructing event theme structures at different granularities; and an enhanced temporal analysis technique for identifying inherent causalities between information blocks. The identification of core information blocks of an event is done using an n-gram based content analysis technique. Frequent word sequences (i.e., n-grams, or key phrases) among a large number of event-related messages are detected in a first stage. Each frequent sequence represents an initial information block. In the second stage, messages that are semantically coherent are merged into the corresponding information blocks. For each one of the remaining messages, messages that do not contain any of the frequent n-gram patterns, a similarity against each core information block is measured by calculating their TF-IDF weights using words that belongs to both. The weighted cosine similarity between each message and each information block allows the merging of messages into the information block with the highest similarity. Messages that belong to a specific "conversation thread" are also merged into the same information block. The construction of hierarchical theme structures is done by applying an incremental (top-down) hierarchical algorithm based on weighted cosine similarity in the previously identified information blocks. Each theme is represented as a tree structure with information blocks as the leaf nodes and subtopics as the internal nodes.

Ozdikis et al. [68] proposed a document expansion based event detection method for Twitter using only hashtags. Their expansion was based on second-order relations, which is also known in NLP as distributional similarity. The event detection technique was based on clustering of hashtags by using the semantic similarities between hashtags. Items (i.e. tweets in this context) are clustered according to their similarity in vector space model using agglomerative text clustering. In their agglomerative clustering implementation, values in tweet vectors, i.e. weights of the corresponding terms for each tweet, are set as TF-IDF values. Cluster vectors are calculated by taking the arithmetic mean of values in tweet vectors in each dimension. The similarity of tweet vectors and cluster vectors is calculated by applying the cosine similarity. Tweets are only added to a cluster in case the similarity of the vectors being above a threshold defined empirically.

With respect on how event detection can work on corpora less structured than newswire releases, Jurgens and Stevens [43] proposed an automatic event detection that aims to identify novel, interesting topics as they are published in blogs. Authors proposed an adaptation of the Random Indexing algorithm [44,87], Temporal Random Indexing, as a new way of detecting events in this media. The algorithm makes use of a temporally-annotated semantic space for tracking how words change semantics and demonstrate how these identified changes could be used to detect new events and their associated blog entries. Based on semantic slice of a single word, which covers all the time periods in which that word has been observed, the detection of events using Temporal Random Indexing is done in three steps: convert the corpus into month long semantic slices; semantic shift are calculated for each word for slices at consecutive timestamps and compared using the cosine similarity. Authors describe changes in angle as a change in a word's meaning, which can signify the presence of an event. Changes in magnitude showed not to be correlated with events; Finally, events are regarded as the selection of the topic words that undergo a significant semantic shift.

Metzler et al. [63] proposed the problem of structured retrieval of historical event information over microblog archives. Unlike all previous work, that retrieves individual microblog messages in response to an event query, they propose the retrieval of a ranked list of historical event summaries by distilling high quality event representations using a novel temporal query expansion technique. Taking a query as input, the proposed microblog event retrieval framework returns a ranked list of structured event representations. This is accomplished through two steps: the timespan retrieval, that identifies the timespans when the event happened; and the summarization step that retrieves a small set of microblog messages for each timespan. Temporal Query Expansion, Timespan Ranking and Timespan Summarization are used in the search task.

Robinson et al. [86] developed an earthquake detector for Australia and New Zealand by monitoring special keywords like "earthquake" and "#eqnz" in Twitter and available geolocation information. Based on the Emergency Situation Awareness (ESA), the earthquake detector monitors Tweets and checks for specific earthquake related alerts. The system uses ESA burst detection methods based on a binomial model to generate an expected distribution of feature occurrences in a given time window. Then a test on the frequency of observed features in fixed-width time-windows against a statistical content model of historical word frequencies is done. In the cases where the historical model of word frequencies does not fit the observed data, an earthquake situation is identified.

Supervised Detection. Controversial events provoke a public discussion in which audience members express opposing opinions, surprise or disbelief. Using social media as a starting point, Popescu and Pennacchiotti [79] addressed the detection of this kind of events, by proposing three alternative regression machine learning models based on Gradient Boosted Decision Trees [31]. Triplets consisting of a target entity, a given time period, and a set of tweets about the entity from the target period, were used. Authors call those triplets a snapshot with the detection task being done in three steps: separation of events and non-event snapshots using a supervised gradient boosted decision trees trained on a manually labeled data set; estimation of a controversy score to each snapshot using an ML regression model; ranking the snapshots according to the controversy score obtained in the previous step. In a successive work, Popescu et al. [80] used additional features with the same framework described earlier to extract events and their descriptions from Twitter. These new features inspired from the document aboutness system Paranjpe [71] allow the ranking of entities in a snapshot with respect to their relative importance to the snapshot.

With the focus on the identification of entertainment event Twitter messages, Benson et al. [15] formulated an approach to the problem as a structured graphical model which simultaneously analyzes individual messages, clusters them according to event, and induces a canonical value for each event property. This technique is able to construct entertainment event records for the city calendar section of NYC.com using a stream of Twitter messages with high precision and acceptable recall. At the message level, the model relies on a conditional random field (CRF) component to extract field values such as the location of the event and artist name. A factor-graph model was used to capture the interaction between each of these decisions. Variational inference techniques allow to make predictions on a large body of messages effectively and efficiently. A seed set of example records constitutes the only source of supervision; alignment between these seed records and individual messages is not observed, nor any message-level field annotation. The output of the model consists of an event-based clustering of messages, where each cluster is represented by a single multi-field record with a canonical value chosen for each field.

By considering a Twitter user as a sensor and tweets as sensory information, Sakaki et al. [88] employed a supervised classification technique to detect specific event types such as earthquakes, typhoons, and traffic jams. Positive events and negative events are classified according to an SVM trained on a manually labelled dataset. Three groups of features are used: statistical features, i.e.: the number of words in a tweet message, and the position of the query word within a tweet; keyword features, the words in a tweet; word context features, the words before and after the query word. The analysis of the number of tweets over time for earthquakes and typhon data revealed an exponential distribution of events. Authors also mentioned that spikes occur on the number of tweets. Subsequently, a probabilistic spatio-temporal model for the target event that can find the center and the trajectory of the event location is produced. The estimation of the earthquake center and typhoon trajectory was done using Kalman filtering and particle filtering. Particle filters outperformed Kalman filter in both cases.

Massoudi et al. [61] presented a model for retrieving microblog posts that is enhanced with textual and microblog specific quality indicators and with a dynamic query expansion model. They used a generative language modeling approach based on query expansion and microblog "quality indicators" to retrieve individual microblog messages. Being the microblogs documents a special type of user-generated content due their limited size, Massoudi et al. [61], enumerated two interesting effects of its limited size: people use abbreviations or change spelling to fit their message in the allotted space, giving rise to a rather idiomatic language; redundancy-based IR methods may not be usable in a straightforward manner to provide effective access to very short documents. To address the first effect, they introduced credibility indicators for blog post search. To overcome the second effect a re-examination of the potential of local query expansion for searching microblog posts is done using a time-dependent expansion flavor that accounts for the dynamic nature of a topic.

Li et al. [56] proposed a domain-specific event detection method based on pre-specified rules called TEDAS. This system detects, analyses, and identifies relevant crime and disaster related events (CDEs) on Twitter. Based on the authors observation that similar types of CDEs share similar keywords, tweets are collected based on iteratively-refined rules (e.g.: keywords, hashtags). Due to the difficulty to manually define a good set of rules, authors adopted the bootstrapping idea to expand the tracking rule set automatically and iteratively. Next, tweets are classified via supervised learning based on content and Twitter-specific features (i.e.: URLs, hashtags, mentions) and CDE-specific features (i.e.: similarity to CDE tweets, time of day with high crime probability, high crime geographical zones). Location information is extracted using both GPS tagged and location information in tweet content. When no location information is present in the tweet, authors predict user's location as the location from his friends or tweets that minimizes the overall distances between locations in his tweets and from his friends. To rank tweets according to their level of importance, authors propose a learning-to-rank approach, which learns a function to assign a score to each tweet, integrating a variety of signals, such as author's credibility and the number of retweets. To predict a tweet's importance precisely, they explored signals from various aspects, including content, user and usage.

Hybrid Detection. Using a set of automatic query building strategies, Becker et al. [12] presented a system for augmenting information about planned events with Twitter messages. Simple query building strategies were used to achieve high precision results in the task of identifying Twitter messages related to a specific event. To improve recall, they employ term-frequency analysis and co-location techniques on the high-precision tweets to identify descriptive event terms and phrases, which are used recursively to define new queries. Additional queries using URL and hashtag statistics from the high-precision tweets for an event are also built. A rule-based classifier is used to select among this new set of queries, and then use the selected queries to retrieve additional event messages. Becker et al. [12] also developed centrality-based techniques for effective selection of quality event content that may, therefore, help improve applications such as event browsing and search. They address this problem with two concrete steps. First, by identifying each event and its associated Twitter messages using an online clustering technique that groups together topically similar Twitter messages. Second, for each identified event cluster, by providing a selection of messages that best represent the event. With the focus on the challenge of automatically identifying user-contributed content for events that are planned across different social media sites, Becker et al. [13] extended and incorporated into a more general approach their developed techniques of query formulation and centrality based approaches for retrieving content associated with an event on different social media sites.

4.2 Unspecified Event Detection

Unspecified event detection systems rely on either on unsupervised or hybrid detection techniques. The following sections describe examples of those two types of systems.

Unsupervised Detection. TwitterMonitor, the trend detection system over the Twitter stream proposed by Mathioudakis and Koudas [62], was also designed to identify emerging topics in real-time. This system also provides meaningful analytics that synthesize an accurate description of each topic. The detection of bursty keywords was done using a data stream algorithm trends are obtained by grouping keywords into disjoint subsets, so all keywords in the same subset appear on the same topic of discussions. Keyword grouping employs a greedy strategy that produces groups in a small number of steps. The system employs context extraction algorithms (such as PCA and SVD) over the recent history of the trend and reports the keywords that are most correlated with it. To identify frequently mentioned entities in trends uses Grapevine's entity extractor [8].

Phuvipadawat and Murata [76] presented a methodology to collect, group, rank and track breaking news using Twitter tweets. Tasks are divided into two stages: story finding and story development: In the story finding, messages are fetched through the Twitter streaming API using pre-defined search queries to get near real-time public statuses. These pre-defined search queries can be messages containing for example, hashtags users often use to annotate breaking news e.g.: #breakingnews and "breaking news" keyword. To accommodate the process of grouping similar messages, an index based on the content of messages is constructed using Apache Lucene. Messages that are similar to each other are grouped together to form a news story. Similarity between messages is compared using TF-IDF with an increased weight for proper noun terms, hashtags, and usernames. A general implementation of linear chain Conditional Random Field (CRF) sequence models, coupled with well-engineered feature extractors was used as the Named Entity Recognition (NER) technique to identify proper nouns. NER was trained on conventional news corpora; In story development, each news story is adjusted with appropriate ranking through a period of time. The final method ranks the clusters of news using a weighted combination of followers (reliability) and the number of re-tweeted messages (popularity) with a time adjustment for the freshness of the message; Phuvipadawat and Murata [76] emphasized that the key aspect to improving the similarity comparison for short-length messages was to put an emphasis on proper nouns.

Traditional first story detection approaches for news media like the one proposed by Allan et al. [5], which was based on the cosine similarity between documents to detect new events that never appeared in previous documents, revealed to be obsolete when used in a real-time event detection method over social data streams. Petrovic et al. [74] being aware of the limitations constraints of classical event detection methods, both in term of speed and efficiency, proposed a constant time and constant space approach to solve this problem. The proposed system [72] achieved over an order of magnitude speedup in processing time in comparison with the a state-of-the-art system on the first story detection task [6]. The author claimed comparable performance event detection on a collection of 160 million tweets. Modern event detection systems face important challenges when dealing with the high-volume, unbounded nature of today social networks data streams. Using an adapted a variant of the Locality Sensitive Hashing methods [33], was able to detect never seen events when a new bucket is created after hashing a new document to calculate its approximate nearest neighbor. In following work, Petrovic et al. [74] evaluated the use of paraphrases [66] and cross stream event detection by combining Twitter data with Wikipedia spikes [67] and Twitter data with traditional newswires sources [75]. In direct comparison with the UMass system [6], Petrovic et al. [74] also concludes that his approximate technique sometimes outperforms the exact technique. The reason for outperforming the exact system lies in the combination of using LSH and the variance reduction strategy.

Long et al. [58] proposed a unified workflow of event detection, tracking and summarization on microblog data composed by three main steps: in the first events from daily microblog posts are detected using clustering of topical words, afterwards related events are tracked by formulating the event tracking task as a bipartite graph matching problem, and finally tracked event chains are summarized for user to better understand what are happening. Summaries are presented using the top-k most relevant posts considering their relevance to the event as well as their topical coverage and abilities to reflecting event evolution over time. Topical words are extracted from messages using word frequency, word occurrence in hashtags, and word entropy. The separation of topical words into event clusters is done using top-down hierarchical divisive clustering on a co-occurrence graph. Authors state that, using any clustering method, their proposed feature selection outperforms, document frequency only and document frequency with entropy. It also stated that top-down hierarchical divisive clustering outperforms both k-means and traditional hierarchical clustering no matter what k to use.

Weng et al. [98] proposed the EDCoW, an event detection algorithm that clusters wavelet-based signals built from the analysis of the text stream in Twitter. The algorithm builds signals for individual words by applying wavelet analysis to the frequency-based raw signals of the words. Then filters away the trivial words by looking at their corresponding signal auto-correlations. Remaining words are then clustered to form events with a modularity-based graph partitioning technique. In a direct comparison with Discrete Fourier Transformation (DFT) approaches [36, 37] that converts the signals from the time domain into the frequency domain, Weng et al. [98] use wavelet transformation to analyses signals in both time and frequency domain. Unlike the sine and cosine used in the DFT, which are localized in frequency but extend infinitely in time, the wavelet transformation allows the identification of the exact time and the duration of a bursty event within a signal. Weng et al. [98] argue why the use of wavelet transformation is, in general, a better choice for event detection, giving as one example an event detection systems using a similar technique on Flickr data [22]. Event detection is performed in four separate steps: Construction of signals for individual words using wavelet analysis. Signal construction is based on time-dependent of document frequency-inverse document frequency (DF-IDF), where DF counts the counts the number of documents containing a specific word, while IDF accommodates word frequency up to the current time step; The detection of events done by grouping a set of words with similar patterns of burst. To achieve this, the similarities between words need to be computed first, by building a symmetric sparse word cross-correlation matrix. This step is called computation of cross-correlation; Applying a modularity-based graph partitioning in the cross-correlation matrix will allow to group co-occurrences of words at the same time. Weng et al. [98] formulated the event detection problem as a graph partitioning problem, i.e. to cut the graph into subgraphs, where each subgraph corresponds to an event, which contains a set of words with high cross-correlation. Finally, the quantification of event significance compute a significance value for each event by summing all the cross-correlation values between signals associated with an event and discounting the significance when the event is associated with too many words.

A lightweight method for event detection using wavelet signal analysis of hashtag occurrences in the Twitter public stream was presented by Cordeiro [23]. In his work hashtags were used to build signals, instead of individual words [98]. The author considered that an abrupt increase in the use of a given hashtag at a given time is a good indicator of the occurrence of an event. Hashtags signals were constructed by counting distinct hashtag mentions grouped in intervals of 5 min. Each hashtag represented a separate time series. The context of the hashtag was kept by concatenating all the text included in documents with mentions to a specific hashtag. Four separate tasks were performed to detect events: representation of each one of the hashtag signals in a time-frequency representation using a continuous wavelet transformations (CWT); Signal pre-processing using Kolmogorov-Zurbenko Adaptive Filters to remove noise; Wavelet peak and local maxima detection using the continuous wavelet transformation; Finally, event summarization was done by applying LDA [18] topic inference to retrieve a list of topics that describes the event.

Li et al. [55] proposed Twevent, a segment-based event detection system for tweets. Authors define a tweet segment as one or more consecutive words (or phrase) in a tweet message. Based on the fact that tweet segments contained in a large number of tweets are likely to be named entities (e.g. Steve Jobs) or some semantically meaningful unit (e.g. Argentina vs. Nigeria), authors refer that a tweet segment often contains much more specific information than any of the unigrams contained in the segment. Where other techniques rely on bursts of terms or topics (unigrams) to detect events, this particular system first detects bursty tweet segments as event segments. Tweets are split into non-overlapping and consecutive segments, this tweet segmentation problem is formulated as an optimization problem with an objective function based on the stickiness of a segment or a tweet by using the generalized Symmetric Conditional Probability (SCP) for n-grams with n greater or equal to 2, supported by statistical information derived from Microsoft Web N-Gram service and Wikipedia. Bursty segments are identified by modeling the frequency of a segment as a Gaussian distribution based on predefined fixed time-window. By considering their frequency distribution and their content similarity, the grouping of event-related segments as candidate events was done using k-Nearest Neighbor graph and a cosine based similarity measure. Each one of the event clusters is regarded as candidate events detected in that time window. Wikipedia is exploited to identify the realistic events and to derive the most newsworthy segments to describe the identified events.

Agarwal et al. [1] model the problem of discovering events that are unravelling in microblog message streams as a problem of discovering dense clusters in highly dynamic graphs. Authors state that the identification of a set of temporally correlated keywords is the starting point to identify an emerging topic. Moreover, they go further and define temporally correlated keywords as keywords that show burstiness at the same time and are spatially correlated, and more specifically keywords that co-occur in temporally correlated messages from the same user. To capture these characteristics, a dynamic graph model that uses the moving window paradigm and is constructed using the most recent messages present in the message stream, was used. An edge between two nodes — representing two keywords — indicates that messages from a user within the recent sliding window involve the respective keywords. A Correlated Keyword Graph (CKG) captures the properties of microblog contents by representing all the keywords, after removing stop words, appearing in the messages in the current window as nodes in an undirected graph. Emerging events are, therefore, identified by discovering clusters in CKG. Clusters of interest are obtained via majority quasi cliques (MQCs). Being the discovering majority quasi cliques an NP-complete problem even for static graphs, authors proposed the use of short cycle property (SCP) of MQCs to make event discovery a tractable and local problem. Because Correlated Keyword Graph is dynamic and not static, efficient algorithms for maintaining the clusters locally even under numerous additions and deletions of nodes and edges were also proposed.

Under the premises that documents that describe the same event contain similar sets of keywords, and graph of keywords for a document collection contain clusters of individual events, Sayyadi et al. [90] proposed an event detection approach that overlays a graph over the documents, based on word co-occurrences. Authors assume that keywords co-occur between documents when there is some topical relationship between them and use a community detection method over the graph to detect and describe events. The method uses two steps: Building of a KeyGraph, by first extracting a set of keywords from documents, then for each keyword calculating the term frequency (TF), document frequency (DF) and the inverse document frequency (IDF). Using keywords with higher occurrences nodes are created in the KeyGraph for keyword. Edges between nodes (keywords) are added if the two co-occur in the same document; Community Detection in KeyGraph, community detection is done removing edges in the graph till communities get isolated. Authors consider that by removing the edges with a high betweenness centrality score, every connected component of the KeyGraph represents a hypothesis about an event, the keywords forming a bag of words summary of the event; Document Clustering, community of keywords are seen as synthetic documents. Original documents are clustered using cosine similarity distance to the keywords synthetic documents. Documents that truly represent events are obtained by filtering keywords synthetic documents with high variance.

Zhao et al. [108] proposed the detection of events by exploring not only the features of the textual content but also the temporal, and social dimensions present in social text streams. Authors define an event as the information flow between a group of social actors on a specific topic over a certain time period. Social text streams are modeled as multi-graphs, where nodes represent social actors, and each edge represents the information flow between two actors. The content and temporal associations within the flow of information are embedded in the corresponding edge. Events are detected by combining text-based clustering, temporal segmentation, and information flow-based graph cuts of the dual graph of the social networks. The proposed method begins with social text streams being represented as a graph of text pieces connected by content-based document similarity. The weight of each word in the text piece is quantified as the TF-IDF, with the content-based similarity being defined as the cosine similarity of the vector representation of each text pieces. Using a graph cut algorithm [60], text pieces are then clustered into a set of topics. The resulting graph then is partitioned into a sequence of graphs based on the intensity along the temporal dimension using the adaptive time series model proposed by Lemire [54]. Each graph in the temporal dimension, for a given topic, represents a communication peak (intensive discussion) that corresponds to a specific aspect or a smaller event. After that, each graph in a specific time window with respect to a specific topic is converted into its dual graph and the dual graph is further partitioned into a set of smaller graphs based on the dynamic time warping [45] based information flow pattern similarity between social actor pairs using graph cut algorithm [60]. Finally, the output of each event will be represented as a graph of social actors connected via a set of emails or blog comments during a specific time period about a specific topic.

Pohl et al. [78] proposed crisis-related sub-event detection using social media data obtained from Flickr and Youtube. Considering the Geo-referenced data an important source of information for crisis management, authors decided to apply a two-phase clustering approach to identify crisis-related sub-events. The method relies on longitude and latitude coordinates of existing data items for sub-event detection. In a pre-processing step each item is therefore represented in two parts: the coordinates, represented by longitude and latitude values; and the terms, extracted from textual metadata fields belonging to a specific item. Term frequency-inverse document frequency (tf-idf) values are also computed. The two-phase clustering consists of the calculation of term-based centroids with a Self-Organizing Map (SOM) Kohonen [47] using the geo-referenced data. In the second phase, the assignment of best fitting data points to the calculated centroids using reassignment and the cosine distance measure is done.

Chen and Roy [22] presents a method to perform event detection from Flickr photos by exploiting the tags supplied by user's annotations. As not every photo represents an event, authors use feature-pivot approaches to detect event-related tags before detecting events of photos. The methods is done in three steps: In Event Tag Detection, the temporal and locational distributions of tag usage are analyzed in order to discover event-related tags using the Scale-structure Identification (SI) approach Rattenbury et al. [83]. A wavelet transform is employed to suppress noise; In Event Generation, by examining the characteristics of the distribution patterns, authors are able to distinguish between aperiodic-event-related and periodic-event-related tags. Event-related tags are clustered such that each cluster, representing an event, consists of tags with similar temporal and locational distribution patterns as well as with similarly associated photos. A density-based clustering method was used (DBSCAN) Ester et al. [28]; In Event Photo Identification, for each tag cluster, photos corresponding to the represented event are extracted.

Corley et al. [24] proposed a conceptual framework for interpreting social media as a sensor network. The system quantifies a baseline from the social sensor measurements. Those baselines provide the expected value at a particular point in time of the volume of social media features fitting some criterion. Using a brute-force approach, they detect aberrations (Events) in the sensor data when an observed value is significantly different from the expected baseline. Signals are built considering the varying time-dependent measures of frequency such as user retweets, term and hashtag usage, and user-specific posts. Measures like the signal magnitude, which is the value of the centered moving average of an indicated time period of that signal, and the social signal noise, defined as the range of counts bounded by the values of two standard deviations above and below the signal magnitude, are used to calculated the signal aberration (or event) is an instance when the social signal exceeds signal noise boundaries. To produce baseline signals for related topics, topic clustering through using the dot product similarity metric between authors and their hashtag usage, over the course of a specified time period, is used.

Tanev et al. [94] described an Information Retrieval approach to link news about events to Twitter messages. The authors also explored several methods for creating event-specific queries for Twitter. They also claim that methods based on utilization of word co-occurrence clustering, domain-specific keywords and named entity recognition have shown good performance. Basic detection of known bi-grams in the input news article is performed using an index of word uni-grams and bi-grams previously calculated. Because each word uni-gram and bi-gram is accompanied by its frequency and the frequency of the co-occurrences with the other uni/bi-grams, the same index is also used to calculate IDF for each term and suggest classes of terms which are used to formulate the queries to Twitter based on the co-occurrence information. Other techniques like word co-occurrences, named entities, domain-specific keywords were used to improved the detection method.

Dou et al. [25] proposed an interactive visual analytics system, LeadLine, that automatically identify meaningful events in news and social media data and support exploration of the events. To characterize events, topic modeling, event detection, and named entity recognition techniques were used to automatically extract information regarding event details. First, text data such as news stories and microblog messages are organized based on topical themes using LDA [18]. An Early Event Detection algorithm is used to identify the temporal scale for events by determining the length and "burstyness" of events.

Supervised Detection. No supervised detection techniques to detect unspecified events were included. No pure supervised event detection systems to detect unspecified events were found in the literature. This fact may be related to the fact that supervised techniques with prior training on ground truth datasets, could not detect unforeseen events in that dataset. Supervised Detection techniques are always used in conjunction with unsupervised techniques (Hybrid Detection) that are being described in the following section.

Hybrid Detection. Sankaranarayanan et al. [89] proposed a news processing system, called TwitterStand, that primarily demonstrates how to use a microblog service (i.e. Twitter) to automatically obtain breaking news from the tweets posted by Twitter users. Since the geographic location of the user as well as the geographic terms comprising the tweets play an important role in clustering tweets and establishing clusters' geographic foci, providing users a map interface for reading this news. This system discards tweets that clearly cannot be news by using a naive Bayes classifier previously trained on a training corpus of tweets that have already been marked as either news or junk. A clustering algorithm based on weighted term vector according to TF-IDF and cosine similarity was used to form clusters of news. The leader-follower clustering [26] algorithm needed to be modified in order to work in an online fashion.

Cataldi et al. [20] use burstiness of terms in a time interval to detect when an event is happening. They proposed a topic detection technique that retrieves in real-time the most emergent topics expressed by the Twitter community. The process begins with the extraction and formalisation of the user-generated content expressed by the tweets as vectors of terms with their relative frequencies; author's authority is calculated by the Page Rank algorithm [69] applied to a directed graph of the active authors based on their social relationships; for each term, its life cycle is modeled according to an aging theory [21] that leverages the user's authority in order to study its usage in a specified time interval; a set of emerging terms is selected by ranking the keywords depending on their life status (defined by an energy value). Supervised term selection relies on a user-specified threshold parameter while the unsupervised term selection relies on an unsupervised ranking model with the cut-off being adaptively computed; finally a navigable topic graph is created which links the extracted emerging terms with their relative co-occurrent terms in order to obtain a set of emerging topics.

Becker et al. [14] explored approaches for analyzing the stream of Twitter messages to distinguish between messages about real-world events and non-event messages. Their approach relies on a rich family of aggregate statistics of topically similar message clusters. Using an incremental, online clustering technique that does not require a priori knowledge of the number of clusters, a task of grouping together topically similar tweets is done. To identify event clusters in the stream, a variety of revealing features is computed using statistics of the cluster messages. Authors used a combination of temporal, social, topical, and Twitter-centric features that must be updated periodically once that they constantly evolve over time. Temporal Features characterize the volume of frequent cluster terms (i.e., terms that frequently appear in the set of messages associated with a cluster) over time. These features capture any deviation from expected message volume for any frequent cluster term or a set of frequent cluster terms. Social Features capture the interaction of users in a cluster's messages. These interactions might be different between events, Twitter-centric activities, and other non-event messages. User interactions on Twitter include retweets, replies, and mentions. Topical Features describe the topical coherence of a cluster, based on a hypothesis that event clusters tend to revolve around a central topic, whereas non-event clusters do not. Twitter-centric features target commonly occurring patterns in non-event clusters with Twitter-centric behavior, including tag usage, and presence of multi-word hashtags. Subsequently, classification via support vector machine (SVM) using the cluster features representation and a previously labeled training set of clusters is done in order to decide whether or not the cluster, and its associated messages, contains event information (i.e.: distinguish between event and non-event clusters).

Ritter et al. [85] proposed TwiCal, an open-domain event-extraction and categorization system for Twitter. The system extract event phrases, named entities, and calendar dates from Twitter by focusing on certain types of words and phrases. Named entities are extracted using a named entity tagger trained on 800 randomly selected tweets, while the event mentions are extracted using a specific Twitter-tuned part-of-speech tagger [84]. The extracted events are classified retrospectively into event types using a latent variable model (LinkLDA [27]) which infers an appropriate set of event types to match the data (via collapsed Gibbs Sampling using a streaming approach [107]), and then classifies events into types by leveraging large amounts of unlabeled data. The approaches used were based on latent variable models inspired on modeling selectional preferences, and unsupervised information extraction.

4.3   Detection Methods

Distinct methods to perform event detection are described in the following sections.

Clustering. Clustering is the most used technique in event detection systems. Different clustering techniques are described in literature, from classical clustering, passing by incremental clustering, hierarchical clustering or graph partitioning techniques, authors see the separation of documents in similar clusters as a valid method to detect events.

Although they require a prior knowledge of the number of clusters, partition clustering techniques such as K-Means, K-median, K-medoid were used by [52]. Clustering of hashtags based on the similarity of documents vectors and cluster vectors using cosine similarity was proposed by [68]. Frequent word sentences (n-grams) using weighted cosine similarity were also used by [35]. The clustering of wavelet signals was proposed in [23] to signals constructed by hashtags occurrences, and using Co-occurrence of words in [98].

With the necessity of grouping continuously arriving text documents, incremental threshold-based clustering approaches need to be used. Examples of this approach are [39] were incremental clustering to the stream of query profiles is proposed and the Locally Sensitive Hashing method proposed by [72] where documents are clustered after applying a dimensional reduction technique. The major drawbacks of these methods are the fragmentation issues and the correct setting for the threshold value.

Table 3. Collection type, corpus size and temporal scope of datasets

Reference	Collection	Corpus size	Temporal scope
Hu et al. [39]	Technorati popular queries	4075 * 15 queries	From 2006-11-08 1 AM to 2008-03-31 10 PM (17 months)
Jurgens and Stevens [43]	Blog articles harvested by BlogLines	15,725,511 blog entries	One year (2006)
Popescu and Pennacchiotti [79]	Twitter streaming API	738,045 Twitter snapshots	July 2009–February 2010
Popescu et al. [80]	Twitter streaming API		
Benson et al. [15]	Twitter streaming API	4.7 Million tweets (5,800 messages)	Three weekends
Lee and Sumiya [52]	Twitter Search API	21,623,947 geo-tagged tweets from 366,556 distinct users	One and a half months (2010/06/04–2010/07/20)
Sakaki et al. [88]	Twitter Search API	49,314 tweets	One month; 2009 Aug. 10 01:00–2009 Oct. 12 18:42
Becker et al. [12]	Twitter Search API	NA	NA
Becker et al. [13]	Last.fm events, EventBrite, LinkedIn events, and Facebook events	NA	May 13, 2011 and June 11, 2011
Massoudi et al. [61]	Twitter Search API	110,038,694 tweets	Nov '09–Apr '10
Metzler et al. [63]	Twitter streaming API	46,611,766 English tweets	July 16, 2010 and Jan 1st, 2011
Gu et al. [35]	Twitter Search API	3.5 million tweets	5 month period
Li et al. [56]	Twitter Search API	1 million of CDE tweets	Two months
Ozdikis et al. [68]	Twitter Search API	388K tweets	March 16, 2012 and March 19, 2012
Sankaranarayanan et al. [89]	Twitter streaming API/Twitter Search API	NA	NA
Cataldi et al. [20]	Twitter streaming API	3 million tweets	13th and 28th of April 2010
Mathioudakis and Koudas [62]	Twitter streaming API	1.2 million per day	NA

(continued)

Table 3. (*continued*)

Reference	Collection	Corpus size	Temporal scope
Phuvipadawat and Murata [76]	Twitter streaming API	NA	February 2010
Petrovic et al. [74]	Twitter streaming API	163.5 million timestamped tweets	Six months (April 1st 2009 to October 14th 2009)
Becker et al. [14]	Twitter streaming API	2,600,000 Twitter messages	February 2010
Long et al. [58]	Sina Microblog API	22 million microblog posts	December 23th, 2010 to March 8th, 2011
Weng et al. [98]	Twitter Search API	4,331,937 tweets	April 13, 2011 till May 13, 2011
Cordeiro [23]	Twitter streaming API	13.651.464 tweets	00:00 of the 10th of November and 23:59 of 18th of November of 2011
Li et al. [55]	Wikipedia/Twitter streaming API	3, 246, 821 articles from wikipedia (30 Jan, 2010)/4, 331, 937 tweets	April 13, 2011 till May 13, 2011
Agarwal et al. [1]	Twitter streaming API	1.3 million	18 h on 29th Feb 2012
Sayyadi et al. [90]	Live Labs's Social Streams platform	18,000 posts	Two months (May and June 2009)
Zhao et al. [108]	Enron Email dataset/Dailykos blog dataset	619,446 messages/249543 blog entries	1998 to year 2002/October 12, 2003 to October 28, 2006
Pohl et al. [78]	Youtube/Flickr	4 datasets (2.039.442, 31.222, 178.274 and 455.700 videos and images)	4 datasets (04–19 May, 22 July, 06–10 Aug and 23–29 Aug)
Chen and Roy [22]	Flickr	7, 405, 135 photos, annotated with 44, 139, 261 tags	Two-year-period starting at Jan 01, 2006, until Dec 31, 2007
Ritter et al. [85]	Twitter streaming API	100 million tweets	November 3rd 2011
Robinson et al. [86]	Twitter streaming API	870 million	September 2011–January 2013
Corley et al. [24]	Twitter streaming API	NA (8.73 TB)	13-June 2011 through 11-March 2013
Tanev et al. [94]	Twitter Search API	NA	NA
Dou et al. [25]	Twitter Search API/CNN news	100,000 tweets/3,130 news articles	Aug 19 to Nov 01 2011/Aug 15, 2011 to Nov 5, 2011

Graph-based clustering algorithms were also used. Hierarchical divisive clustering techniques used on a co-occurrence graph, that connects messages according to word co-occurrences, to divide topical words into event clusters [1,58,94]. Modularity-based graph partitioning techniques are used to form events by splitting the graph into subgraphs each one corresponding to an event [90]. PageRank was used as an alternative to the costly of finding the largest eigenvalue of the modularity matrix [20]. In general hierarchical clustering algorithms do not scale because they require the full similarity matrix. [55] proposed a k-Nearest Neighbour graph of non-overlapping and consecutive document segments based on the generalised Symmetric Conditional Probability (SCP) for n-grams.

Classification. Classification algorithms, commonly used for the detection of specified events, rely mainly on supervised learning approaches. Classification algorithms include naive Bayes [14,89], support vector machines (SVM) [14,88] and gradient boosted decision trees [79,80]. Classifiers are typically trained on a small set of documents collected over a few months or weeks and labeled according event or non-event [14,89], earthquake or non-earthquake [88] and controversial or non-controversial event [79,80]. Usually labeling involves human annotators with domain knowledge and is done manually. Previous filtering of irrelevant messages to increase accuracy is also done, e.g.: [88] filter documents that contains special words like "earthquake".

Dimension Reduction. Dimension Reduction techniques are used in most cases to speed up the event detection methods. They are commonly used in streaming scenarios were very high volumes of documents arrive at very high speeds. Normally they are used in conjunction with other techniques (i.e.: clustering, classification, etc.). Petrovic et al. [72] proposed a first story detection algorithm using Locality-sensitive hashing (LSH). This method performs a probabilistic dimension reduction of high-dimensional data. The basic idea is to hash the input items so that similar items are mapped to the same buckets with high probability (the number of buckets being much smaller than the universe of possible input items). With this improvement the scaling problem of the traditional approaches to FSD, where each new story is compared to all, was overcome by a system that works in the streaming model and takes constant time to process each new document, while also using constant space [65]. The proposed system follows the streaming model of computation where items arrive continuously in a chronological order and are processed, each new one, in bounded space and time.

Early successful approaches such as Latent Semantic Analysis use the Singular Value Decomposition (SVD) to reduce the number of dimensions. Principle Component Analysis (PCA) and SVD event detection techniques were addressed by [62]. Although SVD resulted in significant improvements in information retrieval, their poor performance makes them impractical for use in large corpora. Moreover, the SVD and other forms of PCA must have the entire corpus present at once, which makes it difficult to update space as new words and contexts are added. This is particularly problematic for event detection, as the corpus is expected to grow continuously as new events occur. Random Indexing offers an alternative method for reducing the dimensionality of the semantic space by using a random projection of the full co-occurrence matrix onto a lower dimensional space and was used by [43] as the underlying technique for event detection.

Wavelets Analysis. Weng et al. [98] attempted to solve the event detection in the Twitter online social network by proposing the detection of generic events using signal analysis. The algorithm, called Event Detection with Clustering of Wavelet-based Signals, builds signals for individual words by applying wavelet analysis to the frequency-based raw signals of the words. It then filters away the trivial words by looking at their corresponding signal auto-correlations. The remaining words are then clustered to form events with a modularity-based graph partitioning technique. The algorithm didn't follow the streaming model defined by [65] and is not expected to scale to unbounded text streams. Additionally the authors applied a Latent Dirichlet Allocation algorithm to extract topics from the detected events.

Burstiness Analysis. Analysis of burstiness is also a frequent technique. Analysis if burstiness of special keywords was proposed by [86] while burstiness of topics extracted via LDA was proposed by [25]. Pan and Mitra [70] proposes two event detection approaches using generative models. In the first approach they combined Latent Dirichlet Allocation (LDA) model [18] with temporal segmentation and spatial clustering and afterwards adapted an image segmentation model, Spatial Latent Dirichlet Allocation (SLDA) [96], for spatial-temporal event detection on text.

Other. Hybrid detection approaches are used in techniques composed by more than one step. Supervised classification or detection techniques are commonly used to identify relevant or important documents before performing the unsupervised step (e.g.: clustering) [89]. Other techniques use a factor graph model that simultaneously detects information of events using supervised CRF and then clusters them according to the event type [15]. A temporal query expansion method was proposed by [63] while Generative Language models were proposed by [61].

5   Datasets

Event detection research is hampered by the lack of standard corpora that could be used to evaluate and benchmark systems. Most researchers that work on event detection, often create their ad-hoc corpora to perform the evaluation. Event labelling is typically done by manual inspection or using external sources/systems to mark events, very often are not publicly available, and usually present problems that pass for being: (i) tied to a specific domain application or data source; (ii) they only cover high-volume events ignoring low-volume events; (iii) they do not cover broad range of event types. Table 3 presents all the datasets and respective properties used for the evaluation of each one of the techniques. Through a quick analysis, it can be observed the heterogeneity in terms of source, size and temporal scope of each one of the used corpus.

6 Conclusions

This chapter presents a survey of techniques proposed for event detection in online social networks. The survey also presents an overview of the challenges that event detections techniques face when dealing with today's Online Social Networks data. While some techniques were designed for the detection of specified events (i.e. natural disasters), others were designed to detect events without prior information of the event itself (i.e. unspecified events).

Event detection techniques are classified according to the type of target event into specified or unspecified event detection. Depending on the target application the way data is being analyzed, the techniques are also classified into Online New Event Detection (NED) or in Retrospective Event Detection (RED). Depending on the underlying detection method involved in the event detection, a classification in supervised, unsupervised or hybrid approaches was also done. Depending if the detection method operates at the document or feature domain, the techniques could also be classified in two main categories: Document-pivot techniques and Feature-pivot techniques. A resume of the main detection methods was also provided, clustering methods are the most used in unsupervised detection systems for unspecified event detection. Classification methods are in the basis of most of the supervised methods for specified event detection. Dimension reduction approaches, specially LSH, is used when processing of high volumes of data arriving at very high speeds. Systems based on burstiness analysis are commonly used to monitor trends and changes in behaviour that may indicate the presence of events. In the present survey is also shown that there is a very high variety of applications and sources of data. It was shown that most of the techniques use different datasets and evaluation methods, which makes their direct comparison almost impossible. Some of them also have different event detection objectives and meet specific detection requirements.

Although the extensive literature presented an high degree of maturity of some methods, the event detection problem is still one of the most actives in the research community. The continuous growth and evolving of the Online Social Networks Services is challenging state-of-the-art methods in terms of volume, speed and data diversity. Recent trends in research using approximation methods show that equivalent results were obtained when compared to exact methods in a much more efficient way.

Acknowledgements. This work was supported by national funds, through the Portuguese funding agency, Fundação para a Ciência e a Tecnologia (FCT), and by European Commission through the project MAESTRA (Grant number ICT-2013-612944).

References

1. Agarwal, M.K., Ramamritham, K., Bhide, M.: Real time discovery of dense clusters in highly dynamic graphs: identifying real world events in highly dynamic environments. Proc. VLDB Endow. **5**(10), 980–991 (2012). http://arxiv.org/abs/1207.0138
2. Aggarwal, C.C., Zhai, C.: A survey of text clustering algorithms. In: Aggarwal, C.C., Zhai, C. (eds.) Mining Text Data, pp. 77–128. Springer, New York (2012)
3. Allan, J.: Topic Detection and Tracking: Event-based Information Organization. The Kluwer International Series on Information Retrieval, vol. 12. Springer, New York (2002). http://portal.acm.org/citation.cfm?id=772260
4. Allan, J., Jin, H., Rajman, M., Wayne, C., Gildea, D., Lavrenko, V., Hoberman, R., Caputo, D.: Topic-based novelty detection 1999 summer workshop at CLSP final report (1999). http://old-site.clsp.jhu.edu/ws99/projects/tdt/final_report/report.pdf. Accessed 2 Nov 2013
5. Allan, J., Lavrenko, V., Jin, H.: First story detection in TDT is hard. In: CIKM 2000 Proceedings of the Ninth International Conference on Information and Knowledge Management, pp. 374–381. ACM (2000)
6. Allan, J., Lavrenko, V., Malin, D., Swan, R.: Detections, Bounds, and Timelines: UMass and TDT-3. Information Retrieval, pp. 167–174 (2000). http://maroo.cs.umass.edu/pdf/IR-201.pdf
7. Allan, J., Papka, R., Lavrenko, V.: On-line new event detection and tracking. In: Proceedings of the 21st Annual International ACM SIGIR Conference on Research and Development in Information Retrieval - SIGIR 1998, New York, USA, pp. 37–45 (1998). http://portal.acm.org/citation.cfm?doid=290941.290954
8. Angel, A., Koudas, N., Sarkas, N., Srivastava, D.: What's on the grapevine? In: Proceedings of the 35th SIGMOD International Conference on Management of Data - SIGMOD 2009, p. 1047 (2009). http://portal.acm.org/citation.cfm?doid=1559845.1559977
9. Atefeh, F., Khreich, W.: A survey of techniques for event detection in Twitter. Comput. Intell. (2013). http://doi.wiley.com/10.1111/coin.12017
10. Bampis, E., Jansen, K., Kenyon, C.: Efficient Approximation and Online Algorithms. Springer, Heidelberg (2010). http://www.amazon.com/Efficient-Approximation-Online-Algorithms-Combinatorial/dp/3540322124
11. Barabasi, A.L.: The origin of bursts and heavy tails in human dynamics. Nature **435**, 207 (2005). http://www.citebase.org/abstract?id=oai:arXiv.org:cond-mat/0505371
12. Becker, H., Chen, F., Iter, D.: Automatic identification and presentation of Twitter content for planned events. In: Proceedings of the Fifth International AAAI Conference on Weblogs and Social Media, pp. 655–656 (2011). http://www.aaai.org/ocs/index.php/ICWSM/ICWSM11/paper/download/2743/3198
13. Becker, H., Iter, D., Naaman, M., Gravano, L.: Identifying content for planned events across social media sites. In: Proceedings of the Fifth ACM International Conference on Web Search and Data Mining - WSDM 2012, p. 533 (2012). http://dl.acm.org/citation.cfm?doid=2124295.2124360

14. Becker, H., Naaman, M., Gravano, L.: Beyond trending topics: real-world event identification on Twitter. In: ICWSM, pp. 438–441. Technical Report CUCS-012-11, Columbia University. The AAAI Press (2011). http://www.aaai.org/ocs/index.php/ICWSM/ICWSM11/paper/viewPDFInterstitial/2745/3207

15. Benson, E., Haghighi, A., Barzilay, R.: Event discovery in social media feeds. Artif. Intell. **3**(2–3), 389–398 (2011). http://aria42.com/pubs/events.pdf, http://dl.acm.org/citation.cfm?id=2002472.2002522

16. Berkhin, P.: A survey of clustering data mining techniques. In: Kogan, J., Nicholas, C., Teboulle, M. (eds.) Grouping Multidimensional Data, pp. 25–71. Springer, Heidelberg (2006). http://link.springer.com/chapter/10.1007/3-540-28349-8_2

17. Bifet, A., Kirkby, R.: Data stream mining: a practical approach. Technical report, The University of Waikato, August 2009

18. Blei, D.M., Ng, A.Y., Jordan, M.I.: Latent Dirichlet allocation. J. Mach. Learn. Res. **3**(4–5), 993–1022 (2003). http://www.crossref.org/jmlr_DOI.html

19. Brants, T., Chen, F.: A system for new event detection. In: Proceedings of the 26th Annual International ACM SIGIR Conference on Research and Development in Informaion Retrieval SIGIR 2003, 2002, p. 330 (2003). http://portal.acm.org/citation.cfm?doid=860435.860495

20. Cataldi, M., Torino, U., Caro, L.D., Schifanella, C.: Emerging topic detection on Twitter based on temporal and social terms evaluation. In: Proceedings of the Tenth International Workshop onMultimedia Data Mining, pp. 1–10 (2010). http://dl.acm.org/citation.cfm?id=1814245.1814249

21. Chen, C.C., Chen, Y.-T., Sun, Y., Chen, M.-C.: Life cycle modeling of news events using aging theory. In: Lavrač, N., Gamberger, D., Todorovski, L., Blockeel, H. (eds.) ECML 2003. LNCS (LNAI), vol. 2837, pp. 47–59. Springer, Heidelberg (2003). http://link.springer.com/chapter/10.1007/978-3-540-39857-8_7

22. Chen, L., Roy, A.: Event detection from flickr data through wavelet-based spatial analysis. In: Proceedings of the 18th ACM Conference on Information and Knowledge Management, pp. 523–532 (2009). http://dl.acm.org/citation.cfm?id=1646021npapers2://publication/uuid/8EC6E15D-D958-4A0A-88E5-8D62631BF7C5

23. Cordeiro, M.: Twitter event detection: combining wavelet analysis and topic inference summarization. In: The Doctoral Symposium on Informatics Engineering - DSIE 2012 (2012). http://paginas.fe.up.pt/~prodei/dsie12/papers/paper_14.pdf

24. Corley, C.D., Dowling, C., Rose, S.J., McKenzie, T.: SociAL sensor analytics: measuring phenomenology at scale. In: 2013 IEEE International Conference on Intelligence and Security Informatics, pp. 61–66. IEEE, June 2013. http://iccexplore.ieee.org/lpdocs/epic03/wrapper.htm?arnumber=6578787

25. Dou, W., Wang, X., Skau, D., Ribarsky, W., Zhou, M.X.: LeadLine: interactive visual analysis of text data through event identification and exploration. In: IEEE Conference on Visual Analytics Science and Technology 2012, VAST 2012 - Proceedings, pp. 93–102 (2012)

26. Duda, R.O., Hart, P.E., Stork, D.G.: Pattern Classification, 2nd edn. October 2000. http://dl.acm.org/citation.cfm?id=954544

27. Erosheva, E., Fienberg, S., Lafferty, J.: Mixed-membership models of scientific publications. Proc. Natl. Acad. Sci. U.S.A. **101**(Suppl 1), 5220–5227 (2004)

28. Ester, M., Kriegel, H.P., Sander, J., Xu, X.: A density-based algorithm for discovering clusters in large spatial databases with noise. In: Second International Conference on Knowledge Discovery and Data Mining, pp. 226–231 (1996). http://citeseerx.ist.psu.edu/viewdoc/summary?doi=10.1.1.20.2930

29. Farzindar, A.: Social network integration in document summarization. In: Fiori, A. (ed.) Innovative Document Summarization Techniques: Revolutionizing Knowledge Understanding. IGI-Global, Hershey (2014)
30. Fiscus, J.G., Doddington, G.R.: Topic detection and tracking evaluation overview. In: Topic Detection and Tracking, pp. 17–31 (2002). http://www.springerlink. com/index/T652P42711XW6421.pdf
31. Friedman, J.H.: Greedy function approximation: a gradient boosting machine. Ann. Stat. **29**(5), 1189–1232 (2001). http://citeseerx.ist.psu.edu/viewdoc/ summary?doi=10.1.1.29.9093
32. Fung, G.G.P.C., Yu, J.X.J., Yu, P.P.S., Lu, H.: Parameter free bursty events detection in text streams. In: Proceedings of the 31st International Conference on Very Large Data Bases - VLDB 2005, vol. 1, pp. 181–192 (2005). http://dl.acm.org/ citation.cfm?id=1083616   http://www.scopus.com/inward/record.url?eid=2-s2. 0-33745624002&partnerID=tZOtx3y1
33. Gionis, A., Indyk, P., Motwani, R.: Similarity search in high dimensions via hashing. In: VLDB 1999: Proceedings of the 25th International Conference on Very Large Data Bases, pp. 518–529 (1999). http://portal.acm.org/citation.cfm? id=671516
34. Goorha, S., Ungar, L.: Discovery of significant emerging trends. In: ACM SIGKDD International Conference on Knowledge Discovery and Data Mining (KDD), p. 57 (2010). http://dl.acm.org/citation.cfm?doid=1835804.1835815
35. Gu, H., Xie, X., Lv, Q., Ruan, Y., Shang, L.: ETree: effective and efficient event modeling for real-time online social media networks. In: 2011 IEEE/WIC/ACM International Conferences on Web Intelligence and Intelligent Agent Technology, pp. 300–307. IEEE, August 2011. http://dl.acm.org/citation.cfm?id=2052138. 2052366
36. He, Q., Chang, K., Lim, E.P.: Analyzing feature trajectories for event detection. In: Proceedings of the 30th Annual International ACM SIGIR Conference on Research and Development in Information Retrieval - SIGIR 2007, p. 207 (2007). http://portal.acm.org/citation.cfm?doid=1277741.1277779\n, http://doi. acm.org/10.1145/1277741.1277779
37. He, Q., Chang, K., Lim, E., Zhang, J.: Bursty feature representation for clustering text streams. In: SDM, pp. 491–496 (2007). https://www.siam.org/proceedings/ datamining/2007/dm07_050he.pdf
38. Hounshell, B.: The revolution will be tweeted. Foreign Policy **187**, 20–21 (2011)
39. Hu, M., Sun, A., Lim, E.P.: Event detection with common user interests. In: Proceeding of the 10th ACM Workshop on Web Information and Data Management - WIDM 2008, New York, USA, p. 1. ACM, New York, October 2008. http://dl. acm.org/citation.cfm?id=1458502.1458504
40. Hussein, D., Alaa, G., Hamad, A.: Towards usage-centered design patterns for social networking systems. In: Park, J.J., Yang, L.T., Lee, C. (eds.) FutureTech 2011, Part II. CCIS, vol. 185, pp. 80–89. Springer, Heidelberg (2011). http://dx.doi.org/10.1007/978-3-642-22309-9_10
41. Imran, M., Castillo, C., Diaz, F., Vieweg, S.: Processing social media messages in mass emergency: a survey, July 2014. http://arxiv.org/abs/1407.7071
42. Jo, T., Lee, M.R.: The evaluation measure of text clustering for the variable number of clusters. In: Liu, D., Fei, S., Hou, Z., Zhang, H., Sun, C. (eds.) ISNN 2007, Part II. LNCS, vol. 4492, pp. 871–879. Springer, Heidelberg (2007). http://dx.doi.org/10.1007/978-3-540-72393-6_104

43. Jurgens, D., Stevens, K.: Event detection in blogs using temporal random indexing. In: Proceedings of the Workshop on Events in Emerging Text Types, pp. 9–16 (2009). http://dl.acm.org/citation.cfm?id=1859650.1859652

44. Kanerva, P., Kristofersson, J., Holst, A.: Random indexing of text samples for latent semantic analysis. In: Proceedings of the 22nd Annual Conference of the Cognitive Science Society, vol. 1036, pp. 16429–16429 (2000). http://citeseerx.ist.psu.edu/viewdoc/download?doi=10.1.1.4.6523&rep=rep1&type=pdf

45. Keogh, E.: Exact indexing of dynamic time warping. In: Proceedings of the 28th International Conference on Very Large Data Bases VLDB 2002, pp. 406–417, August 2002. http://dl.acm.org/citation.cfm?id=1287369.1287405

46. Kleinberg, J.: Bursty and hierarchical structure in streams. In: Proceedings of the Eighth ACM SIGKDD International Conference on Knowledge Discovery and Data Mining KDD 2002, vol. 7, no. 4, p. 91 (2002). http://portal.acm.org/citation.cfm?doid=775047.775061

47. Kohonen, T.: The self-organizing map. Proc. IEEE **78**, 1464–1480 (1990)

48. Kontostathis, A., Galitsky, L.M., Pottenger, W.M., Roy, S., Phelps, D.J.: A survey of emerging trend detection in textual data mining. In: Berry, M.W. (ed.) Survey of Text Mining. Springer, New York (2004). http://link.springer.com/chapter/10.1007/978-1-4757-4305-0_9

49. Kumaran, G., Allan, J.: Text classification and named entities for new event detection. In: Proceedings of the 27th Annual International Conference on Research and Development in Information Retrieval, SIGIR 2004, pp. 297–304 (2004). http://portal.acm.org/citation.cfm?doid=1008992.1009044

50. Kwak, H., Lee, C., Park, H., Moon, S.: What is Twitter, a social network or a news media? Categories and subject descriptors. Most **112**(2), 591–600 (2010). http://portal.acm.org/citation.cfm?doid=1772690.1772751

51. Lardinois, F.: Readwritesocial: the short lifespan of a tweet: retweets only happen within the first hour (2010). http://readwrite.com/2010/09/29/the_short_lifespan_of_a_tweet_retweets_only_happen. Accessed 2 Apr 2013

52. Lee, R., Sumiya, K.: Measuring geographical regularities of crowd behaviors for Twitter-based geo-social event detection. In: Proceedings of the 2nd ACM SIGSPATIAL International Workshop on Location Based Social Networks, pp. 1–10 (2010). http://doi.acm.org/10.1145/1867699.1867701

53. Leek, T., Schwartz, R., Sista, S.: Probabilistic approaches to topic detection and tracking. In: Topic Detection and Tracking, pp. 67–83 (2002). http://portal.acm.org/citation.cfm?id=772260.772265

54. Lemire, D.: A better alternative to piecewise linear time series segmentation. In: SIAM Data Mining 2007 (2007)

55. Li, C., Sun, A., Datta, A.: Twevent: segment-based event detection from tweets. In: Proceedings of the 21st ACM International Conference on Information and Knowledge Management - CIKM 2012, New York, USA, p. 155. ACM, New York, October 2012. http://dl.acm.org/citation.cfm?id=2396761.2396785

56. Li, R., Lei, K.H., Khadiwala, R., Chang, K.C.C.: TEDAS: a Twitter-based event detection and analysis system. In: Kementsietsidis, A., Salles, M.A.V. (eds.) 2012 IEEE 28th International Conference on Data Engineering, pp. 1273–1276. IEEE, April 2012. http://dblp.uni-trier.de/db/conf/icde/icde2012.html#LiLKC12, http://ieeexplore.ieee.org/lpdocs/epic03/wrapper.htm?arnumber=6228186

57. Linguistic Data Consortium: TDT 2004: Annotation Manual - version 1.2, 4 August 2004. http://projects.ldc.upenn.edu/TDT5/Annotation/TDT2004V1.2.pdf

58. Long, R., Wang, H., Chen, Y., Jin, O., Yu, Y.: Towards effective event detection, tracking and summarization on microblog data. In: Wang, H., Li, S., Oyama, S., Hu, X., Qian, T. (eds.) WAIM 2011. LNCS, vol. 6897, pp. 652–663. Springer, Heidelberg (2011). http://dl.acm.org/citation.cfm?id=2035562.2035636

59. Makkonen, J., Ahonen-Myka, H., Salmenkivi, M.: Topic detection and tracking with spatio-temporal evidence. In: Sebastiani, F. (ed.) ECIR 2003. LNCS, vol. 2633, pp. 251–265. Springer, Heidelberg (2003). http://citeseerx.ist.psu.edu/viewdoc/summary?doi=10.1.1.1.8469

60. Malik, J.: Normalized cuts and image segmentation. IEEE Trans. Pattern Anal. Mach. Intell. **22**, 888–905 (2000). http://ieeexplore.ieee.org/lpdocs/epic03/wrapper.htm?arnumber=868688

61. Massoudi, K., Tsagkias, M., de Rijke, M., Weerkamp, W.: Incorporating query expansion and quality indicators in searching microblog posts, pp. 362–367, April 2011. http://dl.acm.org/citation.cfm?id=1996889.1996936

62. Mathioudakis, M., Koudas, N.: TwitterMonitor: trend detection over the twitter stream. In: Proceedings of the 2010 International Conference on Management of Data - SIGMOD 2010, p. 1155. ACM, New York (2010). http://portal.acm.org/citation.cfm?id=1807306, http://portal.acm.org/citation.cfm?doid=1807167.1807306

63. Metzler, D., Cai, C., Hovy, E.: Structured event retrieval over microblog archives. In: Proceedings of the 2012 Conference of the North American Chapter of the Association for Computational Linguistics: Human Language Technologies, pp. 646–655 (2012). http://www.aclweb.org/anthology/N12-1083

64. Mohd, M.: Named entity patterns across news domains. In: BCS IRSG Symposium: Future Directions in Information Access (FDIA), pp. 1–6 (2007). http://www.mendeley.com/research/named-entity-patterns-across-news-domains/

65. Muthukrishnan, S.: Data streams: algorithms and applications. Found. Trends Theoret. Comput. Sci. **1**(2), 117–236 (2005). http://www.nowpublishers.com/product.aspx?product=TCS&doi=0400000002

66. Osborne, M., Lavrenko, V., Petrovic, S., Osborne, M., Lavrenko, V.: Using paraphrases for improving first story detection in news and Twitter. In: Proceedings of the 2012 Conference of the North American Chapter of the Association for Computational Linguistics Human Language Technologies, pp. 338–346. The Association for Computational Linguistics (2012). http://www.aclweb.org/anthology/N12-1034

67. Osborne, M., Petrovic, S., McCreadie, R., Macdonald, C., Ounis, I.: Bieber no more: first story detection using Twitter and Wikipedia. In: Proceedings of TAIA 2012 (2012)

68. Ozdikis, O., Senkul, P., Oguztuzun, H.: Semantic expansion of hashtags for enhanced event detection in Twitter. In: The First International Workshop on Online Social Systems (WOSS 2012) (2012)

69. Page, L., Brin, S., Motwani, R., Winograd, T.: The PageRank citation ranking: bringing order to the web. World Wide Web Internet Web Inf. Syst. **54**, 1–17 (1998). http://ilpubs.stanford.edu:8090/422

70. Pan, C.C., Mitra, P.: Event detection with spatial latent dirichlet allocation. In: Proceedings of the 11th Annual International ACM/IEEE Joint Conference on Digital Libraries, vol. 20, pp. 349–358 (2011). http://portal.acm.org/citation.cfm?id=1315460

71. Paranjpe, D.: Learning document aboutness from implicit user feedback and document structure. In: Proceeding of the 18th ACM Conference on Information and Knowledge Management, p. 365 (2009). http://dl.acm.org/citation.cfm?id=1645953.1646002

72. Petrovic, S., Osborne, M., Lavrenko, V.: Streaming first story detection with application to twitter. In: Proceedings of NAACL (2010). http://citeseerx.ist.psu.edu/viewdoc/download?doi=10.1.1.170.9438&rep=rep1&type=pdf

73. Petrovic, S.: Real-time event detection in massive streams. Ph.D. thesis, University of Edinburgh (2012). http://homepages.inf.ed.ac.uk/s0894589/petrovic-thesis.pdf

74. Petrovic, S., Osborne, M., Lavrenko, V.: Streaming first story detection with application to twitter. In: HLT-NAACL, pp. 181–189. The Association for Computational Linguistics (2010)

75. Petrovic, S., Osborne, M., McCreadie, R., Macdonald, C., Ounis, I., Shrimpton, L.: Can Twitter replace newswire for breaking news? In: 7th International AAAI Conference on Web and Social Media (ICWSM) (2013)

76. Phuvipadawat, S., Murata, T.: Breaking news detection and tracking in Twitter. In: 2010 IEEE/WIC/ACM International Conference on Web Intelligence and Intelligent Agent Technology, WI-IAT 2010, vol. 3, pp. 120–123. IEEE, August 2010. http://ieeexplore.ieee.org/lpdocs/epic03/wrapper.htm?arnumber=5616930

77. Li, P., Burges, C.J.C., Wu, Q.: McRank: learning to rank using multiple classification and gradient boosting. In: Advances in Neural Information Processing Systems 20, Proceedings of the Twenty-First Annual Conference on Neural Information Processing Systems, Vancouver, British Columbia, Canada, 3–6 December 2007. http://www.researchgate.net/publication/221619438_McRank_Learning_to_Rank_Using_Multiple_Classification_and_Gradient_Boosting, http://citeseerx.ist.psu.edu/viewdoc/summary?doi=10.1.1.143.6630

78. Pohl, D., Bouchachia, A., Hellwagner, H.: Automatic identification of crisis-related sub-events using clustering. In: Proceedings - 2012 11th International Conference on Machine Learning and Applications, ICMLA 2012, vol. 2, pp. 333–338 (2012)

79. Popescu, A.M., Pennacchiotti, M.: Detecting controversial events from Twitter. In: Proceedings of the 19th ACM International Conference on Information and Knowledge Management - CIKM 2010, New York, USA, p. 1873 (2010). http://dl.acm.org/citation.cfm?id=1871751, http://portal.acm.org/citation.cfm?doid=1871437.1871751

80. Popescu, A.M., Pennacchiotti, M., Paranjpe, D.: Extracting events and event descriptions from Twitter. In: Proceedings of the 20th International Conference Companion on World Wide Web - WWW 2011, New York, USA, p. 105. ACM, New York, March 2011. http://dl.acm.org/citation.cfm?id=1963192.1963246

81. Raimond, Y., Abdallah, S.: The event ontology (2007). http://motools.sf.net/event

82. Rajaraman, A., Ullman, J.D.: Mining of Massive Datasets. Cambridge University Press, Cambridge (2012). http://www.amazon.de/Mining-Massive-Datasets-Anand-Rajaraman/dp/1107015359/ref=sr_1_1?ie=UTF8&qid=1350890245&sr=8-1

83. Rattenbury, T., Good, N., Naaman, M.: Towards automatic extraction of event and place semantics from flickr tags. In: Proceedings of the 30th Annual International ACM SIGIR Conference on Research and Development in Information Retrieval, SIGIR 2007, 103 pages (2007)

84. Ritter, A., Clark, S., Mausam, Etzioni, O.: Named entity recognition in tweets: an experimental study. In: Proceedings of the 2011 Conference on Empirical Methods in Natural Language Processing, pp. 1524–1534 (2011). http://www.aclweb.org/anthology/D11-1141

85. Ritter, A., Etzioni, O., Clark, S.: Open domain event extraction from Twitter. In: Proceedings of the 18th ACM SIGKDD International Conference on Knowledge Discovery and Data Mining - KDD 2012, p. 1104 (2012). http://dl.acm.org/citation.cfm?id=2339530.2339704

86. Robinson, B., Power, R., Cameron, M.: A sensitive Twitter earthquake detector. In: WWW 2013 Companion - Proceedings of the 22nd International Conference on World Wide Web, pp. 999–1002 (2013). http://www.scopus.com/inward/record.url?eid=2-s2.0-84893039051&partnerID=tZOtx3y1

87. Sahlgren, M.: Vector-based semantic analysis: representing word meaning based on random labels. In: ESSLI Workshop on Semantic Knowledge Acquistion and Categorization (2002). http://www.sics.se/~mange/papers/VBSA_Esslli.ps

88. Sakaki, T., Okazaki, M., Matsuo, Y.: Earthquake shakes Twitter users: real-time event detection by social sensors. In: Proceedings of the 19th International Conference on World Wide Web, pp. 851–860. ACM (2010). http://dl.acm.org/citation.cfm?id=1772690.1772777

89. Sankaranarayanan, J., Samet, H., Teitler, B.E., Lieberman, M.D., Sperling, J.: TwitterStand: news in tweets. In: Proceedings of the 17th ACM SIGSPATIAL International Conference on Advances in Geographic Information Systems - GIS 2009, New York, USA, vol. 156, p. 42 (2009). http://portal.acm.org/citation.cfm?id=1653781, http://portal.acm.org/citation.cfm?doid=1653771.1653781

90. Sayyadi, H., Hurst, M., Maykov, A., Livelabs, M.: Event detection and tracking in social streams. In: Proceedings of International Conference on Weblogs and Social Media (ICWSM), pp. 311–314 (2009). http://www.aaai.org/ocs/index.php/ICWSM/09/paper/viewFile/170/493

91. Schonfeld, E.: Techcrunch: mining the thought stream (2009). http://techcrunch.com/2009/02/15/mining-the-thought-stream. Accessed 9 July 2013

92. Snowsill, T., Nicart, F., Stefani, M., De Bie, T., Cristianini, N.: Finding surprising patterns in textual data streams. In: 2010 2nd International Workshop on Cognitive Information Processing, pp. 405–410, June 2010. http://ieeexplore.ieee.org/lpdocs/epic03/wrapper.htm?arnumber=5604085

93. Strassel, S.: Topic Detection & Traking (TDT-5) (2004). http://www.ldc.upenn.edu/Projects/TDT2004

94. Tanev, H., Ehrmann, M., Piskorski, J., Zavarella, V.: Enhancing event descriptions through Twitter mining. In: Sixth International AAAI Conference on Weblogs and Social Media, pp. 587–590 (2012). http://www.aaai.org/ocs/index.php/ICWSM/ICWSM12/paper/view/4631\n, http://www.aaai.org/ocs/index.php/ICWSM/ICWSM12/paper/view/4631/5065

95. Baldwin, T., Paul Cook, M., Baldwin, T., Cook, P., Lui, M., Mackinlay, A., Wang, L.: How noisy social media text, how diffrnt social media sources? In: Proceedings of IJCNLP 2013, pp. 356–364 (2013). http://citeseerx.ist.psu.edu/viewdoc/summary?doi=10.1.1.385.1683

96. Wang, X., Grimson, E.: Spatial latent Dirichlet allocation. In: Advances in Neural Information Processing Systems, vol. 20, pp. 1–8 (2007). http://people.csail.mit.edu/xgwang/papers/STLDA.pdf

97. Wang, X., Zhai, C., Hu, X., Sproat, R.: Mining correlated burstytopic patterns from coordinated text streams. In: Proceedings of the 13th ACM SIGKDD International Conference on Knowledge Discovery and Data Mining, pp. 784–793 (2007). http://dl.acm.org/citation.cfm?id=1281276\npapers2://publication/uuid/A6A05DF5-1873-4DC4-BAB3-F73712691FCA

98. Weng, J., Yao, Y., Leonardi, E., Lee, F.: Event detection in Twitter. Development **98**, 401–408 (2011). http://www.aaai.org/ocs/index.php/ICWSM/ICWSM11/paper/download/2767/3299

99. Wikipedia: Facebook — Wikipedia, the free encyclopedia (2013). http://en.wikipedia.org/w/index.php?title=Facebook&oldid=548760277. Accessed 7 Apr 2013

100. Wikipedia: Google+ — Wikipedia, the free encyclopedia (2013). http://en.wikipedia.org/w/index.php?title=Google%2B&oldid=548920007. Accessed 7 Apr 2013

101. Wikipedia: Linkedin — Wikipedia, the free encyclopedia (2013). http://en.wikipedia.org/w/index.php?title=LinkedIn&oldid=549175950. Accessed 7 Apr 2013

102. Wikipedia: Twitter — Wikipedia, the free encyclopedia (2013). http://en.wikipedia.org/w/index.php?title=Twitter&oldid=549164139. Accessed 7 Apr 2013

103. Yang, C.C., Shi, X., Wei, C.P.: Discovering event evolution graphs from news corpora. IEEE Trans. Syst. Man Cybern. Part A Syst. Hum. **39**, 850–863 (2009)

104. Yang, Y., Carbonell, J., Brown, R., Pierce, T., Archibald, B., Liu, X.: Learning approaches for detecting and tracking news events (1999)

105. Yang, Y., Pierce, T.T., Carbonell, J.G.: A study of retrospective and on-line event detection. In: SIGIR 1998: Proceedings of the 21st Annual International ACM SIGIR Conference on Research and Development in Information Retrieval, Melbourne, Australia, 24–28 August 1998, pp. 28–36. ACM, New York (1998). http://portal.acm.org/citation.cfm?doid=290941.290953

106. Yang, Y., Zhang, J., Carbonell, J., Jin, C.: Topic-conditioned novelty detection. In: Proceedings of the Eighth ACM SIGKDD International Conference on Knowledge Discovery and Data Mining - KDD 2002, pp. 688–693 (2002). http://dl.acm.org/citation.cfm?id=775047.775150

107. Yao, L., Mimno, D., McCallum, A.: Efficient methods for topic model inference on streaming document collections. In: Proceedings of the 15th ACM International Conference on Knowledge Discovery and Data Mining, 2009, vol. 4, p. 937 (2009). http://portal.acm.org/citation.cfm?doid=1557019.1557121

108. Zhao, Q., Chen, B., Mitra, P.: Temporal and information flow based event detection from social text streams. In: Proceedings of the 22nd National Conference on Artificial Intelligence - AAAI 2007, vol. 2, pp. 1501–1506. AAAI Press (2007). http://www.aaai.org/Papers/AAAI/2007/AAAI07-238.pdf

Detecting Events in Online Social Networks: Definitions, Trends and Challenges

Nikolaos Panagiotou, Ioannis Katakis$^{(\boxtimes)}$, and Dimitrios Gunopulos

Department of Informatics and Telecommunications,
National and Kapodistrian University of Athens,
Panepistimioupolis, Ilisia, 15784 Athens, Greece
{n.panagiotou,katak,dg}@di.uoa.gr

Abstract. Event detection is a research area that attracted attention during the last years due to the widespread availability of social media data. The problem of event detection has been examined in multiple social media sources like Twitter, Flickr, YouTube and Facebook. The task comprises many challenges including the processing of large volumes of data and high levels of noise. In this article, we present a wide range of event detection algorithms, architectures and evaluation methodologies. In addition, we extensively discuss on available datasets, potential applications and open research issues. The main objective is to provide a compact representation of the recent developments in the field and aid the reader in understanding the main challenges tackled so far as well as identifying interesting future research directions.

Keywords: Event detection · Social media · Stream processing

1 Introduction

The Web 2.0 era brought a lot of revolutionary changes in the way World Wide Web content is generated and utilized. Social media and online Social Networks are nowadays the most widely used services along with search engines. Data generated from Web 2.0 activity are of great *value* since they reflect aspects of real-world societies. Moreover, data are *easily accessible* since they can be collected through web-crawlers or public APIs. These two qualities constitute the main motivation for researchers studying online social networks.

The range of novel data analysis applications is impressive. A prominent technique, known as 'sentiment analysis', analyzes user opinions in order to extract the expressed emotion about products [14,35,75], services, or even political figures [80]. Marketing in particular found a perfect fit since now businesses are able to analyse a large volume of public data and identify trends [55], influential profiles [18], experts [33] or to provide personalized advertisements and documents [20]. From another perspective, social scientists study knowledge cascades [32], information propagation [32] or community dynamics [49]. In health care, researchers have been able to track and predict diseases like influenza [76]

© Springer International Publishing Switzerland 2016
S. Michaelis et al. (Eds.): Morik Festschrift, LNAI 9580, pp. 42–84, 2016.
DOI: 10.1007/978-3-319-41706-6_2

and identify disorders such as depression [28]. The list is incomplete and expands rapidly.

From all the above applications, the task of event detection stands out due to its complexity and social impact. Broadly speaking event detection is the problem of automatically identifying *significant incidents* by analysing social media data. Such events can be a concert, an earthquake or a strike.

Most approaches tackle event detection similarly to a clustering problem. Clustering can be performed on the textual features of users' messages (Topic Clustering) or on their spatio-temporal attributes (Spatio-Temporal clustering). Some of the identified clusters correspond to real events while others are just groups of similar messages. The identification of the *event clusters* is often tackled with scoring functions or machine learning classifiers [12]. Some approaches utilize novelty tests [66] while others focus on sentiment peaks [83] and keyword bursts [1]. A common element in many methods is a change detection component necessary to identify that 'something happened out of the ordinary'. Change is detected through statistical analysis of the messages' content or the network's structure (e.g. an increasing number of new connections in the social graph). There are many more lines of research in event detection. The most prominent ones are organized and discussed in the following sections.

The purpose of the article is to provide a categorization of existing approaches in order to let the reader easily grasp the motivation, basic steps and issues of each group of algorithms. In brief, the contribution of this article can be summarized into the following points:

- It provides definitions of numerous concepts related to event detection from Web 2.0 data. These definitions aim at formalizing the problem by disambiguating fuzzy concepts. Additionally, they allow a common terminology that will aid in presenting the state of the art under the same framework.
- The state of the art is identified, organized and discussed.
- Open issues and potential future research directions are presented.

The remainder of this article is organized as follows. In Sect. 2 recently introduced definitions of event detection are presented. In Sect. 3, an overview and a taxonomy of event detection approaches is given. Section 4 presents the algorithms in more detail emphasizing on intuition, main advantages and disadvantages. Section 4 outlines architectures utilized in relevant systems for efficient event detection. After that, in Sect. 5, we review a large number of Event Detection applications. Next, in Sect. 6 we summarize the evaluation procedures (protocols, datasets, metrics) that are utilized in evaluating the algorithms and we comment on the obtained results. Finally, the paper concludes with a discussion on related problems and open issues.

2   Research Challenges and Requirements

There are numerous research challenges inherent in event detection. In this section we discuss the ones that differentiate this task from other well known problems. Hence, we justify why off-the shelf data and text mining approaches are not suitable for tackling event detection.

Volume and Velocity. Data from social media come in great volume and velocity. Therefore, algorithms should be online and scalable in memory and computational resources. High data volume makes batch processing computationally infeasible. Data structures like Count-Sketch [19], randomized data structures such as Bloom [38] and Bloomier [50] filters, sampling methods [54] and streaming algorithms are often used in real-time streaming applications. The authors in [2] use the Count-Min Sketch [24] data structure to improve the efficiency of the Content Summary clustering algorithm they propose. Osborne et al. in [66] use a hashing function to calculate neighbours in constant time and [43] uses simple *inverse document frequency* (IDF) scores in order to avoid document-to-document comparisons and to reduce the number of computations. Most of the related work aim in building online systems capable of processing high rate streams such as the Twitter Sample stream (1%) or even the Firehose stream (100%) [56].

Real-Time Event Detection. Events should be identified as soon as possible, especially when the approach is intended to be used in critical applications like emergency response. In this case, methods for event detection should be evaluated not only in terms of Precision and Recall but also in terms of how fast they can identify a specific type of event. In [56], the authors offer a detailed description of the real-time elements of their approach and comment on advantages and disadvantages of making the process parallel.

Noise and Veracity. It is only natural that user generated information is characterized by noise. Social media are filled with spam messages, advertisements, bot accounts that publish large volumes of messages, hoaxes, as well as internet memes [45]. Another obstacle is that textual information in social media is very limited. Users usually publish very short messages a fact that makes off-the-shelf Text Mining and NLP methods unsuitable.

Feature Engineering. Selecting the most suitable features to utilize in supervised or unsupervised learning components is not a trivial task. Textual representations such as Term-Document matrices are not sufficient. As many researchers have observed, there are specific characteristics that appear in event related messages. These features could be content-based attributes such as TF-IDF scores, number of tags and emoticons or structural features like the number of followers (Twitter) or friends (Facebook). Supervised approaches mostly focus on content features in order to train classifiers such as Naive Bayes or Support Vector Machines. Many researchers have concluded that the utilization of the correct feature-set is very crucial for the event detection process. As an example, Becker et al. [11] presents a comparison between structural features and Term-Document matrices. They conclude that the combination of textual and non-textual features lead to a statistically significant gain in Precision.

Evaluation. Algorithm evaluation sets difficult to overcome obstacles. Unfortunately, availability of event detection datasets is very limited [57]. The TDT5[1] dataset is used by many researchers such as [43,66] in order to evaluate Precision. However, the nature of this data is quite different (topic detection and tracking) and hence it serves only as last resort. Results obtained from TDT5 could significantly vary from those obtained from Twitter or Facebook. TDT5 comes from news-wire articles and contains well formed high quality text. On the other hand, social media text has unique textual characteristics including abbreviations, use of slang language and misspellings. A public dataset gathered from social media sources is very important since it could be used to train supervised classifiers and also evaluate the algorithms in terms of Precision or Recall. Since such dataset is not available most research teams create their own corpora that are manually annotated [12,57,66] with a small number of events. This fact makes the results subjective to sample bias and also hinders comparative experiments.

3   Definitions and Context

The lack of a formal definition for the problem of event detection initiates a lot of issues since the problem is multi-dimensional and many aspects are not obvious. Up until now there were some individual efforts towards defining specific sub-problems. We begin this section by presenting such definitions of tasks that relate to event detection or similar problems. Next, we propose definitions that extend and unify the ones that appear in the literature.

According to the Topic Detection and Tracking (TDT)[2] project [3], an event is "something that happens at specific time and place with consequences". The consequences may motivate people to act in social media and hence the events will be reflected in network activity (e.g. large number of tweets on Twitter, new groups on Facebook and new videos on YouTube). Aggarwal et al. [2] provides a definition of *News Event* as "something that happens at specific time and place but is also of interest to the news media". That is, apart from making an impact to the Web 2.0 world, an event should also affect conventional news media. In [12] the authors state that "an event is a real-world occurrence e with a time period T_e and a stream of Twitter messages discussing the event during the period T_e". Their definition has a Twitter scope and is related to an increased amount of messages in a time window. However, it could be applied to other platforms that operate as a stream of documents.

McMin et al. [57] defines *event* as "something *significant* that happens at specific time and place". The authors state that something is significant when it is discussed by the news media. This is quite similar to the definition of [2]. Weng et al. in [89] state that an event is "a set of posts sharing the same topic and words within a short time". Abedelhaq et al. [1] state that events stimulate people to post messages but in a substantial geographic space. This is connected

---

[1] http://www.itl.nist.gov/iad/mig/tests/tdt/resources.html.

[2] http://www.itl.nist.gov/iad/mig/tests/tdt/tasks/fsd.html.

to *localized event detection* (local events versus global events). In [1] the authors define *localized events* as "events with a small spatial extent". Boettcher et al. in [15] state that an event is "an occurrence or happening restricted on time". This definition differentiates *Real World* events from *Virtual* events. Virtual events are restricted within the limits of the online world. Examples of such cases are memes, trends or popular discussions. Discussions are considered events since people are active in social media because of them. Nevertheless, they do not correspond to a real world incident. Wang et al. [87] define *Social Events* as events among people when the one is an acquaintance of the other.

A slightly different definition that also includes the concept of *social* event is given in [26]. The authors identify four event categories: *local non-social*, *local social*, *global social* and *global*. A social event is an event that involves participants that have been together again in another situation. An example of such a *social* event is a conference where the participants have attended the same conference in the past. Finally, Popescu et al. [69] describe the "event snapshots" idea as a tuple $s = (e, \Delta_t, tweets)$. The tuple consists of a set of *tweets* that are correlated with an entity e for a time period Δ_t. This definition could be mostly considered for celebrity-related events like popular actors or singers. An alternative definition that considers sentiment information, is given in [83]. The authors define the task of event detection as: "The identification of those messages that alter significantly and abruptly the emotional state of a large group of people."

It is clear that the aforementioned definitions are not always consistent with each other. Some of them require the events to happen in a specific geographical region while others do not take into account the space dimension. Other articles state that the time the events take place should be finite and short in duration. Such a definition is not applicable for Global Events that may concern communities for weeks.

Event Types. In the literature we come across the following types of events:

- *Planned:* Events with a predefined time and location (e.g. a concert).
- *Unplanned:* Events that are not planned and could happen suddenly (e.g. a strike, an earthquake).
- *Breaking News:* Events connected to breaking news that are discussed in conventional news media (e.g. the result of the elections in Greece discussed by the global press).
- *Local:* Events limited to a specific geographical location. The event impacts only this area (e.g. a minor car accident).
- *Entity Related:* Events about an entity (i.e. a new video clip of a popular singer).

Table 1 summarizes the range of the different event types in terms of space and time. It also reports in which media these events are more probable to be observed in.

The rest of this section presents definitions that unify and extend the ones proposed in the literature. They are based on the observation that *events* can be identified by analysing *actions* of *accounts* in the online social network (OSN).

Table 1. Different type of events and their properties

Event type	Time duration restrictions	Geographical distribution	Observable in
Planned	High	Medium	Social media, news media, event portals
Unplanned	Low	High	News media
Breaking news	High	Low	News media
Global	Low	Low	News media, online sources
Local	High	High	Local media, online sources
Entity	High	Low	News media, blogs

Definition 1. Account *(p): An agent that can participate (i.e. perform actions) in a social network after following a registration procedure.*

Accounts can be operated by individuals, groups of people or computational agents (bots). Accounts usually maintain a *profile* in the OSN.

Definition 2. Content object *(c): A textual or binary object that is published or shared via the social network (e.g. text, image, video).*

Definition 3. Action *(a): Depending on the social network, an action, a, can be either: (i) a post of new content (e.g. a new tweet), (ii) an interaction with another profile (e.g. a new follower, a friend request, etc.), (iii) an interaction with another user's content (e.g. a retweet, or a "like").*

It is obvious that some of these actions can be *observable* or *un-observable* by agents that are not connected with the action-generating accounts. In the task of event detection, we are interested in a *set of N actions $A^e = \{a_i, \ldots, a_N\}$* that are correlated with (or caused by) the event e. Such a set of actions has also a temporal definition $T_{A^e} = [t_{(A^e, start)}, t_{(A^e, end)}]$. The actions that an event produces are most of the times in a different time window compared with the actual event, i.e. $t_{(e, start)} < t_{(A^e, start)}$ and $t_{(e, end)} \neq t_{(A^e, end)}$.

A^e is the ideal, ground truth set that contains *all* actions that event e has caused. This set of actions comprises the effect of the event and it is the only source of information that a computational agent can analyse in order to "sense" the event. By analysing A^e, location and actors can be inferred.

Definition 4. Event (e): *In the context of online social networks, (significant) event e is something that causes (a large number of) actions in the OSN.*

Intuitively, the importance of an event can be measured by the mass of the actions that it generates. It is implied that, in event detection, we are interested in significant events. Naturally, significant events can have global or local character.

The textual *representation* or *summary* $R(e)$ of an event could be a title heading or a set of keywords. An event is linked with a specific time frame $T_E = [t_{(e,start)}, t_{(e,end)}]$ (duration of the event). An event is sometimes correlated with a set of *involved actors* I^e and a location Loc^e.

Definition 5. Event detection in an online Social Network: *Given a stream of actions A_n of the OSN n, identify a set of real-world events and provide some of the following information:*

(a) the (textual) representation of the event $R(e)$,
(b) a set of actions that relate to this event $A^e \subset A_n$,
(c) a temporal definition of the set of actions

$$T_A^e = [t_{(A^e,start)}, t_{(A^e,end)}]$$

(d) a location loc_e that is correlated with the event,
(e) the involved actors I^e.

Currently, approaches presented in the literature only provide some of the above information. This is totally acceptable in some applications.

In other words, the problem of event detection could be defined as: "Given a stream of actions A_n in an online social network n identify all tuples $E = \{e_1, \ldots, e_M\}$", where M is the number of events and

$$e_i = \prec R(e_i), A^{e_i}, T_A^{e_i}, loc_{e_i}, I^{e_i} \succ$$

4   Organization of Methods

We present an organization of Detection approaches under two perspectives. We firstly organize methods according to the technique they utilize (clustering, first story, etc.) (Sect. 5). Then we organize approaches according to whether they are looking for New or Past events or whether they are operating off-line or online (Sect. 6). Details of each algorithm are presented in the following section. An earlier overview of Twitter specific event detection approaches can be found in [9]. Although most work on event detection is using Twitter data, we describe techniques on other sources as well (Youtube, Flickr, etc.). Furthermore, we provide a hierarchical organization of the methods and emphasize on architectural issues, evaluation procedures, dataset availability and dataset labeling.

4.1   Taxonomy

In this section we present a taxonomy of the related work based on the fundamental data mining techniques that they utilize (clustering, outlier detection, classification, etc.). An illustration of the taxonomy can be seen in Fig. 1. Details, as well as more references will be presented in the next section.

Most event detection algorithms tackle the problem, at least in a first stage, as a Stream *clustering* task. The identified clusters are organized into "event-clusters" or "non-event-clusters". This assignment can be either supervised or unsupervised. In the unsupervised case, a scoring function is used that is usually based on features extracted from the clusters. In the supervised case, a classifier is trained either using textual features, structural features, or both. The advantage of the supervised approach is that the classifiers automatically "learn" the task based on historical cases. On the other hand, a training set should be available and the classifiers must be retrained periodically.

A different approach targets at the detection of *anomalies* in the content of the network. The idea is to first model the content in normal circumstances and then detect outlying messages. The first step is to build language models capturing term usage from historical data. When, for example, a group of terms demonstrates increased usage, then this is considered an indication of an event. Typically, sentiment information is utilized along with the assumption that significant deviations in sentiment indicate events.

An alternative strategy is to use novelty scores on incoming messages. Novelty scores are mostly used in the *First Story Detection* (FSD) problem. FSD is usually applied on news streams and aims at detecting the first story about an event by examining a set of 'neighbour' (i.e. similar) documents. That is, if a message is significantly different from its nearest neighbours, it is considered novel and indicative of a new event. Auxiliary sources of information like Wikipedia are exploited in order to identify evidence for the detected events.

Some methods, focus on events concerning *specific* topics such as a music band. After the messages that talk about these topics are identified, the algorithms detect anomalies. For example, in [94], authors find events about the NFL 2010–2011 games. [7] detects increases in flu-related messages while the TEDAS system [52] focuses on crime-related and disastrous events. An issue with these approaches is that the topic should be known a priori and other event types will not be identified. We refer to this category as *topic specific* event detection.

According to Topic Detection and Tracking task (TDT[3]) the two main approaches of event detection are *Document Pivot* and *Feature Pivot*. In document pivot techniques, clustering is used to organize documents according to their textual similarity and neighbours are identified through direct comparison. These approaches were mainly used in TDT challenges. However, they are not directly applicable to social media like Twitter or Facebook. The first issue is that not all documents are related to events (e.g. memes) as it is assumed in the TDT challenge. A second problem is that Document Pivot techniques require batch processing and are not scalable to large amounts of data.

Feature Pivot techniques focus on event topics that were previously unseen or growing rapidly. Many Feature Pivot techniques focus on burst detection. Bursts could be defined as term or sentiment deviations. Kleinberg et al. [44] define a finite state automaton to detect bursts in documents streams, while [31] model words as a binomial distribution in order to detect bursts. Similarly in the Twitter Monitor system [55] a streaming algorithm called Queue Burst is used in order to detect bursts on the Twitter stream. In general, Feature Pivot techniques focus on change and burst detection of text features. Most algorithms

---

[3] http://www.itl.nist.gov/iad/mig//tests/tdt/1998/.

from TDT that are applied in social media are mainly Feature Pivot algorithms. We will discuss Pivot algorithms in more detail in the following sections.

4.2   NED vs. RED and Online vs. Offline

Even from the TDT era (see previous section), two important categories of event detection were identified. These were the retrospective event detection (RED) and the new event detection (NED). RED focuses mostly on identifying previously unknown events from historical collections [92]. NED targets at events from live streams in real time [5]. RED mines historical data in order to detect events that were not previously known. There is no time constraint since the events already happened in the past and their identification could not support decision making. On the other hand, NED has online nature involving real time event detection with the goal to support crucial decision making (e.g. in emergency situations).

Most articles related to event detection in social media focus on online event detection. Online event detection aims at deciding if a message is about an event as soon as it arrives without the need of time consuming batch processing. The offline event detection algorithms require complex procedures that can not be used in real-time processing. They are useful mostly for retrospective event analysis where execution time is not a major requirement. Some hybrid approaches have an online part that is used for real-time analysis of messages and an offline part that post-processes data. Such an offline part may be the training of a classifier. Typical examples of online approaches are those that cluster messages and then use scoring functions to decide which clusters are event clusters. Approaches like [12] require offline components since the classifier used for the cluster categorization requires training.

5   Event Detection Methods

Following the taxonomy presented in the previous section, we present here representative methods from each category in more detail. Initially the clustering based approaches are presented including supervised and unsupervised scoring techniques. Then, approaches based on anomalies, such as keyword bursts, are discussed. After that, First Story Detection approaches inspired by TDT are presented. The Section concludes with methods that focus on detecting specific events.

5.1   Clustering Based Event Detection

Event detection approaches from social media streams is often faced using clustering of messages. After that, the clusters are classified as "Event-Related" or "Non-Event-Related" (see Fig. 1). This assignment could be resolved with supervised or unsupervised learning. In the supervised case, clusters are classified using *a learning algorithm* such as Naive Bayes or Support Vector Machine based on a

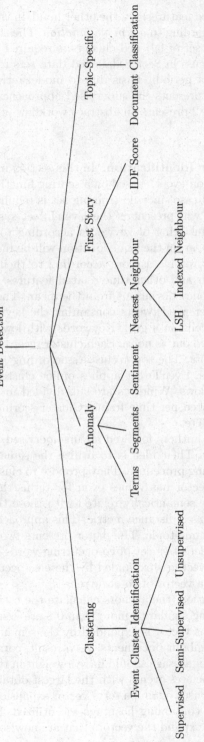

Fig. 1. An abstract taxonomy of event detection approaches.

certain group of extracted features. On the other hand, in the unsupervised case, clusters are classified according to *a scoring function*. The main difference is that in the supervised case a set of labelled clusters is required in order to train the classifier. As we will discuss in Sect. 8, labelled data sets require large amounts of annotation effort on a periodic basis due to model retraining requirements. The rest of this Section presents the supervised approaches and continues with the unsupervised. Figure 2 presents an abstract workflow of the clustering based event detection.

Unsupervised Cluster Identification. In this section methods that identify clusters as "event" or "non-event" based on a scoring function will be presented. This is the unsupervised case since no training set is required.

The first system we will present is the EvenTweet system [1]. EvenTweet is based on an initial clustering of keywords according to their spatial signature. Keywords that appear on the same location will be included into the same cluster. These keywords receive a score according to their level of burstiness, their spatial distribution and other time-related features. Burstiness is calculated according to frequency deviations from the mean. The spatial signature is calculated using geo-referenced tweets containing the keywords and it is fixed on a set of pre-defined cells on a grid. Keywords with low burstiness and high spatial entropy are filtered out as noise. Each cluster receives a score equal to the sum of its keywords' score. The top-k clusters according to their score are the candidate event Clusters. EvenTweet applies online clustering by dividing the stream into sliding windows. Windows are sub-divided into time frames. Keywords' scores are calculated per time frame. Cluster scoring is updated when a new time-frame is complete.

Similarly, in [60] the authors followed an unsupervised approach for detecting events from Twitter. Their idea is to utilize the semantic relationships of terms during the clustering procedure. They propose to cluster expanded TF-ID vectors. An expanded vector has values even for terms that are missing from the document if they are semantically related with those that are present. The cosine similarity is used as a distance metric. This approach leads to clusters of tweets that discuss the same topic. The paper presents two semantic expansion methods. The first one detects a set of co-occurring words from a static corpus and then the document vector is expanded by these co-occurrences. The second one treats each word as a vector of co-occurrences.

Using this representation, the authors calculate the cosine similarity among all vectors. Vectors having similarity more than 0.8 are assumed to be semantically related. A document vector is expanded by the semantically related terms being present at the neighbour documents. As a result, correlated words even if they do not appear in the message should have a weight in the expanded vectors. Finally, the method associates events with the largest obtained clusters.

A similar method, focused again on term vector expansion, is followed in [61]. In this case only tweets containing hashtags are utilized. Hashtag correlations are exploited in order to expand the vectors that are now solely contain hashtag information. The results are improved in comparison to [60] especially in how fast the events are identified. Such a result is not expected since document (word) information is discarded. Nevertheless, on specific event types it turns out that hashtag features alone are sufficient.

On [47] the authors describe an event detection method that is based on topic clustering of the tweets. The clustering involves both textual and social features such as the unique users that posted messages about the event. They give a definition of *user diversity* within a cluster as the entropy of its users. The more users a cluster has the higher the diversity. Then they formally define the event detection method as an optimization problem where the goal is to both maximize the documents similarity as well as the user diversity and prove that it is NP-hard. As a result, they use an approximate time efficient and one-pass online clustering algorithm in order to cluster tweets topically. Then the clusters are periodically checked for their user diversity and those with a diversity more than a threshold are identified as event clusters.

The TwEvent system [51] implements the idea of using tweet segments (N-grams) instead of unigrams. Segment extraction is based on Wikipedia corpus and the Microsoft N-gram service[4]. Segments are selected according to their appearances on historical data using a "cohesiveness" metric formally defined on the article. Thus, only coherent segments are considered while the rest are filtered out. The segment extraction algorithm has linear complexity. In the next step, they approximate the frequencies of the segments and detect the bursty ones. Candidate event segments are identified based on burstiness and number of unique authors. Then, candidate segments are clustered using a modification of the Jarvis Patric algorithm [41]. According to this algorithm, two segments result in the same cluster if one is the nearest neighbour of the other. The similarity between two segments s_a and s_b is based on a time-indexed sliding window W consisting of m parts. The similarity $sim_t(s_a, s_b)$ is defined in the following Equation (Eq. 1)

$$sim_t(s_a, s_b) = \sum_{m=1}^{M} w_t(s_a, m) w_t(s_b, m) Sim(T_t(s_a, m), T_t(s_b, m)) \tag{1}$$

$Sim(T_t(s_a, m), T_t(s_b, m))$ is the similarity of the tweets concatenation containing segments s_a and s_b during the sub-window m. $T_t(s_a, m)$ is the concatenation of tweets containing the segments s_a during the sub-window m. $Sim(T_t(s_a, m), T_t(s_b, m))$ is the similarity of the concatenated documents $T_t(s_a, m)$ and $T_t(s_b, m)$ is extracted from the associated segments s_a and s_b respectively. This similarity is based on the cosine similarity of the TF-IDF vectors. Weight $w_t(s_a, m)$ equals to the ratio of tweets containing s_a to tweets that do not. A problem with this approach is that the computation complexity of the Jarvis Patric algorithm is $O(n^2)$. However, the authors state that the algorithm should perform well for a small number of tweets. A score to each cluster is assigned according to the number of segment appearances in Wikipedia. The top clusters are classified as event related clusters.

---

[4] http://research.microsoft.com/en-us/collaboration/focus/cs/web-ngram.aspx.

[30], similarly to [52], considers Twitter as a social sensor where the users provide valuable information for the authorities. The authors use their system to detect flood events analyzing tweets from Germany for a period of eight months. They follow a visual analytics approach in order to present flood-related Tweet messages on the map. Two interesting approaches are presented. The first one is based on increased local tweet activity. The second assumes that similar messages appearing in nearby locations can possibly refer to disastrous events taking place in that area.

For their first approach (increased spatial twitter activity), the authors divided Germany into a number of cells using Voronoi Polygons. All these cells are associated with a normal activity' characterized by a mean and a standard deviation. When increased activity is detected an event alarm is triggered.

The second approach, at first, filters out irrelevant messages (not related to floods) using a dictionary approach. The OPTICS [6] density based clustering algorithm is utilized in order to identify similar messages. Validation of clusters is achieved through the use of external sources. If the tweets contained in a cluster correspond to a news story then the cluster is identified as an event cluster. The authors conclude that the second approach is more effective than the first one.

The authors of [2] utilize content and social features of the Twitter network in order to detect events. Following a similar path to many of the aforementioned approaches, clustering comprises the first step of the method. Clustering is topic based since textual features are analyzed. However, an important difference with other approaches is that the clustering takes into account user profiles. The algorithm is named Cluster Summary (CS). Cluster centroids consist of two parts: (i) the *content* summary which is a term-frequency matrix and (ii) the *user* summary which is a user-frequency matrix. The distance metric of the clustering algorithm is a linear combination of the two summaries:

$$Sim(S_i, C_i) = \lambda * SimS(S_i, C_i) + (1 - \lambda) * SimC(S_i, C_i) \qquad (2)$$

Content similarity $SimC$ is based on TF-IDF [21] and utilizes the cosine distance. The structure similarity $SimS$ depends on how many users the tweet and the cluster have in common. The authors associate each tweet with all the followers of the author of the tweet.

In the same article, the authors state that the user summary of a cluster can be represented using a randomized counting data structure called Count-Min Sketch [23]. The Count-Min Sketch could be used to approximate the user frequency using constant amounts of memory. Count-Min Sketch is a data structure that overestimates the counters of an element . A possible drawback of the Count-Min Sketch for such applications is that its error rate increases with time [23].

Once a cluster is formed then its size is periodically checked according to its recent history. If the cluster growth over two consecutive sliding windows is more than a predefined threshold the cluster is identified as an event.

Supervised Cluster Identification. In this group of approaches, decision about event clusters is made through supervised machine learning classifiers. The classifiers take advantage of textual features as well as other attributes that are usually domain dependent.

In [86] the authors initially clustered the tweets according to their spatio-temporal information using a set of predefined rules. The clustering algorithm presented is simple, online and fast. The clusters formed are considered event-candidate clusters. In the next step, textual and non-textual features are extracted from the clusters in order to train a classifier. The top extracted features according to individual feature evaluation are the following:

- Unique authors
- Word overlap
- Number of mentions
- Unique coordinates
- Number of fourthsquare[5] posts

For example, the more unique coordinates or fourthsquare posts within a cluster, the more likely this is an event cluster. Three classifiers are trained using a manually annotated dataset: a Decision Tree, a Naive Bayes classifier and a Multilayer Perceptron. The classifiers are compared using only textual features against using both textual and non-textual features. The result showed a statistically significant improvement when all features are used. This is an indication that in some cases the text itself is not enough.

Another algorithm based on message clustering followed by supervised classification is described in [12]. The authors, as in similar approaches, set the requirement of not knowing apriori the number of clusters and they use an online threshold-based clustering method. The documents are presented as vectors that are TF-IDF weighted using a bag-of-words approach. The clustering algorithm is simple as in [86]. When a new point has a distance less than a threshold from the nearest centroid it is added to that cluster, otherwise a new cluster is created. Then, features are extracted from the clusters in order to train the classifier. The features are topical, temporal, social and Twitter specific. Temporal features may describe deviations on the volume of common terms as well as changes on their usage frequency. Social features capture interactions among users. Topic based features capture the thematic coherence of the cluster. Twitter specific features are often present in non-event clusters, for example Twitter tagged conversations that do not correspond to a real-world event (e.g. the hashtag #ff "Follow Friday"). When the feature extraction is complete, a Support Vector Machine is trained and compared against a Naive Bayes classifier using only textual features. The conclusion is that the manually extracted features provide a very important advantage over the baseline method using only text features. The clustering step is fast and online. However, training the Support Vector Machine is computationally demanding involving parameter tuning and is prone to over-fitting especially when the training set is relatively small.

---

[5] https://foursquare.com/.

The event detection method presented above, originally described in [12], is revisited in [70] in order to boost the clustering procedure. The authors state that the original online clustering algorithm takes into account just textual information in the form of TF-IDF vectors. They then propose the usage of two new features from the similarity function used in the clustering. The first feature originates from the parsing of URLs within the document, favoring documents with the same or similar URL. The intuition behind that is the fact that documents that contain the same URL should result into the same cluster since the shared URL indicate that the documents correspond to the same event. The second feature is called "Bursty Vocabulary". This is a set of keywords per cluster that exhibit bursty behaviour identified trough a computationally inexpensive outlier test on consecutive sliding windows. The frequency of the identified bursty keywords during the next sliding window is estimated and used for the assignment of new documents to the cluster. The two features presented so far are textual. However, the authors suggested that temporal features should be used also during the document allocation to clusters. They propose the usage of a Gaussian attenuator, highly similar to the one presented in [73], that takes into account the time of the latest cluster document and the time of the document to be assigned. This temporal feature is embedded to the clustering similarity function and penalizes inactive clusters. The above improvements not only supply more textual information to the clustering algorithm but also exploit temporal information resulting into a textual-temporal method.

Another interesting system is described in [73]. TwitterStand targets at detecting tweets that relate to *Breaking News*. In contrast with other approaches, it does not utilize information extracted only from the Twitter API. Their data originate from tweets of the top-2000 users with the most tweets, the 10 % of the public tweets, the Twitter Search API[6] and an API named BirdDog that receives tweets from a large number of Twitter users.

A Naive Bayes classifier is built in order to classify tweets as "news-tweets" or "junk-tweets". Their approach of classifying tweets obtained from keyword based searches is similar to [52]. The classifier is trained on a static corpus consisting of tweets labelled as "junk" or "news". A smaller dynamic corpus is exploited to periodically update the classifier. This corpus consists of tweets related to news reported by conventional media. "News" tweets are clustered into topics. The clustering algorithm used is called leader-follower [29] and allows content and temporal clustering. Regarding content similarity required by the clustering algorithm, the TF-IDF weighted vectors of the tweets are utilized. The similarity metric is a variant of the cosine similarity containing a temporal factor. Content similarity between document d and cluster c is defined as:

$$\delta(d, c) = \frac{TFV_t \cdot TFV_c}{||TFV_t|| \cdot ||TFV_c||} \tag{3}$$

where TFV_t and TFV_c are the term vector of the tweet (TF-IDF weighted) and the cluster centroid respectively. In order to capture time the similarity metric

---

[6] https://dev.twitter.com/docs/api/1.1/get/search/tweets.

is expanded with a Gaussian attenuator (Eq. 4). T_t and T_C are the tweet time and the time of the latest tweet respectively. When the clustering is complete the system presents the clusters on a map by estimating the location of tweets using a text-based geo-tagging technique.

$$\hat{\delta}(d,c) = \delta(d,c) * e^{\frac{-(T_t - T_c)}{2\sigma^2}} \qquad (4)$$

EventRadar [15] follows a similar idea in order to detect localized events. A term vector is created for each tweet at pre-processing. Then, unigrams, bigrams and trigrams are extracted and the algorithm examines if in a recent history H there are tweets that contain these n-grams. If these tweets are close in space and time, they are considered as event candidates. DBSCAN is utilized for clustering the tweets. A Logistic regression classifier is trained in order to reveal which of the clusters are real events. The features used for the classification task include two Boolean variables related to the tweets locations and keywords. In addition, they include the number of tweets containing relevant keywords for a period of seven days as a feature for the classification. The final result is a list of events described by a set of keywords as well from a set of representative tweets.

The work in [69] aims at recognizing controversial events. These are events where users express opposing opinions. The authors use the idea of an *entity snapshot* as the sum of the *tweets* related to an entity e published during a period Δ_t. It is defined as a triple $s = (e, \Delta_t, tweets)$. Some of these snapshots are about real-events related to an entity e and are called *event snapshots*. Snapshots are created from entities gathered from Wikipedia, tweets referring directly (using the @ symbol) or indirectly to these entities published during Δ_t. The next step of the method is to train a Gradient Boosted Decision Tree to classify snapshots as event or non-event snapshots. Features from Twitter as well as from external sources, including sentiment information, are utilized. For each of the detected events a regression model outputs a controversy score. These models mainly use textual features extracted from annotated samples as well as from a sentiment lexicon and a controversy lexicon derived from Opinion Finder[7] and Wikipedia pages.

Semi-supervised Cluster Identification. In [39] a semi-supervised approach is utilized since the labelling of Tweet clusters, involving thousands of tweets, is a quite time consuming task. The system tracks events in news media, extract keywords, and labels tweets that contain similar keywords as 'events tweets'. These tweets are able to propagate their label to related tweets using the social structure of Twitter. Social ties such as message re-tweets, mentions and hashtags are used to propagate labels. Using a reputable seed the authors are able to obtain a training set and propagate the observed labels. The next step is to construct the wavelet signal for every term that appears in the tweets. For these signals auto-correlation is calculated and common words or words appearing every day are filtered out (e.g. high auto-correlation). For example the hashtag

---

[7] http://mpqa.cs.pitt.edu/opinionfinder/.

"#ff" (follow Friday) appearing every Friday would be filtered out. For the resulting set of words a cross-correlation matrix is calculated. This matrix is presented as a graph in order to apply graph partitioning [89]. This way word groups are created and tweets are clustered to these word groups according to their content. For the classification of the clusters as event or non-event, a Support Vector Machine is trained using TF-IDF weighted document term features. Named entities are removed from the terms in order to avoid over-fitting issues. The event-clusters identified are then spatially grouped according to the tweets' geo-locations. For tweets that do not contain geo-location, the location is propagated from related tweets using social ties. In the end, the system provides a visualization of the event clusters on a map. We should note that such an approach could be highly valuable in cases where a reputable seed is available allowing label propagation without the need of manual annotation.

Clustering Approaches Summary. Table 2 presents an overview of the clustering-based approaches used for event detection. The second column notes the clustering algorithm that is utilized in the approach. The third and forth column present the features and the similarity metric that are exploited. Lastly, "scoring" indicates how the event cluster identification is achieved (supervised or unsupervised). Table 3 presents an overview of the supervised approaches, along with features and classifiers used.

Table 2. A summary of the clustering approaches used for event detection.

References	Clustering type	Features	Similarity metric	Scoring
[1]	Keyword clustering	Spatial	Cosine similarity	Unsupervised
[52]	Topic	Segments	Content similarity	Unsupervised
[60,61]	Tweet clustering	Expanded TF-IDF vectors	Cosine similarity	Unsupervised
[86]	Spatio-temporal	Spatio-temporal	Rule based distance	Unsupervised
[12]	Topic	TF-IDF vectors	Cosine similarity	Supervised
[30]	Spatio-temporal density	Spatio-temporal	Distance threshold	Unsupervised
[73]	Content-temporal	TF-IDF vectors, Temporal	Modified cosine similarity	Unsupervised
[39]	Topic clustering	Term vectors	Overlapping tweet terms	Supervised
[39]	Term clusters, spatial clusters	TF-IDF vectors	Cross-correlation	Semi-Supervised

5.2   Anomaly Based Event Detection

The methods of this section follow the path of identifying *abnormal* observations. Examples include: unexpected word usage in the last time window, irregular

Table 3. A summary of the supervised approaches used for event detection.

References	Features	Algorithms
[12]	Temporal, topical, social	Naive Bayes, SVM
[86]	Textual, spatial, temporal	Decision Tree, Neural Net
[73]	Textual	Naive Bayes
[15]	Textual	Logistic Regression
[69]	Social (internal and external), textual	Gradient Boosted Decision Tree

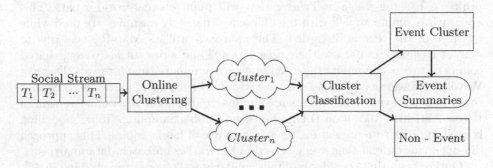

Fig. 2. An example of the general event detection approach using stream clustering. The online clustering component groups tweets that are close in space. The Cluster Classification module uses a supervised or an unsupervised method in order to classify the clusters as event cluster. For each of the event clusters a summary is extracted using different summarization methods.

spatial activity, or a distribution of emotion that is different from the average. The approaches discussed in this section track the social stream and raise an alert for an event candidate when an anomaly is observed.

The methods presented in [82,83] focus on identifying events from social media using a sentiment analysis. The main idea is that users will respond to an event in order to express their opinion causing this way fluctuations in the sentiment levels. According to the authors, when an event happens it affects the emotional state of a group of people that are close to the event. In the proposed system, users are initially clustered according to their geographic locations and their messages are aggregated over sliding windows. For each sliding window, sentiment sensors are responsible for specific regions (e.g. a sensor per city or district). The emotions of each region are analysed over four sentiment classes. When a significant deviation in sentiment levels is detected, an event alarm is triggered. The system is compared against the EdCow [89] system that uses keyword count deviations instead of sentiment information. Authors of [83] report that TwitInsight outperforms EdCow on its capability to detect events. In addition, TwitInsight is much faster since it does not depend on computationally expensive procedures and is able to run in real-time.

A strong point of the above work is the sentiment level outlier detection method. The authors of [83] assume that the sentiment level distribution is *unknown amd changes* overtime. The idea is to dynamically estimate the *Probability Density Function* (PDF) in a streaming fashion and use it for detecting outlying sentiment levels. The tool used for this methodology originates from [78] where a streaming estimation technique for unknown Probability Density Functions (PDF) is utilized based on kernels and dynamic sampling. The resulting PDF is used in order to perform non-parametric density outlier detection.

Authors in [22] use the Discrete Wavelet Transformation in order to detect peaks on hashtag usage in Twitter that will point on real-world events. This approach is similar to [89] with the difference that only hashtags are used while the rest of the text is discarded. The approach utilizes Map-Reduce jobs to extract hashtags and create their time series. Time series consist of aggregated counts of tweets containing the hashtag over five minute time intervals. Discrete Wavelet Transformation is used in order to detect *bursts* of hashtags since that could indicate events. The events are summarized using a fast online version of Latent Dirichlet Allocation (LDA) based on Gibb's Sampling. Topic modelling is used in order to represent events as a mixture of latent topics. This approach did not focus on real-time event detection but rather on batch data analysis.

Outlier tests that consider hashtags similarly to [22] are presented in [25, 46]. Hashtags are commonly used to indicate *topics* but some times correspond to real *breaking news events*. Moreover, in some cases, they represent "memes" or "virtual events". The authors in [25] extracted content features from hashtag including "frequency instability", "meme characteristics" and "authors entropy". They classified hashtags as "Advertisements", "Miscalculation", "Breaking News" and "Memes". Their method is able to discriminate breaking news from meme-hashtags regardless of language. Similarly in [46], hashtags are considered to be associated with event or with memes. The authors extracted hashtag features like the number of words used with a hashtags, the number of replies a tweet with a hashtag is getting, number of URLs, and more. Using these features and a training-set they utilize a set of supervised classifiers including Random Forests and Support Vector Machines. According to their report, discrimination between event-hashtags and meme-hashtags is successful with 89.2 % accuracy.

Watanabe et al. [88] built the Jasmine system in order to detect *local* events in real time. They used the streaming Twitter API to collect tweets from Japan. A location database is created from messages posted on Forthsquare[8]. This database is utilized in order to geo-tag tweets not including location information. The approach is simple and fast. Based on geo-tagged tweets, popular places are identified using a hashing algorithm called "geo-hash". According to this algorithm, close locations result into the same hashing bucket. From the most popular places, keywords that describe the localized event are extracted. Jasmine could lead to an interesting mobile application that detects local parties or concerts instead of larger scale events like earthquakes.

---

[8] https://foursquare.com/.

Focusing on a different social network, in [85] the authors describe their app-roach for the MediaEval Benchmark 2012[9]. This dataset contained 167 thou-sands images from Flickr and the challenge is to find (a) technology events that took place in Germany, (b) soccer events that took place in Hamburg and (c) demonstrations and protests that took place in Madrid. The research team used some preprocessing techniques involving removal of common words and text cleaning. They also used the Google Translate API[10] in order to translate non-English words. Based on the image description text, they classified pictures based on their TF-IDF vectors. As for the pictures with no textual information, user profile information is utilized. Since the challenge required topic-specific event detection, Latent Dirichlet Allocation is utilized in order to extract topics. For the event detection task they used peak detection on the number of photos assigned to each topic. If a topic received more photos than expected, an event is identified for this topic. Since this approach requires computationally expensive procedures such as LDA, it is not easily applicable for high rate streams.

5.3 First Story Detection

The authors of [66] tackle the problem of detecting the first story about a news event. This problem is known as *first story detection (FSD)* and is equivalent to the problem of new event detection (NED) (see Sect. 4). A common approach to solving the FSD problem is to calculate for every document in the corpus the distance to their nearest neighbours [5]. If this distance is larger than a threshold, this document is considered novel and a "First Story". This unsupervised app-roach is extended in [48] to a supervised method using as features the distance, the entity overlap as well as the term overlap utilizing a SVM classifier.

Osborne et al. in [66] suggests that the conventional approach described by [5,48] will not scale for streaming data and therefore proposes a more efficient approach. Nearest neighbour calculation is computationally intensive. Even fast nearest neighbour algorithms such as KD-Trees and Indexing-Trees [93] will not scale in the case of large and fast social streams. The authors propose the usage of a hashing technique in order to detect the nearest neighbour. It is called *Locality Sensitive Hashing (LSH)* [77] and is a hashing scheme that provides an approximate nearest neighbour in constant time. LSH hashes documents to buckets. If the documents are similar, they are hashed into the same bucket. However, the approach is randomized and errors may occur. In order to reduce the variance of the neighbour errors, [66] uses multiple LSH data structures. That is, a document is hashed to multiple buckets, one per LSH data structure, using different hashing families. The neighbour computation similarly involves the exploration of all these buckets.

In order to bound the memory requirements as well as the number of com-putations per incoming document they restrict the maximum size of a LSH bucket and the computations performed within it by discarding old invaluable documents. Similar approaches are used in [53] in order to bound the memory

---

[9] http://www.multimediaeval.org/mediaeval2012/.
[10] https://cloud.google.com/translate/docs.

consumption using careful delete operations. This system detects the First Story about an event and then new documents that are similar are linked together in order to create *event threads*. Threads are document sets that represent the event. This is similar to the clustering approaches we discussed in the previous sections. The Threads are created and then presented in a sorted list according to their size, number of users and thread entropy (expressed as the distribution of terms). They compare their results to [5] and they suggest that event detection performance is similar. However, the efficiency of [66] is improved from [5] achieving constant memory and computation time per document resulting in a streaming FSD solution.

Petrovic et al. in [59] observed that the FSD system of [66] had low Precision due to many false positives. This is something expected since most tweets are not about real-world events. The authors use two streams in order to detect events. They constructed a stream of Wikipedia page views using Wikipedia logs[11]. The method in [66] is ranking the event threads according to their entropy in order to identify the top-k events. In [59] the ranking is modified in order to take into account Wikipedia page views that are related to the event. For every detected thread, the Wikipedia stream is checked for outlying behaviour (i.e. an unexpectedly large numbers of views) in pages that had a similar title. A very important drawback of this method is that Wikipedia Stream lags on average two hours behind the Twitter Stream causing problems for real time event detection. This is explained by the fact that users initially discuss the event topic on the social platform and then some of them may visit the relevant Wikipedia page. An overview of the system is presented in Fig. 3.

Osborne et al. [67] extends [66] in order to cope with "tweet paraphrases". That is, the feature vector is extended with synonyms of existing terms. This approach may remind to the reader the method described in [60] where the term vectors are semantically expanded. In [67] the authors used online sources in order to create a list of paraphrases while in [60] the authors computed term co-occurrences from a static Twitter corpus. The idea of [67] is to use a term-to-term matrix Q in order to exploit term synonyms. Using this matrix the similarity of

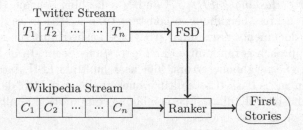

Fig. 3. Overview of the approach presented in [59]. A Wikipedia page-view stream is utilized in order to validate events detected by the FSD algorithms

---

[11] http://meta.wikimedia.org/wiki/Data_dumps#Content.

two tweet vectors x and y, is computed as:

$$Sim(x, y) = y^T Q x \tag{5}$$

Since such a similarity computation requires intensive matrix-vector multiplications they use a heuristic to calculate the inner product faster by using the square root matrix of Q. Since the square root matrix computation is also expensive $O(n^3)$ they approximate $Q^{\frac{1}{2}}$ as if Q is a sparse matrix. Using the heuristics their system is only 3.5 times slower from the original FSD system [66]. The improvement in the Precision of the system is significant achieving higher Precision than the state of the art UMASS system [4].

A similar approach that uses Locality Sensitive Hashing for event detection is followed in [42]. The approach is quite similar to [66]. The authors utilize Twitter posts and Facebook messages. They use LSH in order to group messages into buckets. Their algorithm works in two phases. In the first one, new events (first stories) from both sources (Twitter, Facebook) are independently identified and stored. In the second phase, first stories are hashed into buckets and the corresponding messages are stored as 'event messages'.

The authors in [65] applied an FSD system on Twitter and on a news feed that is referred as Newswire. The performance of FSD is evaluated on both Twitter and the newswire. In an additional experiment, the 27 events originally detected in [66] are used in this work in order to clarify which of the events will be present in both media. They found that almost all events appeared on Twitter and newsWire. However, the events appeared in different time points in the two streams. Events related to sports usually appear faster on Twitter since users post about them while they happen. On the other hand, on events related to breaking news, the newswire stream had a minor advantage.

An efficient first story detection method is presented in [43]. The authors focus on new event detection using a novelty score that is based on term-usage. The main goal of this approach is to detect novel documents avoiding the computation of distances among similar documents. Such an approach is very useful since neighbor computations in the TF-IDF weighted vector space could be computationally intensive for a large corpus. The proposed algorithm uses the Inverse Document Frequency (IDF) per keyword as a novelty score component. Each document receives a novelty score that is the sum of its terms' IDF weights. That is, a document is considered novel if its terms are novel. If the novelty score is above a threshold the document is detected as event related. In the same work, the authors suggest also the probabilistic IDF (pIDF) as a scoring function. Given a term q taken from a corpus C, pIDF is defined as:

$$pIDF(q, C) = log \frac{N - df_q}{df_q} \tag{6}$$

df_q is the frequency of the term q among the documents in the corpus C and N is the size of the corpus C. The probabilistic IDF violates a set of rules about a scoring function since it can take negative values. The authors state that this is beneficial since it penalizes documents if they contain common terms. Notably, score calculation using IDF weights [43], is invariant of the corpus size and the complexity of processing a document d is $O(|d|)$. Where $|d|$ is the number of terms used in the document. However, it should be noted that since social media such as Twitter consist of extremely dynamic content, the IDF scores should be periodically updated in order to reflect accurately the content of the stream.

5.4   Topic Specific Event Detection

The methods presented so far aim at identifying events that could be of any type. This section, presents algorithms that target at identifying and tracking events of a specific predefined type.

One of these efforts is the TEDAS system that is described by Li et al. in [52]. TEDAS is built for recognizing criminal and disastrous events such as tornadoes, floods or law-breaking evidence. The system collects tweets using the Twitter API and returns tweets related to crime and disaster using topic related keywords. Tweets are captured using an initial keyword seed that is predefined by the authors. This seed is expanded according to co-occurrences with keywords from the received tweets. In other words, the system looks for paraphrases or for semantically linked terms. Since not all tweets containing these keywords are about crimes and disasters, a classifier is trained using Twitter features such as use of hashtags, mentions and some predefined pattern-features in the content. Such a pattern feature is the presence of time and location in a tweet. The system clusters all crime and disaster (CDE) tweets according to their spatial information and presents them to a map. The high-level description of the TEDAS system can be seen in Fig. 4.

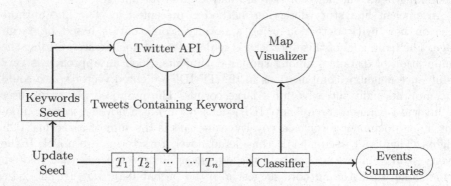

Fig. 4. The TEDAS system. An initial seed with crime/disaster keywords in applied to the Twitter API in order to collect relevant tweets. Based on the result set, the keyword list is expanded. Each tweet is classified as event or non-event, and in the second case it is presented to the map.

Another approach that targets events of specific type is that of Sakaki et al. [72]. This work focuses on earthquakes. Initially, tweets containing at least one earthquake-related query-term are collected. Assuming that these tweets actually talk about earthquakes the authors train a Support Vector Machine on these data. For the classification task, they used tweet statistics (#words in a tweet), textual features (terms of a tweet) and context features (keywords before or after the query term). The goal is to detect the location and the trajectory of the event. The geo-tagging of tweets is exploited to detect the location of the event. Finally, Kalman Filters and Particle Filters aid in identifying the trajectory of the event.

Packer et al. [62] expand a seed of keywords related to a topic using external structured information. In order to collect tweets about a topic (e.g. a music band), they use RDF structured information to identify related entities. For example, many music bands have an entry in DBPedia[12], a large RDF database based on Wikipedia. Therefore, entities related to the band (e.g. band members) could easily be extracted. These additional entities are used to extract tweets that refer to the topic. Events related to a topic are identified according to the number of the times the corresponding entities are mentioned in tweets. According to the experimental evaluation the usage of the external sources gave a boost in the event detection performance. Furthermore, the authors observed an important correlation between the actual time period of the event and the time the related entities are mentioned in Twitter. This observation suggests that users usually tweet during an event.

The work presented in [94] focuses on detecting sports-related events. The case study is the National Football League games of the 2010–2011 season. Using a lexicon-based heuristic the authors collect relevant tweets. For identifying events, they propose a sliding adaptive window-based method. If the ratio of relevant tweets in the second half of the window is larger than a predefined threshold then that is an indicator that something is happening. The window size is adapted when the tweet-ratio(of relevant tweets) highly deviates from that of the previous window. The algorithm is able to detect game related events such as *touchdowns* and *interceptions*. The idea of using an adaptive sliding window is quite interesting since it will enable capturing events of different magnitudes.

The approach proposed in [7] targets at detecting influenza incidents using Twitter. Similarly with above, flu-related keywords are utilized in order to collect a number of potentially relevant tweets. A classifier is then trained in order to filter tweets that are not relevant. The classifier is built on bag-of-words text features. By considering the output of the classifier (relevant tweets), a time series is created based on flu-related tweets count. In order to evaluate their results, the authors compared their methods to Infection Disease Surveillance reports from clinics and to Google trends[13]. They conclude that this type of events can be tracked via Twitter and detected before Google trends and even before a potential break out. A similar approach is suggested in [34] where search engine queries are utilized instead of social media messages.

---

[12] http://dbpedia.org/About.
[13] http://www.google.com/trends/.

Medvet et al. [58] focus on keywords that suddenly received increased popularity. Their algorithm monitors words related to a predefined topic. Whenever these words demonstrate an increased frequency - compared with their history, are identified as candidate event keywords. The most recent tweets containing these terms are classified according to sentiment (positive, negative or neutral). Tweets from these three classes are then used to generate a summary for the event. This application could be very useful for brand related events and also for market research software in order to track product feedback.

6   Architecture

From a computational point of view, an apparent obstacle in social media analysis is Big Data management. Real time detection in Web 2.0 data requires algorithms that efficiently scale in space and time. Extreme data volumes impel researchers and engineers to consider distributed environments. Inevitably many of the papers discussed in Sect. 5 focus on architectural aspects and suggest frameworks suitable for real-time social media analysis. Methods that are focused on smaller data volumes without intensive computations however are able to perform in real-time even with the usage of a single machine.

The frameworks proposed in the literature recently can be organized in the following categories:

- *Multi-Component: Single Machine or Distributed.* The system consists of many components, each of which is responsible for a different task. Many times, the components that considered to be a bottleneck, are replicated on multiple machines in order to increase throughput if possible.
- *Data Stream Topologies:* Multiple nodes are responsible for different tasks. These approaches utilize a stream topology that is suitable for scaling with high-rate data input. A common configuration for this case is Storm[14] along with a NoSQL database like MongoDB.

6.1   Architectures of Multiple Components

In this section we provide an overview of systems built for event detection that utilize a multiple components structure running on a single or multiple computers.

In [86] the authors focus on identifying events in *real-time*. They utilize a sample of the Twitter stream that produces 3 million tweets per day. The core of the system is a MongoDB[15] database. MongoDB is suitable for storing data such as tweets in JSON format. The choice of this type of database is supported by the fact that MongoDB supports geo-spatial and temporal indices. Another important feature is that it can easily scale in number of instances. In case data rate increases (e.g. due to Twitter increased popularity) MongoDB would deploy

---

[14] http://storm.incubator.apache.org/.
[15] https://www.mongodb.org/.

additional machines. The system consists of multiple components. These are a Twitter Fetcher, a Cluster Creator, a Cluster Updater and a Cluster Scorer. Initially a single machine is used, however modules such as the Cluster Scorer could be easily replicated on more than one machine to handle increased load. This architecture is presented in Fig. 5.

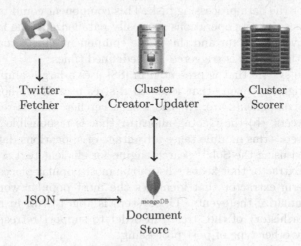

Fig. 5. The architecture of [86]. The four components are the Twitter Fetcher, the Cluster Creator-Updater and the Cluster Scorer. The cluster scorer is represented as a set of servers since it could be replicated.

In [22] the author established a MongoDB database to store Twitter data. The choice of MongoDB is justified by the requirement of having the pre-processing done by Map-Reduce jobs. MongoDB supports Map-Reduce jobs as javascript functions. Map-Reduce as a pre-processing engine is an intuitive choice given that tasks like noise filtering and natural language processing are computationally demanding. The dataset consisted of 1.7 million tweets per day. However, the method did not target at run-time processing since it only considered Retrospective event detection. The dataset is processed in batch steps that involved computationally expensive procedures like Discrete Wavelet Transformation (DWT) and Latent Dirichlet Allocation (LDA). The main contribution of this work is that it provides an insight on the capabilities of map-reduce for efficiently preprocessing textual information. MongoDB as well as other NoSQL databases such as CouchDB attracted the interest of the research community due to their document storage and scaling capabilities.

Abdelhaq et al. [1] focused on identifying local events in a streaming fashion. The system is implemented as a plug-in for the JOSM[16] framework. Similarly to [86] it consists of a number of modules. The Tweets Repository module is responsible for gathering tweets using the Twitter API. The Buffer module keeps in

---

[16] https://josm.openstreetmap.de/.

main memory the last window of tweets. The window is indexed according to time frames in order to ensure quick access to the data. In addition, this module maintains a table with word-count statistics calculated from the streams' history. This component results in large amounts of memory requirements. Sketch randomized data structures could be a solution in such cases and be applied for keeping the word-count table in main memory. The Content Processor module is responsible for the data processing task. This component could be distributed on more machines since its operations are easily parallelized. The last module is the Localized Event Detector and this is the component that actually performs the event detection and it is triggered at predefined times.

The Jasmine system that is presented in [88] uses a large sample of Twitter (15 % of the original stream) that leads to a stream of 15 million tweets per day. The basic components are: (a) the Tweet Fetcher, that is responsible for downloading tweets, (b) the Geotag allocator that is responsible for assigning locations to tweets - this module takes advantage of a locations database where places are stored using the Solr[17] search engine for efficient text search, (c) the Popular Place Extractor that keeps a list of the most popular places, and finally, (d) the Key Term extractor that identifies the most popular words in tweets in order to summarize the events. The system is able to run in real-time but also maintains a history of the stream in order to support retrospective event detection or any other type of post processing.

An overview of the system is presented in Fig. 6.

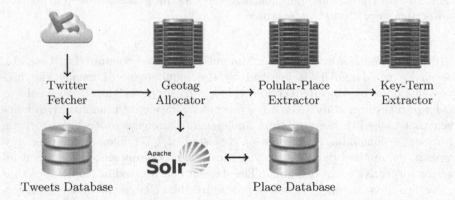

Fig. 6. The architecture of Jasmine System [88]. The first component is the Twitter Fetcher that receives tweets from the 15 % of the Twitter stream and stores them in a database. The geotag allocator geotag the tweets using the places database. The popular place component keeps in memory the most popular places. Finally, the Key-Term extractor extracts key-terms for localized events.

The TwitterStand system presented in [73] used 4 different sources of information: (a) Twitter Gardenhose (deprecated privileged Twitter API providing

---

[17] https://lucene.apache.org/solr/.

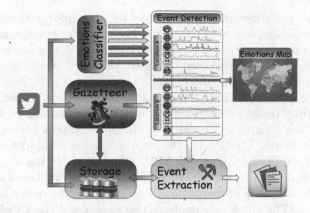

Fig. 7. The architecture of the TwInsight system [84].

10 % sample of tweets), (b) the BirdDog service (deprecated API for receiving posts from up to 200, 000 Twitter users), (c) a 2000 user stream and (d) a keyword stream of 2000 terms. The first component of the system is responsible for collecting the tweets from the sources. The next component is a fast Naive Bayes classifier with the purpose to filter out "junk" from "news" tweets. The classifier is not trained on a static corpus but it is updated from a dynamic one. This justifies the choice of the Naive Bayes classifiers since it has a low update computational cost. Another benefit is that its simplicity makes it tolerant to increased data volumes. The third component is the Clusterer that performs topic clustering using textual features. The Clusterer depends on the classifier component since it clusters only the tweets that are classified as "news". The last module is the Geo-Tagging Component that groups together the topic-clustered tweets according to their geo-location. The above system is represented as a graph of connected stream processing units, each of them receiving the output of the previous one, defining a processing topology. Thus, it is straightforward to think that it could be implemented from a stream processing framework such as Apache Storm[18] and distributed on multiple processing engines if necessary. Units that act as bottlenecks can be enhanced with more cores.

Another system structured in multiple-components is presented in [84]. Two approaches are presented. The first one uses Twitter data while the second one exploits mobile information. Figure 7 presents the architecture of the approach operating on Twitter data. This system is utilized by the TwInsight system which is presented in [83]. The first layer consists of the Twitter feed fetcher. In the second layer, an emotions classifier, a Gazetteer and a storage component are included. The Gazetteer assigns geo-locations to tweets and users utilizing an algorithm presented in [81]. This part can be a bottleneck for the system and therefore can be replicated to multiple machines. Emotions are assigned to tweets using a Machine Learning classifier. The storage component contains a database that stores the tweets with their extracted meta-data from the previous

---

[18] https://storm.apache.org/.

two components including emotion as well as location information. Finally, the resulting tweets are processed in the third layer by the event extractor. This component performs the event detection and provides a summarization of the detected events. A visualization component on the third layer is responsible for providing sentiment level information on a map.

The purpose of INSIGHT's[19] Twitter Intelligent Sensor Agent (ISA) is to detect in real-time traffic or flood related incidents in the city of Dublin. The architecture of the Twitter-ISA consists of multiple components similar to [83]. The first component is a Tweet fetcher responsible for gathering topic related tweets through the Twitter Filtered API[20]. This enables tracking of specific users, keywords and locations in order to collect a decent number of topic related tweets. Since the majority of tweets do not include location information, a Geo-tagger is utilized. The Geotagger analyzes the tweets and checks whether there are references to places. If this is the case, it assigns coordinates to tweets using Open Street Maps[21] and a Lucene[22] index following the method described in [27]. The resulting tweets are forwarded to the Text Classifier component that identifies tweets that talk about traffic or flood incidents. All identified event-related tweets are stored to a MongoDB database for further analysis. The bottleneck of the system are the Geotagger and the Text Classifier components. However, these components are easily replicated on multiple machines each of them handling a different sub-stream of the original stream without any impact on the effectiveness of event detection. The architecture of the Twitter-ISA is presented in Fig. 8.

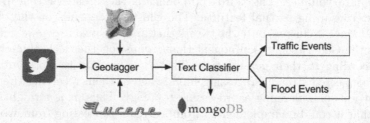

Fig. 8. The architecture of the Twitter-ISA of the INSIGHT system.

6.2   Data Stream Topologies

The authors in [56] suggested a distributed framework for high-volume data streams like the Twitter Firehose (nearly 100 % of the Twitter feed). They propose a Storm topology in order to enable parallel and distributed computations

---

[19] http://www.insight-ict.eu/.

[20] https://dev.twitter.com/streaming/reference/post/statuses/filter.

[21] https://www.openstreetmap.org.

[22] https://lucene.apache.org/.

implementing the first story detection algorithm described in [66]. A key compo-nent of this algorithm is finding the nearest neighbor of a document. The basic idea of the distributed streaming topology is to divide the Twitter stream into sub-streams without reducing the evidence that could aid event detection.

The first topology layer is the Vectorizer that converts the tweets to the vector space using a bag-of-words approach. The next layer is the "Hashing" that dis-tributes the tweets on different processing units. The authors suggest the usage of Locality Sensitive Hashing (LSH) in order to partition the documents into mul-tiple LSH-buckets, with similar content, belonging on Storm Bolts. The intuition behind the multiple-bucket partitioning, using multiple LSH data-structures, is to reduce the nearest neighbour error caused by LSH and is described in more detail in [66]. Each document is sent to multiple bolts involving extra com-munication cost but reducing the LSH error. Those bolts belong to the Local Distance layer where each of them reports the nearest neighbour to a document identified from its buckets. In the next layer named Global Distance, the nearest neighbors are aggregated and the one with the smallest distance from the new document is selected. Documents who have distance more than a threshold from the global nearest neighbour are sent to the "K-Means clustering" layer that performs online clustering, the rest are discarded. Each of the formed clusters may correspond to a real world event. The storm topology is presented on Fig. 9.

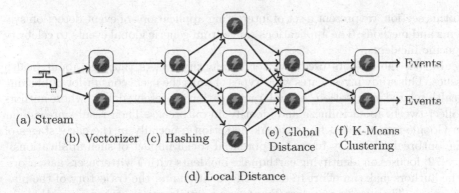

Fig. 9. The storm topology with multiple bolts per component. The (d) local distance layer is allocated the most cores since it is the most computationally intensive.

They found that the fastest layer is, as expected, the Vectorizer. The slowest one is the Local Distance. This is explained by the fact that this bolt imple-ments a nearest neighbour computation involving similarity calculations on high dimensional vectors. It is important to measure the slowest layer in order to allocate cores where it is necessary. The authors suggest that the throughput of the system scales linearly as more and more cores are added to the right bolt layers. After investigating the number of cores that will be needed in order to process the entire Twitter Firehose (5000 tweets/s) the authors concluded that 70 cores or 9 8-core machines will suffice.

6.3   Summary

In this section, system architectures that are capable of performing real-time event detection are presented. Most of them focus on the Twitter sample stream and the Twitter Garden hose with almost 10 million tweets per day. For managing and mining the sample stream (1 % of Twitter) a distributed modular architecture is required.

Notably, the 1 % of the Twitter stream on March of 2014 was about 8 million tweets per day. This volume is similar to the volume of the Gardenhose API (10 % of Twitter) provided some years before. This fact confirms the growth of the Twitter usage over the years. If one is willing to process the entire Twitter stream she needs to turn to architectures like Storm and to use the appropriate number of processing units. For a smaller dataset such as a stream of 1 million tweets per day a single machine with high amounts of main memory and processing power will be able to process the data at run-time. It is important to note that not all modules involved in the methods presented can be replicated. This is due to the fact that not all algorithms can be parallelised without affecting the quality of results (e.g. [66]). A summary of the above approaches that contains the stream rate and the approach architecture is presented on Table 4.

7   Applications

In this section, we present a set of interesting applications of event detection systems and methods. The applications range from generic global events to celebrity specific incidents.

In [36] the authors used Twitter to identify tweets that are about health issues. This study investigates what types of links the users consult for publishing health related information. A similar application is presented in [7] where authors collect tweets about Influeza and identify flu outbreaks. Their results are similar to Google-trends based flu outbreak detection especially in the early stages of the outbreak. It is easy to see the potential social impact of such applications.

[72] focuses on identifying earthquake incidents with Twitter users as sensors. The authors make an effort to detect the location and the trajectory of the phenomenon. The system monitors Twitter and emails citizens when an earthquake

Table 4. The size of corpus and frameworks used in the papers presented in this Section.

Reference	Data volume	Frameworks
[86]	3 M/Day	MongoDB Spatial-temporal indexes, horizontally Scaling
[1]	Twitter sample stream 10 M/day	JOSM
[88]	15 M/day	Apache Solr, geo-hash
[73]	Above 15 M/day	-
[56]	Gardenhose and Firehose	Storm Topology

is detected. The response time of the system is proved to be quite fast, similar to the Japan Meteorological Agency. In [30] the authors detect flood events in Germany providing visual information on the map. The TEDAS system [52] targets Crime and Disaster incidents by identifying where and when they happened. A map visualization of tweets is available. Flickr and Youtube are utilized in [68] where the goal is to detect content related to an emergency. The above systems help the authorities in detecting real-time incidents as well as in extracting useful information after the event.

Another set of approaches focused on finding global important events for a given time period. These include [12, 43, 66]. [66, 67] emphasize on finding the first story about a new event (new event detection). Such approaches are valuable since they can aid in identifying unexpected events.

Medvent et al. [58] focused on detecting events related to specific brands. They focused on three major brands: Google, Microsoft and Apple. Examples of such events are the release of a new product like the new iPad or Microsoft's Security Essential software. In order to achieve the desired outcome, the authors study the sentiment of the tweets. These techniques are utilized for marketing purposes. A similar approach is presented in [69] where events of controversial sentiment are targeted. Automatic identification of controversies is very useful in order to track and manage a large number of discussion groups.

Noettcher et al. [15], developed an Android application that finds local events given a specific geographic area. The application is able to provide summaries to the users. The Jasmine system detects local events for the user according to the desired size of event and the number of users attending it. It presents summaries and some important tweets per event in order to provide with a short description. This group of applications could support mobile users looking for "happenings" near by.

In the area of sports analytics, the EvenTweet system [1] could detect the start time and the location of football matches for UEFA 2012. The system described in [94] was able to detect National Football League events of the 2010–2011 season. The events include touchdowns, interceptions and goals.

8   Evaluation

Event Detection in social media is a relatively new and rather complex problem, especially when it comes to evaluating the suggested approaches. Most authors have to evaluate their algorithms during a period where important global events take place. This will enable the validation of their techniques. Another approach is to insert artificially event tweets into the stream. In this section a review of the most common evaluation practises is presented. Experimental set-up, utilized metrics, and obtained results are presented. Moreover we will cover strategies for labeling the data and provide links to publicly available datasets.

8.1 Dataset Labelling

Many approaches constructed a dataset using the Twitter API. From the collected tweets they create clusters of messages and label these clusters either using the most important words of the cluster or the centroid of the cluster. Usually, more than one annotators are used and the agreement between them is measured using Cohen's Kappa [90]. Only the annotations with high agreement are used in the most cases while the low agreement annotations are discarded since they are considered noise.

The authors in [2] followed a supervised approach on classifying event-candidate clusters as event or non-event. Their system required a training set for the supervised classifier used by their method. In order to assemble such a dataset they initially extracted 1000 clusters using their one pass online clustering algorithm. Then they manually labelled these clusters. 319 clusters are labeled as events whereas 681 are labelled as non-events suggesting that, as expected, the two classes are slightly unbalanced.

In a similar fashion the authors in [12] tracked the same problem. Sharing the supervised classification idea with [86] they required a training set consisting of example clusters marked as event and non-event. They manually annotated clusters, but these clusters are carefully selected. Instead of annotating the clusters or annotating a random subset, they restricted the cluster selection to the top-20 fastest growing clusters per hour. The assumption behind this is that usually a cluster that suddenly increases in size, will be an event cluster. For the testing set they sample randomly clusters per hour from the whole cluster pool in order to depict the real balance between "event" and "no-event" clusters. The annotators labelled the clusters as "real world event", "Twitter centric activity", "non-event" and "ambiguous". Two annotators provided judgments and Cohen's Kappa is used in order to measure the agreement. The clusters used for the training are 504, favoring the event class due to the careful cluster selection. The test-set consisted only of 300 clusters.

The authors in [66] used the Edinburgh Fist Story Detection (FSD) Corpus. A simply modified version of this dataset will be presented in the next sections. From this dataset they created threads of messages using their threading algorithm described in the same paper. The threading algorithm links related documents according to their textual distance and creates clusters of similar documents. Authors divided the stream in sliding windows and then for every sliding window they extracted the fastest growing threads(clusters) and manually labelled them. The top 1000 fastest growing threads from a sliding window of 100, 000 threads are labelled using two annotators and using Cohen's Kappa coefficient. Clearly, it is important to note that again the reason why the fastest growing clusters are selected for annotation is in order to favor the event clusters similarly to [12]. Otherwise, the dataset would contain only a very small proportion of clusters labeled as "event".

An alternative approach is described in [57]. There, the wisdom of the crowd is utilized through Amazon's Mechanical Turk.

They selected candidate clusters of events using the LSH [66] algorithm and the Cluster Summary algorithm [2]. They also utilized the Wikipedia Events Portal in order to receive event clusters as well as tweets about the detected events. Then the crowd is used to determine if the cluster tweets are about the event. In addition some clever heuristics are used in order to increase the annotators agreement and also to filter out low-quality annotations. On the same time, these heuristics provided the annotators motivation to continue their high quality work.

8.2   Evaluation Metrics and Results

In this section, we provide an overview of the results obtained by various studies presented in previous sections. Given the fact that there is an absence of shared datasets, a direct comparison is impossible. However, the following metrics serve as indicators of performance in various problems. Furthermore we present information on the metrics used in each case.

Many authors decided to test the performance of their algorithms on the TDT5 dataset. This dataset contains news articles extracted from traditional news media and was widely used for the TDT challenge. Naturally, results will diverge in a Twitter dataset since the two information sources are different in many ways. For example, the streaming FSD algorithm [66] demonstrated much better performance on the TDT5 dataset in comparison to a Twitter dataset.

The most common evaluation metrics originate from fields like Information Retrieval and Natural Language Processing. Typical examples are Precision, Recall and the F-Measure in terms of the detected events. In some cases, Accuracy is reported whereas in others the number of detected events is used as an indicator of effectiveness. The latter might be misleading in cases of unbalanced classes such as event detection. At this point, we will define some of the basic metrics frequently used in the literature. Precision is defined in Eq. 7. *Actual Events* (or *True Positives*) is the number of times that the algorithm detected an event and it is actually an event. *Recall* shows the percentage of the actual events that the system is able to identify (see Eq. 8). F-Measure is the harmonic mean of Precision and Recall and it is defined in Eq. 9.

Many approaches can achieve high Recall but with limited Precision due to the large number of False Positives. This is one of the reasons that additional 'filtering' techniques are utilized before or after the core approach. Some methods use ranked versions of the above evaluation metrics such as Precision at k ($P@K$). Such metrics allow the evaluation of methods that can provide an ordered list of predicted events.

$$Precision = \frac{\text{Number of Actual Events Detected}}{\text{Number of Detected Events}} \tag{7}$$

$$Recall = \frac{\text{Number of Actual Events Detected}}{\text{Number of Actual Events}} \tag{8}$$

$$F\text{-}Measure = \frac{2 * Precision * Recall}{Precision + Recall} \tag{9}$$

The authors in [12] decided to use a manually labelled dataset for the evaluation of their system. Their approach utilized classifiers that are compared based on the F_1 measure. SVMs outperformed the Naive Bayes classifier (0.837 over 0.702).

In [66] the authors evaluate their system in terms of average Precision. This decision is straightforward since labels are available only for the detected events (True Positives). The authors evaluate their first story detection system on the top-k event stories ranked according to different scoring functions. The best obtained results in terms of Precision at k ($P@k$) is 34.0 %.

The same event detection system is tested in [67] where Wikipedia is utilized in order to re-rank the detected events. The authors observed that the use of Wikipedia provided an improvement in Precision. However, as the authors comment, Wikipedia causes a two-hour delay in event detection in comparison with the original approach [66].

The authors in [59] evaluated their system on the same dataset [66]. They conclude that their approach outperformed UMASS [4] while the detection time is only 3.5 times slower than the method in [66]. However, the benefits of the approach are not so clear in Twitter data as they are in FSD data (TDT5).

[43] utilized the corpus of [59]. They used Detection Trade-off (DET) curves for evaluating the effectiveness of the approach. DET curves display the ratio of *Miss Probability* to *False Alarm Probability*). The conclusion is that their system outperformed all baseline approaches included in the experiments. On top of that, a significant improvement is observed in execution time.

In [86] the authors evaluated three classifiers in terms of Precision, Recall and F-Measure. They used a manually labelled dataset by 10-fold cross-validation. Their best performing classifier is a Pruned Decision Tree with a F_1 score of 0.857. They also experimented with the impact of the content and non-content features. The authors observed a statistically significant improvement when all types of features are considered. One could note however that these results deviate from the ones reported in [12,67]. This gap demonstrates the effect of the dataset in an experimental evaluation.

The inventors of TwEvent [51] provided with their own definitions of Recall and Duplicate Event Rate (DER) that are the metrics used in their evaluation. Recall is defined as the number of detected events while DER is the ratio of duplicate events found. DER is useful in order to penalize multiple alerts on the same event. They compared the approach against the EdCow system [89] and concluded that TwEvent achieved an important improvement in terms of Recall (75 over 13 detected actual events). An improvement is also observed in Precision (86.1 % over 76.2 %). The DER metric of TwEVent system and EdCow is 16.0 % and 23.1 % respectively.

Similarly, Popescu et al. [69] utilized a manually labelled dataset that consists of 800 labelled events using two human annotators. Three alternative approaches are compared on Precision at k ($P@k$). The so-called *blended* model performed best with 0.9 Precision at rank-1 and 0.80 Precision at rank-4. The Area Under Curve (AUC) of the three models suggests that they have good discriminative power in comparison to baseline algorithms. A final note is that the performance differences between the three systems is not statistically significant.

8.3 Available Datasets

The Edinburgh FSD corpus[23] was created in order to test the method in [66]. The dataset contains $51,879,318$ tweet IDs. The content of the tweets is removed due to Twitter's terms of use. In order to take advantage of the dataset one has to use the Twitter API to download the messages that correspond to the tweet IDs. The authors identified 27 topics in the data. 3034 tweets are labelled according to the procedure described in [59]. This dataset was created for detecting first stories. However, it is suitable (and was utilized) for general event detection tasks.

A dataset that is not Twitter specific but is useful for event detection evaluation is the NewsWire dataset[24]. The dataset contains links to news articles. The articles contain a timestamp and a relevance value to some of the aforementioned 27 topics [65]. The dataset contains 47751 links to articles.

The dataset of the MediaEval challenge is also available and can be utilized for event detection. The dataset consisted of Flickr images and 1.327 videos from YouTube with their metadata. Another one consists of Instagram pictures instead of Flickr images. The labelled part of the dataset was created using human annotators. This dataset contains 8 event types. These are music events, conferences, exhibitions, fashion shows, protests, sport events, theatrical/dance events and other events.

9 Related Problems

Trend detection is a highly related task to event detection and is commonly applied to social media (e.g. Twitter trending topics) and News portals (e.g. Yahoo News). Many trend detection methods like [13,55] are similar to feature-pivot event detection techniques. In these methods, a keyword burst identification is a core element. Similarly to event detection, scalability for high volumes of data is a major concern.

Information diffusion [37] is another problem that shares many similarities with event detection. Twitter [71,91] and Facebook [10] have been extensively studied on how information flows inside the network. Information diffusion examines the impact of the network structure, which users are influential or why some content becomes viral.

'Event Detection' is a term commonly used in video/image analysis and computer vision [40,79,95]. In this case the goal is to identify in a video feed an incident - usually of specific type. Similar efforts have been observed in image and video streams in social networks like Instagram, Flick and YouTube [63,64,85].

Other domains for event detection emerge as new information sources become available. Mobile and Urban data are now in abundance in smart cities. Hence,

---

[23] Available at http://demeter.inf.ed.ac.uk/cross/docs/fsd_corpus.tar.gz.

[24] Available at http://demeter.inf.ed.ac.uk/cross/docs/Newswire_Events.tar.gz.

data analysis and event processing techniques as well as complete streaming frameworks are exploited in order to identify incidents in the streets of a city [8,16,17,74]. Data sources that are utilized in such cases are SCATS[25] data (traffic volume information) or vehicle data like public transport data (e.g. GPS location of buses moving around the city). The INSIGHT project develops a system that targets at identifying disastrous events from city data.

10   Conclusion and Open Challenges

In this paper we presented an overview of the most recent techniques for detecting events in online social networks. This is an area of research that emerged during the last years, in parallel with the growth of user participation in social networks. In this overview, we made an effort to organize the most important research lines as well as their results. Furthermore we focused on the architecture element of such systems. Due to large volumes of data, state-of-the-art data stream and database frameworks had to be utilized. Finally we discussed how the evaluation is being executed in event detection and mentioned the most common evaluation metrics and datasets used. We believe that this survey will benefit researchers in the field as well as practitioners working in commercial applications that exploit social network applications.

The problem of event detection is a very challenging one. The definition of the problem in Sect. 3, suggests that there are many dimensions to it. It is not sufficient to detect that something happened, in other words, detect anomalies. Event detection requires the automatic answering of what, when, where, and by whom. After reporting on the most recent efforts in the area, it is clear that no method addressed all of these questions. Therefore, there is a lot of space for improvement towards this direction.

Another challenge that has to be addressed is the lack of public datasets. Privacy issues along with Social Network companies' terms of use hinder the availability of shared data. This obstacle, is of great significance since it relates to the repeatability of experiments and comparison between approaches. It is not hard to observe that most approaches focus on the Twitter platform. This is of course due to the usability and accessibility of the Twitter API. However, a research area that depends on a single data source, as interesting as it is, entails many risks. Nonetheless, it is expected that as new media sources emerge, event detection will remain significant and challenging.

Acknowledgments. This work is funded by the projects EU FP7 INSIGHT (318225), GGET Thalis DISFER and GeomComp.

References

1. Abdelhaq, H., Sengstock, C., Gertz, M.: EvenTweet: online localized event detection from twitter. Proc. VLDB Endow. **6**(12), 1326–1329 (2013)

---

[25] http://en.wikipedia.org/wiki/Sydney_Coordinated_Adaptive_Traffic_System.

2. Aggarwal, C.C., Subbian, K.: Event detection in social streams. In: SDM, pp. 624–635. SIAM/Omnipress (2012)
3. Allan, J.: Introduction to topic detection and tracking. In: Allan, J. (ed.) Topic Detection and Tracking, pp. 1–16. Springer, New York (2002)
4. Allan, J., Lavrenko, V., Malin, D., Swan, R.: Detections, bounds, and timelines: Umass and TDT-3. In: Proceedings of Topic Detection and Tracking Workshop, pp. 167–174 (2000)
5. Allan, J., Papka, R., Lavrenko, V.: On-line new event detection and tracking. In: Proceedings of the 21st Annual International ACM SIGIR Conference on Research and Development in Information Retrieval. ACM (1998)
6. Ankerst, M., Breunig, M., Kriegel, H., Sander, J.: OPTICS: ordering points to identify the clustering structure. ACM SIGMOD Rec. **28**, 49–60 (1999)
7. Aramaki, E., Maskawa, S., Morita, M.: Twitter catches the flu: detecting influenza epidemics using Twitter. In: Proceedings of the Conference on empirical methods in natural language processing, pp. 1568–1576 (2011). http://dl.acm.org/citation.cfm?id=2145600
8. Artikis, A., Weidlich, M., Schnitzler, F., Boutsis, I., Liebig, T., Piatkowski, N., Bockermann, C., Morik, K., Kalogeraki, V., Marecek, J., et al.: Heterogeneous stream processing and crowdsourcing for urban traffic management. In: EDBT, pp. 712–723 (2014)
9. Atefeh, F., Khreich, W.: A survey of techniques for event detection in twitter. Computat. Intell. **31**, 132–164 (2013)
10. Bakshy, E., Rosenn, I., Marlow, C., Adamic, L.: The role of social networks in information diffusion. In: Proceedings of the 21st International Conference on World Wide Web, pp. 519–528. ACM (2012)
11. Becker, H., Iter, D., Naaman, M., Gravano, L.: Identifying content for planned events across social media sites. In: Proceedings of the Fifth ACM International Conference on Web Search and Data Mining, WSDM 2012, p. 533 (2012)
12. Becker, H., Naaman, M., Gravano, L.: Beyond trending topics: real-world event identification on twitter. In: Proceedings of the Fifth International AAAI Conference on Weblogs and Social Media (ICWSM 2011), pp. 1–17 (2011)
13. Benhardus, J., Kalita, J.: Streaming trend detection in twitter. Int. J. Web Based Commun. **9**(1), 122–139 (2013). http://inderscience.metapress.com/index/906V117647682257.pdf
14. Bifet, A., Frank, E.: Sentiment knowledge discovery in twitter streaming data. In: Pfahringer, B., Holmes, G., Hoffmann, A. (eds.) DS 2010. LNCS, vol. 6332, pp. 1–15. Springer, Heidelberg (2010)
15. Boettcher, A., Lee, D.: EventRadar: a real-time local event detection scheme using twitter stream. In: 2012 IEEE International Conference on Green Computing and Communications, pp. 358–367, November 2012
16. Boutsis, I., Kalogeraki, V.: Privacy preservation for participatory sensing data. In: 2013 IEEE International Conference on Pervasive Computing and Communications (PerCom), pp. 103–113. IEEE (2013)
17. Boutsis, I., Kalogeraki, V., Gunopulos, D.: Efficient event detection by exploiting crowds. In: Proceedings of the 7th ACM International Conference on Distributed Event-Based Systems, pp. 123–134. ACM (2013)
18. Cha, M., Haddadi, H., Benevenuto, F., Gummadi, P.K.: Measuring user influence in twitter: the million follower fallacy. In: ICWSM 2010, pp. 10–17 (2010)
19. Charikar, M., Chen, K., Farach-Colton, M.: Finding frequent items in data streams. Theort. Comput. Sci. **312**(1), 3–15 (2004)

20. Chen, K., Chen, T., Zheng, G., Jin, O., Yao, E., Yu, Y.: Collaborative personalized tweet recommendation. In: Proceedings of the 35th International ACM SIGIR Conference on Research and Development in Information Retrieval, pp. 661–670. ACM (2012)

21. Chowdhury, G.: Introduction to Modern Information Retrieval. Facet Publishing, London (2010)

22. Cordeiro, M.: Twitter event detection: combining wavelet analysis and topic inference summarization. In: Doctoral Symposium on Informatics Engineering, DSIE (2012). http://paginas.fe.up.pt/prodei/dsie12/papers/paper_14.pdf

23. Cormode, G., Muthukrishnan, S.: An improved data stream summary: the count-min sketch and its applications. Theoretical Computer Science **55**(1), 58–75 (2005). http://linkinghub.elsevier.com/retrieve/pii/S0196677403001913

24. Cormode, G., Muthukrishnan, S.: What's hot and what's not: tracking most frequent items dynamically. Theoretical Computer Science **30**(1), 249–278 (2004). http://portal.acm.org/citation.cfm?d=1061318.1061325

25. Cui, A., Zhang, M., Liu, Y., Ma, S., Zhang, K.: Discover breaking events with popular hashtags in twitter. In: Proceedings of the 21st ACM International Conference on Information and Knowledge Management, CIKM 2012, p. 1794 (2012). http://dl.acm.org/citation.cfm?d=2396761.2398519

26. Daly, E.M., Geyer, W.: Effective event discovery: using location and social information for scoping event recommendations. In: Proceedings of the Fifth ACM Conference on Recommender Systems, pp. 277–280. ACM (2011)

27. Daly, E.M., Lecue, F., Bicer, V.: Westland row why so slow?: fusing social media and linked data sources for understanding real-time traffic conditions. In: Proceedings of the 2013 International Conference on Intelligent User Interfaces, pp. 203–212. ACM (2013)

28. De Choudhury, M., Gamon, M., Counts, S., Horvitz, E.: Predicting depression via social media. In: AAAI Conference on Weblogs and Social Media, vol. 2 (2013)

29. Duda, R.O., Hart, P.E., Stork, D.G.: Pattern Classification. Wiley, New York (2012)

30. Fuchs, G., Andrienko, N., Andrienko, G., Bothe, S., Stange, H.: Tracing the German centennial flood in the stream of tweets: first lessons learned. In: Proceedings of the Second ACM SIGSPATIAL International Workshop on Crowdsourced and Volunteered Geographic Information, GEOCROWD 2013, pp. 31–38. ACM, New York (2013). http://doi.acm.org/10.1145/2534732.2534741

31. Fung, G.P.C., Yu, J.X., Yu, P.S., Lu, H.: Parameter free bursty events detection in text streams. In: Proceedings of the 31st International Conference on Very Large Data Bases, pp. 181–192. VLDB Endowment (2005)

32. Galuba, W., Aberer, K.: Outtweeting the twitterers-predicting information cascades in microblogs. In: Proceedings of the 3rd Conference on Online Social Networks (2010). http://static.usenix.org/events/wosn10/tech/full_papers/Galuba.pdf

33. Ghosh, S., Sharma, N., Benevenuto, F., Ganguly, N., Gummadi, K.: Cognos: crowdsourcing search for topic experts in microblogs. In: Proceedings of the 35th International ACM SIGIR Conference on Research and Development in Information Retrieval, pp. 575–590. ACM (2012)

34. Ginsberg, J., Mohebbi, M.H., Patel, R.S., Brammer, L., Smolinski, M.S., Brilliant, L.: Detecting influenza epidemics using search engine query data. Nature **457**(7232), 1012–1014 (2009)

35. Go, A., Huang, L., Bhayani, R.: Twitter sentiment analysis. Nature **17**, 1–6 (2009)

36. Goot, E.V.D., Tanev, H., Linge, J.: Combining twitter and media reports on public health events in medisys. In: Proceedings of the 22nd International Conference on World Wide Web Companion, pp. 703–705. International World Wide Web Conferences Steering Committee (2013). http://dl.acm.org/citation.cfm?id=2488028
37. Gruhl, D., Guha, R., Liben-Nowell, D., Tomkins, A.: Information diffusion through blogspace. In: Proceedings of the 13th International Conference on World Wide Web, pp. 491–501. ACM (2004)
38. Guo, D., Wu, J., Chen, H., Yuan, Y., Luo, X.: The dynamic bloom filters. IEEE Trans. Knowl. Data Eng. **22**(1), 120–133 (2010)
39. Hua, T., Chen, F., Zhao, L., Lu, C., Ramakrishnan, N.: STED: Semi-Supervised Targeted Event Detection (2013). people.cs.vt.edu, http://people.cs.vt.edu/ramakris/papers/kdddemo13_sted.pdf
40. Itti, L., Baldi, P.: A principled approach to detecting surprising events in video. In: IEEE Computer Society Conference on Computer Vision and Pattern Recognition, CVPR 2005, vol. 1, pp. 631–637. IEEE (2005)
41. Jarvis, R.A., Patrick, E.A.: Clustering using a similarity measure based on shared near neighbors. IEEE Trans. Comput. **100**(11), 1025–1034 (1973)
42. Kaleel, S.B.: Event Detection and trending in multiple social networking sites. In: Proceedings of the 16th Communications and Networking Symposium. Society for Computer Simulation International (2013)
43. Karkali, M., Rousseau, F., Ntoulas, A., Vazirgiannis, M.: Efficient online novelty detection in news streams. In: Lin, X., Manolopoulos, Y., Srivastava, D., Huang, G. (eds.) WISE 2013, Part I. LNCS, vol. 8180, pp. 57–71. Springer, Heidelberg (2013)
44. Kleinberg, J.: Bursty and hierarchical structure in streams. In: Proceedings of the Eighth ACM SIGKDD International Conference on Knowledge Discovery and Data Mining, KDD 2002, p. 91 (2002). http://portal.acm.org/citation.cfm?d=775047.775061
45. Knobel, M., Lankshear, C.: Online memes, affinities, and cultural production. In: Knobel, M., Lankshear, C. (eds.) A New Literacies Sampler, pp. 199–227. Peter Lang, New York (2007)
46. Kotsakos, D., Sakkos, P., Katakis, I., Gunopulos, D.: # tag: meme or event? In: 2014 IEEE/ACM International Conference on Advances in Social Networks Analysis and Mining (ASONAM), pp. 391–394. IEEE (2014)
47. Kumar, S., Liu, H., Mehta, S., Subramaniam, L.V.: From tweets to events: exploring a scalable solution for twitter streams. arXiv preprint, arXiv:1405.1392 (2014)
48. Kumaran, G., Allan, J.: Using names and topics for new event detection. In: Proceedings of the Conference on Human Language Technology and Empirical Methods in Natural Language Processing, pp. 121–128. Association for Computational Linguistics (2005)
49. Kwak, H., Lee, C., Park, H., Moon, S.: What is Twitter, a social network or a news media? In: Proceedings of the 19th International Conference on World Wide Web, pp. 591–600. ACM (2010)
50. Levenberg, A., Osborne, M.: Stream-based randomised language models for SMT. In: Proceedings of the 2009 Conference on Empirical Methods in Natural Language Processing, vol. 2. Association for Computational Linguistics (2008)
51. Li, C., Sun, A., Datta, A.: Twevent: segment-based event detection from tweets. Proceedings of the 21st ACM International Conference on Information and Knowledge Management (2012). http://dl.acm.org/citation.cfm?id=2396785

52. Li, R., Lei, K.H., Khadiwala, R., Chang, K.C.C.: TEDAS: a twitter-based event detection and analysis system. In: 2012 IEEE 28th International Conference on Data Engineering, pp. 1273–1276, April 2012

53. Luo, G., Tang, C., Yu, P.S.: Resource-adaptive real-time new event detection. In: Proceedings of the 2007 ACM SIGMOD International Conference on Management of Data, SIGMOD 2007, p. 497 (2007)

54. Manku, G.S., Motwani, R.: Approximate frequency counts over data streams. In: Proceedings of the 28th International Conference on Very Large Data Bases (2002). http://dl.acm.org/citation.cfm?id=1287400

55. Mathioudakis, M., Koudas, N.: Twittermonitor: trend detection over the twitter stream. In: Proceedings of the 2010 International Conference on Management of Data, pp. 1155–1157 (2010). http://dl.acm.org/citation.cfm?id=1807306

56. McCreadie, R., Macdonald, C.: Scalable distributed event detection for Twitter. In: 2013 IEEE International Conference on Big Data, 6–9 January 2013. IEEE (2013)

57. McMinn, A.J., Moshfeghi, Y., Jose, J.M.: Building a large-scale corpus for evaluating event detection on twitter. In: Proceedings of the 22nd ACM International Conference on Conference on Information and Knowledge Management, CIKM 2013 pp. 409–418 (2013). http://dl.acm.org/citation.cfm?doid=2505515.2505695

58. Medvet, E., Bartoli, A.: Brand-related events detection, classification and summarization on twitter. In: 2012 IEEE/WIC/ACM International Conferences on Web Intelligence and Intelligent Agent Technology, pp. 297–302 (2012). http://ieeexplore.ieee.org/lpdocs/epic03/wrapper.htm?arnumber=6511900

59. Osborne, M., Petrovic, S.: Bieber no more: first story detection using Twitter and Wikipedia. In: Proceedings of the Workshop on Time-Aware Information Access, TAIA (2012)

60. Ozdikis, O., Senkul, P., Oguztuzun, H.: Semantic expansion of tweet contents for enhanced event detection in twitter. In: 2012 IEEE/ACM International Conference on Advances in Social Networks Analysis and Mining, pp. 20–24 (2012). http://ieeexplore.ieee.org/lpdocs/epic03/wrapper.htm?arnumber=6425790

61. Ozdikis, O., Senkul, P., Oguztuzun, H.: Semantic expansion of hashtags for enhanced event detection in Twitter. In: Proceedings of the 1st International Workshop on Online Social Systems (2012). http://www.cs.ubc.ca/welu/woss2012/papers/1-ozdikis.pdf

62. Packer, H.S., Samangooei, S., Hare, J.S., Gibbins, N., Lewis, P.H.: Event detection using Twitter and structured semantic query expansion. In: Proceedings of the 1st International Workshop on Multimodal Crowd Sensing, CrowdSens 2012, p. 7 (2012)

63. Papadopoulos, S., Schinas, E., Mezaris, V., Troncy, R., Kompatsiaris, I.: The 2012 social event detection dataset. In: Proceedings of the 4th ACM Multimedia Systems Conference, pp. 102–107. ACM (2013)

64. Papadopoulos, S., Troncy, R., Mezaris, V., Huet, B., Kompatsiaris, I.: Social event detection at MediaEval 2011: challenges, dataset and evaluation. In: MediaEval (2011)

65. Petrovic, S., Osborne, M.: Can twitter replace newswire for breaking news. In: Proceedings of the Seventh International AAAI Conference on Weblogs and Social Media 2011 (2013)

66. Petrovic, S., Osborne, M., Lavrenko, V.: Streaming first story detection with application to twitter. In: Proceedings of the NAACL (2010)

67. Petrović, S., Osborne, M., Lavrenko, V.: Using paraphrases for improving first story detection in news and Twitter. In: Proceedings of the 2012 Conference of the North American Chapter of the Association for Computational Linguistics: Human Language Technologies, pp. 338–346 (2012)

68. Pohl, D., Bouchachia, A., Hellwagner, H.: Automatic sub-event detection in emergency management using social media. In: Proceedings of the 21st International Conference Companion on World Wide Web, WWW 2012 Companion, p. 683 (2012). http://dl.acm.org/citation.cfm?d=2187980.2188180

69. Popescu, A.M., Pennacchiotti, M.: Detecting controversial events from twitter. Proceedings of the 19th ACM International Conference on Information and Knowledge Management, CIKM 2010, p. 1873 (2010). http://portal.acm.org/citation.cfm?d=1871437.1871751

70. Psallidas, F., Becker, H., Naaman, M., Gravano, L.: Effective event identification in social media. IEEE Trans. Comput. **36**(3), 42–50 (2013). http://sites.computer.org/debull/A13sept/p42.pdf

71. Romero, D.M., Meeder, B., Kleinberg, J.: Differences in the mechanics of information diffusion across topics: idioms, political hashtags, and complex contagion on twitter. In: Proceedings of the 20th International Conference on World Wide Web, pp. 695–704. ACM (2011)

72. Sakaki, T., Okazaki, M., Matsuo, Y.: Earthquake shakes Twitter users: real-time event detection by social sensors. In: Proceedings of the 19th International Conference on World Wide Web (2010). http://dl.acm.org/citation.cfm?id=1772777

73. Sankaranarayanan, J., Samet, H.: Twitterstand: news in tweets. In: Proceedings of the 17th ACM SIGSPATIAL International Conference on Advances in Geographic Information Systems. ACM (2009). http://dl.acm.org/citation.cfm?id=1653781

74. Schnitzler, F., Liebig, T., Mannor, S., Morik, K.: Combining a Gauss-Markov model and Gaussian process for traffic prediction in Dublin city center. In: Proceedings of the Workshop on Mining Urban Data at the International Conference on Extending Database Technology (2014, to appear)

75. Sharma, J., Vyas, A.: Twitter sentiment analysis. Indian Institute of Technology (2010, unpublished). http://home.iitk.ac.in/jaysha/cs365/projects/report.pdf)

76. Signorini, A., Segre, A.M., Polgreen, P.M.: The use of Twitter to track levels of disease activity and public concern in the US during the influenza A H1N1 pandemic. IEEE Trans. Comput. **6**(5), e19467 (2011)

77. Slaney, M., Casey, M.: Locality-sensitive hashing for finding nearest neighbors [lecture notes]. IEEE Trans. Comput. **25**(2), 128–131 (2008)

78. Subramaniam, S., Palpanas, T., Papadopoulos, D., Kalogeraki, V., Gunopulos, D.: Online outlier detection in sensor data using non-parametric models. In: Proceedings of the 32nd International Conference on Very Large Data Bases, pp. 187–198. VLDB Endowment (2006)

79. Tang, K., Fei-Fei, L., Koller, D.: Learning latent temporal structure for complex event detection. In: 2012 IEEE Conference on Computer Vision and Pattern Recognition (CVPR), pp. 1250–1257. IEEE (2012)

80. Tumasjan, A., Sprenger, T.O., Sandner, P.G., Welpe, I.M.: Predicting elections with twitter: what 140 characters reveal about political sentiment. In: ICWSM 2010, 178–185 (2010)

81. Valkanas, G., Gunopulos, D.: Location extraction from social networks with commodity software and online data. In: 2012 IEEE 12th International Conference on Data Mining Workshops pp. 827–834. http://ieeexplore.ieee.org/lpdocs/epic03/wrapper.htm?arnumber=6406525

82. Valkanas, G., Gunopulos, D.: Event detection from social media data. IEEE Transactions on Computers **36**(3), 51–58 (2013)
83. Valkanas, G., Gunopulos, D.: How the live web feels about events. In: Proceedings of the 22Nd ACM International Conference on Conference on Information and Knowledge Management, CIKM 2013, pp. 639–648. ACM, New York (2013). http://doi.acm.org/10.1145/2505515.2505572
84. Valkanas, G., Gunopulos, D., Boutsis, I., Kalogeraki, V.: An architecture for detecting events in real-time using massive heterogeneous data sources. In: Proceedings of the 2nd International Workshop on Big Data, Streams and Heterogeneous Source Mining Algorithms, Systems, Programming Models and Applications, BigMine 2013, pp. 103–109 (2013). http://dl.acm.org/citation.cfm?d=2501221.2501235
85. Vavliakis, K.N., Tzima, F.A., Mitkas, P.A.: Event detection via LDA for the MediaEval2012 SED task. In: MediaEval, pp. 5–6 (2012)
86. Walther, M., Kaisser, M.: Geo-spatial event detection in the twitter stream. In: Serdyukov, P., Braslavski, P., Kuznetsov, S.O., Kamps, J., Rüger, S., Agichtein, E., Segalovich, I., Yilmaz, E. (eds.) ECIR 2013. LNCS, vol. 7814, pp. 356–367. Springer, Heidelberg (2013)
87. Wang, Y., Sundaram, H., Xie, L.: Social event detection with interaction graph modeling. In: Proceedings of the 20th ACM International Conference on Multimedia (2012). http://dl.acm.org/citation.cfm?id=2396332
88. Watanabe, K., Ochi, M., Okabe, M., Onai, R.: Jasmine: a real-time local-event detection system based on geolocation information propagated to microblogs. In: Proceedings of the 20th ACM International Conference on Information and Knowledge Management, pp. 2541–2544 (2011). http://dl.acm.org/citation.cfm?id=2064014
89. Weng, J., Lee, B.: Event detection in twitter. In: ICWSM (2011)
90. Wood, J.M.: Understanding and computing Cohen's kappa: a tutorial. WebPsychEmpiricist. Web J. (2007) http://wpe.info/
91. Yang, J., Counts, S.: Predicting the speed, scale, and range of information diffusion in twitter. In: ICWSM 2010, pp. 355–358 (2010)
92. Yang, Y., Pierce, T., Carbonell, J.: A study of retrospective and on-line event detection. In: Proceedings of the 21st Annual International ACM SIGIR Conference on Research and Development in Information Retrieval, pp. 28–36. ACM (1998)
93. Zhang, K., Zi, J., Wu, L.G.: New event detection based on indexing-tree and named entity. In: Proceedings of the 30th Annual International ACM SIGIR Conference on Research and Development in Information Retrieval, SIGIR 2007, p. 215 (2007). http://portal.acm.org/citation.cfm?d=1277741.1277780
94. Zhao, S., Zhong, L.: Human as real-time sensors of social and physical events: a case study of twitter and sports games. arXiv preprint, arXiv:1106.4300, 1–9 June 2011. http://arxiv.org/abs/1106.4300
95. Zhong, H., Shi, J., Visontai, M.: Detecting unusual activity in video. In: Proceedings of the 2004 IEEE Computer Society Conference on Computer Vision and Pattern Recognition, CVPR 2004, vol. 2, pp. II–819. IEEE (2004)

Why Do We Need Data Privacy?

Volker Klingspor[(✉)]

Bochum University of Applied Sciences, 44801 Bochum, Germany
volker.klingspor@hs-bochum.de
http://www.hochschule-bochum.de/fbw/personen/klingspor.html

Abstract. In recent years, various socio-political debates and scandals
have raised old and new questions regarding data protection that, among
other things, will also lead to new European legislation initiatives. However relevant each issue may be, there is far too little discussion in the
public which potentials, be it positive or negative, exist with the possibility of combining data from different sources. In this article I want to
give a non-exhaustive overview of the manner in which such information
about everyone of us is collected today, before I discuss the social risks
this may entail. I close the article with some theses outlining a path that
helps to protect the rights of freedom of the citizens despite the extensive collection and analysis of data (My heartfelt thanks goes to Edward
Sodmann for proofreading this text. He required tons of hours to generate
something from my text, that can be understood at all.).

Keyword: Data privacy

1   Current Trends

Information processing is increasingly integrated into everyday consumer electronics. While the average Internet user is worried about his data collected by
Amazon or posted on Facebook, information about him is being collected and
interpreted extensively in much more private and sensitive areas. Examples are:

1.1   Communication Data

WhatsApp, Twitter, and Facebook not only save the data we post about ourselves, be it the already quite personal daily joggs or photos of their loved ones.
Hardly anyone is aware that a personality profile with information about their
marital status (marriage, children, etc.), education, consumer behavior, etc. can
be determined solely from the use of language, without actually communicating
this information [19]. So we unconsciously give many more details about us than
we realize at the time of communication.

Email is a highly decentralized form of electronic communication. There are
many providers or one can even host an email server oneself. Thus, there is no
easy way to access all emails of a person. It requires a relatively deep intrusion

© Springer International Publishing Switzerland 2016
S. Michaelis et al. (Eds.): Morik Festschrift, LNAI 9580, pp. 85–95, 2016.
DOI: 10.1007/978-3-319-41706-6_3

into the network infrastructure (as the NSA has shown us) to analyze email communications on a large scale. Moreover, it is very easy to create different email accounts for different purposes. This makes it even more difficult to create a comprehensive profile of an email user.

The conventional forums allow for a certain anonymity. Forums are typically themed: I am interested in motorcycles and diving, so I can log into two completely independent forums using different pseudonyms. But again, it is of course possible to use the IP address to link the various pseudonyms to the same person.

The communication via social networks and communication platforms such as Facebook, Twitter and WhatsApp has influenced the ability to collect data dramatically. Social networks thrive on the fact that they are, on the one hand, people-oriented, and, on the other hand, forming communication monopolies. Only if most of the communication is done via a single platform, the platform makes sense. Facebook, Twitter and WhatsApp are successful because *almost all* use the same platforms – similar platforms such as the VZ platforms have been forced out of the market. However, the fewer platforms exist, the more they know about us.

No one wants to have to post the same information on different platforms just to reach different friends. Then one could just use email. But this monopoly pushes towards a centralization of data. More so since the purchase of WhatsApp by Facebook, and (after a cooling-off period) the subsequent pooling of data from the two platforms, Facebook knows a great many details of billions of people: their lives, their preferences, their behavior, their whereabouts, and, in particular, their social networks.

A Facebook user profile can draw fairly accurate conclusions about the sexual orientation, ancestry, religion, political attitude, personality, intelligence, well-being, age, gender, relationship status, and the drug abuse of the user. All this information can be obtained indirectly, without asking the user any questions whatsover [19]. With Facebook knowing twenty percent of the world population, this is no trifle matter!

1.2   Photos and Videos

While public video surveillance is subject to strict conditions and hence the data protection is essentially ensured, countless pictures and videos are posted on all kinds of communication platforms in the so-called private sector. Furthermore, ubiquitous smartphone photography and filming with action cams is steadily increasing. Each pastime is now captured on video without considering the people that may unknowingly be in the background. The fact that, unlike photography and classic filmmaking, is especially critical concerning action cams because no conscious image design takes place. While image design usually tries to have as little distraction in the picture as possible, action cams record everything that happens to be in front of the lens.

This is particularly problematic since we have little impact on the publication of pictures in which we happen to be coincidentally. While we can ignore

Facebook and the like (albeit considered as old-fashioned or nutters), we can only object to the publication of photos if we become aware of a photo, and if we are more than a random accessory in the background. While this used to be no problem, as long as the photos were developed on paper and only shown in the circle of friends, digital images quite often appear on the Internet.

Photos and videos are no highly critical information as such. Linking names to photos and videos leads to a critical conflict situation in the near future. The more photos are tagged, the better the face recognition method of Google, Facebook, etc. will perform. In turn, images that have not yet been tagged, can be assigned to specific people. Similarly, as the search for buildings with a photo is available on Google, it is a likely scenario that in a few years one can find the names of the people who are on a photo or video. In addition, it will be feasible to find pictures or movies with non-public persons on the net – possibly only because they appear randomly somewhere in the background of a picture or video.

Since the photos and videos are often marked with metadata such as date and location, more information, such as motion profiles, can be extracted or extrapolated.

1.3 Smartphones

In 2014, worldwide about 1.85 billion people use smartphones [16]; in Germany there are approximately 41 million people [15]. Apps on these devices are easily able to collect very detailed profiles about their users – without them knowing it. A look at the permissions of the first six matches for Android flashlight apps reveals that these can all access the Internet, and five of the six apps require additional permissions. The app with the most permissions, for example, can read the device status and device ID, change system settings, retrieve running applications, take pictures and videos, as well as read, modify, and delete arbitrary files on the device.

For an app that is supposed to only turn on the LED, it's a lot of rights! This begs the suspicion that this app is to produce not only light in the darkness of smartphone owners, but also light in the dark of the supplier by consistently uploading information about smartphone usage so that the supplier is able to analyze the behavior of the user. Ironically, this app is rated above average, so that it is installed probably more often than other, perhaps less curious apps.

Unfortunately, the user of an Android device is not empowered to withdraw special rights of an installed app (unless he roots his smartphone). Apple with their iOS is far ahead: once an app wants access to resources of the smartphone, the user is asked in advance.

1.4 Internet of Things

More and more everyday items are based on electronic control modules. Thus, for efficient energy use one's own home becomes a networked, intercommunicating system. Devices will cooperate in the future in order e.g. to adapt the current electricity consumption to the power generation status or to weather conditions. Heaters will automatically adjust to the habits of the inhabitants.

And of course, these devices must always communicate with the owners. Either explicitly through concrete instructions ("the washing should be done by 16:30"), or by recording the behavior ("if it is recognized that no one is at home on Tuesdays, the heater control can respond accordingly"). Most times, increased comfort is made possible by the extensive collection of data. If one can control blinds and light via a smart home functionality via the Internet, a lot can be recorded detailing the course of our days.

Again, of course: as long as the data remain stored locally, and do not leave their homes, this is not problematic for the time being. However, it is already being planned for the detection of current consumption that information about the load behavior in households are stored centrally in order to facilitate the control of power plants and grids. Fortunately, this issue is still in the hands of the legislature, so anonymization and aggregation of data are regulated [7]. In the consumer sector, however, the authorities cannot respond to each new product with new legislation. And here a technology invades our households whose potential goes far beyond the Orwellian fantasies. Today's television sets can be controlled by gestures or voice commands. This is especially concerning when the TV can be switched on via gestures or language because this requires that a camera or microphone is continuously in receiving mode. TV sets recognize who sits in front of them and report the appropriate user to the Internet services so that the viewer can use these services without prior log-in. At least the manufacturer Samsung stipulates that in the terms that the recorded data may be sent to third parties [9].

Another example of the increasing networking of everyday objects is the rising number of equipment in vehicles with communication modules. Today, all modern cars record not only functional errors but also parameters of driving behavior. High-end vehicles are fitted with communication modules that are able to transmit these data online to the manufacturer. Together with the driver recognition via electronic key or seating positions information can be generated from the speakerphone's address data or from the GPS data that go far beyond mere motion profiles. Without knowing a concrete study, I am sure that the mood of the driver can be recognized by the accelerator and brake protocol. In contrast to the opportunities that arise here, the possibilities of black boxes for detection of driver behavior as they are offered by commercial truck insurers are less threatening.

1.5   Linking Information

To date, information substantially exists as data islands. Every company, every forum, every game portal, each social media site collects data about customers or users. For some of these providers, we may mask our identity with pseudonyms. Once we enter into a business transaction, this is generally not possible. In addition, Facebook in particular expects and reviews the application with a real name, although this is actually not necessary. Research from 2006 shows that in an individual case it is possible to relate different, even anonymous user profiles with each other, matching it to a specific person. The best-known example is the assignment of a IMBD profile to a video rental customer's account [13]. While in this example, only a single person was de-anonymized, whole subnets of the Flickr network could be de-anonymized in [12]. In 2000, Latanya Sweeney showed that 87 % of all Americans could be uniquely identified using only three bits of information: ZIP code, birth date, and sex [17]. One year before, she identified and related the anonymized medical records of the governor of Massachusetts [6].

Meanwhile various approaches to de-anonymize people are developed by combining distributed data. Researchers can given credit that they want to demonstrate to users the possibilities and to suggest a less generous attitude with their data. Nevertheless, it is expected that within a few years, techniques will be on the market that link user and search profiles on a large scale. Information which is regarded as confidential and anonymous by Internet users can be assigned to them anyway after all.

1.6 Pre-crime Detection

Data Mining techniques are now being used for the prevention of crime and terror. "The Future Attribute Screening Technology project (FAST) system has the capability to monitor physiological and behavioral cues without contact. That means capturing data like the heart rate and steadiness of gaze of passengers about to board a plane. The cues are then run through algorithms in real-time to compute the probability that an individual is planning to commit a crime" [10].

In Memphis, Tennessee (USA), data analysis has been used to preventively to monitor locations where potential offenses may be committed since 2005. In 2010, the police and IBM celebrated with a decline of serious crime by more than 30 %, including a 15 % reduction in violent crimes since 2006 [11]. The reason for the decline is, according to the publication, that particular gang disputes could be detected early. Unfortunately, it is not clear from text to what extent people were arrested preventively, and, conversely, to what extent honest persons preemptively avoid areas where they can potentially be arrested.

2 Why Is Privacy Important?

2.1 Loss of Autonomy and Freedom Rights

"I have nothing to hide!" This sentence is heard often at discussions about data protection and in particular on data that is to be given to law enforcement authorities. I would like to emphatically disagree.

"Privacy describes the extent to which a person other people are permitted to enter one's own world" [4,18]. We constantly negotiate the limits of our privacy and our voluntary disclosure with others but also with ourselves. The scope of one's privacy is very individual, and may even vary at different times. We need an open mental sphere to evolve. Young people need privacy to learn to think for themselves and to act self-confidently. The less a teenager has trust in his privacy, the less he will dare to act contrary to the norms of his peer group, and the more uniform and less confident he will be. But adults need privacy to personally and professionally develop, too. Sabine Trepte formulated as a benefit of privacy among others the autonomy to break social norms and to experiment with new behaviors and thoughts [18].

The German Federal Constitutional Court ruled in its landmark ruling on the census [1] that if someone is not reasonably certain how his personal information is used and shared, he can be inhibited in its freedom, to plan and decide in a self-determined way. People who do not know what information is held on them, will try to behave as inconspicuously as possible. There is the likelihood that central fundamental rights are waived without being aware.

The protection of privacy not only means that "thoughts are free", but also that I may share my thoughts in a safe room with the people that I trust. For this reason, the inviolability of the home is an essential and fundamental right. Equally important is the inviolability of communication. Of course, not every conversation, every chat, every posting is equally confident, nor any confidentiality is of equal importance. But I must be able in a figurative sense "to close the door" at all times in order to monitor accurately who participates in a conversation.

In academic literature, scholars distinguish three dimensions of privacy [14]: *Informational privacy* refers to the fact that my data will not be public unless I want to. *Decisional privacy* describes the right to be protected in decisions and actions from unwanted external influences. *Local privacy* describes the protection against the entry of other into private rooms and areas.

And only those that respect the privacy of a person, respect him as an autonomous person in the sense that he has the freedom to live his life independently and to seek his own happiness [14]. Conversely, secret knowledge about other people gives institutions power over these, which can lead to changes in behavior and behavioral adaptations. In this sense, to understand the ruling of the Constitutional Court: this power through secret knowledge threatens the freedom of expression and freedom of assembly, and thus central fundamental rights.

Unfortunately, many Internet users already resigned: *"It is already too late, they already know everything about me"*. Many people who do not agree with the data collection use this as an excuse and keep using the new communications media. I agree with that in so far as all information that we have disclosed via social media services, etc., cannot be taken back, i.e. cannot be deleted. This certainly does not mean that we should continue feeling unconcerned. The more information that is available about us, the easier it will be in the future to associate this information with other strings of information, and finally to generate complex personality profiles.

The sooner we begin to be careful with our data, the better. The human quality to forget or to put into perspective what is said over time is just an important tool for personal development, such as the aforementioned privacy.

Only trusting the fact that details about my actions and statements over time disappear from the memories, allows me to freely discuss. Unfortunately, the Internet and data collection companies do not forget what we have posted!

The safest option for confidentiality is the boycott of as many Internet-based services as possible. This way less or no information will be disclosed, of course, and cannot be misused. And although I certainly advocate a more restrictive use of the Internet: this path sooner or later leads to an exclusion from the peer groups that communicate via social-media platforms. If I'm not on Facebook, I do not get the information provided exclusively via this platform. Also note that not even the complete Internet abstinence really solves the problem; this strategy will only delay the collection of data and cannot really prevent it, either. Data are now collected at every traditional purchase, every contract, and any travel booking.

To illustrate this more vividly: you search old classmates for a class reunion? Check with a credit agency such as Bürgel or Creditreform. They know the complete move history, including any change of name of any bank client. Just give as reason for the request "business contacts", and – a couple of euros poorer – you know the current address of your old class mate and, in addition, you can verify if he is able to pay his bill himself? [8].

2.2   Transition from a State of Law to a State of Prevention

A constitutional state is different from a preventive state in that the former does not preemptively act against potential offenders – only the charge of an offence may lead to a penalty. Preventive actions always lead to a restriction of the freedom of action – especially for innocent citizens.

Not without reason will the establishment of any public video surveillance be discussed intensively – particularly from the people that have nothing to hide. There is a very fine line between security and upholding fundamental rights. In a constitutional state we are not able to prevent all crimes, nor will we be able to solve every crime because of the given legal means. However, the rulings of the Federal Constitutional Court (e.g. concerning data retention) seem to suggest that this tightrope walk works quite well – at least in Germany.

2.3   Desolidarization Society

Machine Learning techniques as well as traditional statistical techniques are often used for risk assessment. However, automatic risk assessment can cause certain members of the population to be unreasonably excluded from a fair treatment that is based on facts and not extrapolation. If scoring functions are used already, they should produce accurate results in every case. The well-known example of someone receiving no credit because he lives in the wrong area indicate the poor scoring functions. However, I do not understand why trained and experienced bank branch managers no longer have the power to overrule the scoring result. The individual applicant may have sufficiently good proof that ultimately speaks for getting his credit.

In the insurance industry a more accurate risk assessment leads to a loss of solidarity. The more closely I can determine the risk of insurance, and the more accurately I can set the premium on this risk, the more the idea of solidarity insurance is lost. In almost superfluous insurances such as the legal protection insurance (for "normal" individuals) this might be a trifle.

When, however, the risk assessment excludes participation from society, we lose a fundamental principle of our state. Consider, for example, the car insurance sector: the fact that the premium depends on the type of vehicle is unproblematic – cars that carry lower premiums can be an option. Since the current premium determination has started considering whether the parents also own a car, and how old the driver is, the premiums for liability insurance can easily vary from case to case by a factor of three to four. I can even drive as prudently and cautiously as I want to – I may be charged higher premiums just because the statistics speak against me and I may not be able to afford a car. A corresponding analysis in the field of health insurances might point at an even worse scenario for a nation.

In my view, we are losing the basic idea of an insurance. Their idea is to spread the risk as much as possible, to give everyone the same opportunities. If, due to individual misconduct, premiums rise modestly, that is certainly acceptable. But to exclude people from insurance policies putting up high premiums without letting them have an influence on the design of the premium does not correspond to my idea of our society.

3   Is Data Collection and Analysis to Demonize Generally?

Data analysis is just a technique, and is not bad in itself. In many situations, including those that will perhaps be viewed by the population as questionable data analysis carries overall positive aspects.

The quality of Google search results bases on a very detailed analysis of user behavior. The fact that almost always a relevant hit is already on the first results page, reflects the quality of this analysis. The extrapolation quality of the on-screen advertising is very high, too. There is, of course, the self-interest of Google because the ads usually must be paid only if they are clicked. You do not get anything for nothing – without collecting and analyzing data Google search works just not as well (the older readers among us remember surely still with horror at the previous attempts to adjust the keywords as closely as possible to the search target, to aid search engines to find something useful).

And I love the assistance provided by many vendors displaying similar products. I can browse through music, and come across new, hitherto unknown pieces. Invoices are archived for years by the manufacturers and may be accessed by the customer at any time. There is no need for him to file away the bills at home in folder. As long as the data is saved by the providers, there is little danger from them. Scary is the fact that much more data is collected about customers than is required for the sale. Amazon e.g. gathers intensive data about the reading habits of the users of its e-book reader [3].

And also in the scoring of credits, there are positive aspects of a systematic data analysis. The denial of a loan is not only to protect the financial lender against losses, but also very much the borrower against over-indebtedness. The more accurate the scoring process the better I can grant credits to those that do not overburden themselves with this loan. It is a pity, however, if the scoring algorithm is not able to justify the classification, and if, at the same time, the employees of the lenders are completely exempted of the freedom to lend based on personal assessments and counter-scoring results.

There are many other data analysis applications from which citizens benefit more or less directly. The analysis of traffic flows may optimize traffic control and planning, so that road users get faster and more energy-efficiently to the destinations. The analysis of the movement patterns of people in shops or at events such as conferences can lead to improvements in the placement of objects (sale items, break counters, . . .).

In this sense, there is certainly an unlimited number of useful applications of data analysis that can help to cope with the challenges that we have to face because of limited and increasingly costly resources. But how should society deal with the conflict between benefits and risks?

4   Conclusion

Big Data cannot be stopped, and in many areas data analysis is really helpful. It also seems to point out that with the larger amounts of data the statistical errors become less relevant – that are so used to analyze more data, the better the results are.

However, we need an intense social discourse about the ways our data is handled. We need empowered and educated citizens who are aware of the dangers and consequences of data collection. Citizens who actually read the privacy statements of companies or the permissions of apps, and discard applications when they have doubts about the legality of the use. Citizens who understand the principles and apply encryption techniques, and quit service providers who do not offer any encoding. Citizens who consider carefully what information they publish about themselves. At the same time, we need citizens who critically evaluate information gathered from the Internet and that rely on different sources of information, and that are always aware of the existence of fake identities and propaganda dressed up as information.

For these citizens, however, we need a better infrastructure to ensure their data are safe. Easy to use encryption programs that allow local data and data stored in cloud storage to be automatically protected by encryption. Communication systems, including in particular email, encrypting messages end-to-end, and thereby allowing a simple but faithful key exchange. Although this software already exists, it is still underused, which is, in my view, due to complicated use for IT laymen. It would be desirable if pictures, videos and other documents could be provided with an expiry date, after which they can no longer be seen. This would ensure that missteps in the identification process, particularly of young people would not permanent part of their public life.

It is absolutely certain that we also need new laws. The Federal Constitutional Court took very clear position in 1983 with the census ruling ensuring the right to informational self-determination. All European countries are now following the European Data Protection Directive, published in 1994 by having adopted country-specific data protection laws. All these laws do no longer work effectively in the global world, as they are binding only for those providers who are headquartered in the EU. Those who do not want to keep to European law, need only to open their businesses in Tonga (or the US), and can then handle data in an unrestricted way. Furthermore, the penalties for offenses against the privacy law are too low to deter potential perpetrators. Since 2012, a new European data protection regulation has been discussed, in particular addressing the two points mentioned above [2]. First, the privacy regulation is applicable as soon as *product* is offered in Europe, regardless of where this occurs. Then, European citizens are protected against suppliers from countries that have weak privacy rights. On the other hand, a significant increase in the potential penalties is planned so that this right can be enforced.

We need to ensure that people will not be constantly put into a box. The larger the amount of data that can be used to analyze, the better purely statistical correlations work, and the less analyses that are based on models are used [5]. Since we lack the explanations for automatically-made decisions, a bank employee is unable to overrule the scoring decision because he cannot identify an inconsistency between decision and request. I believe we should have the right to oppose purely statistical assessment procedures that force the seller to disclose the reasons for a refusal.

Still less can it be permissible to criminalize people on the basis of statistical analyses. In a constitutional state an offender still must have actually committed a crime before he is indicted. Each statistics-based crime prevention that determines and monitors potential criminals leads away from the rule of law towards a prevention and surveillance state.

And finally, we need more learning techniques (and other techniques); which ensure a very early real anonymization of the data collected. Research by e.g. the Fraunhofer Institute for Intelligent Analysis and Information Systems shows, that an analysis of pedestrian flows is possible just on the basis of aggregated data that cannot not be de-anonymized [20].

References

1. BVerfGE 65, 1; Az. 1 BvR 209, 269, 362, 420, 440, 484/83 (1983)
2. Proposal for a regulation of the european parliament and of the council on the protection of individuals with regard to the processing of personal data and on the free movement of such data (general data protection regulation) (2012). http://eur-lex.europa.eu/legal-content/EN/TXT/PDF/?uri=CELEX:52012PC0011&from=EN

3. Alter, A.: Your E-Book is reading you. Wall Street J. (2012). http://www.wsj. com/articles/SB10001424052702304870304577490950051438304
4. Altman, I.: Privace: a conceptual analysis. Environ. Behav. 8(1), 7–29 (1976)
5. Anderson, C.: The end of theory: the data deluge makes the scientific method obsolete. Wired 16(7) (2008)
6. Anderson, N.: "Anonymized" data really isn't - and here's why not. Ars Technica (2009). http://arstechnica.com/tech-policy/2009/09/ your-secrets-live-online-in-databases-of-ruin/
7. Bräuchle, T.: Die datenschutzrechtliche Einwilligung in Smart Metering Systemen - Kollisionslagen zwischen Datenschutz- und Energiewirtschaftsrecht. In: Pödereder, E., Grunske, L., Schneider, E., Ull, D. (eds.) INFORMATIK 2014: Big Data - Komplexität meistern, pp. 515–526. Lecture Notes in Informatics, Gesellschaft für Informatik (2014)
8. Dambeck, H.: Verschollene Mitschäler: Bärgel und Creditreform retten Klassentreffen. Spiegel online (2006). http://www.spiegel.de/netzwelt/web/ verschollene-mitschueler-buergel-und-creditreform-retten-klassentreffen-a-452447. html
9. Eikenberg, R.: Spion im Wohnzimmer - Privacy und Sicherheit bei Internet-fähigen TVs. c't (4) (2014)
10. Erickson, M.: Pre-crime detection system now being tested in the U.S. (2012). http://bigthink.com/think-tank/ pre-crime-detection-system-now-being-tested-in-the-us
11. IBM: Memphis police department reduces crime rates with IBM predictive analytics software (2010). http://www-03.ibm.com/press/us/en/pressrelease/32169.wss
12. Narayanan, A., Shi, E., Rubinstein, B.I.P.: Link prediction by de-anonymization: How we won the kaggle social network challenge. CoRR abs/1102.4374 (2011). http://arxiv.org/abs/1102.4374
13. Narayanan, A., Shmatikov, V.: How to break anonymity of the netflix prize dataset. CoRR abs/cs/0610105 (2006). http://arxiv.org/abs/cs/0610105
14. Rössler, B.: Der Wert des Privaten. suhrkamp taschenbuch (2001)
15. Statista: Anzahl der Smartphone-Nutzer in Deutschland in den Jahren 2009 bis 2014. http://de.statista.com/statistik/daten/studie/198959/umfrage/ anzahl-der-smartphonenutzer-in-deutschland-seit-2010
16. Statista: Prognose zum weltweiten Bestand an Smartphones von 2008 bis 2017. http://de.statista.com/statistik/daten/studie/312258/umfrage/ weltweiter-bestand-an-smartphones/
17. Sweeney, L.: Simple demographics often identify people uniquely. In: Data privacy working paper. Carnegie Mellon University (2000)
18. Trepte, S.: Privatsphäre aus psychologischer Sicht. In: Schmidt, J.H., Weichert, T. (eds.) Datenschutz - Grundlagen, Entwicklungen und Kontroversen, pp. 59–66. Bundeszentrale für politische Bildung (2012)
19. Tufekci, Z.: Ein Datensatz mit X. The European (3) (2013). http://www. theeuropean.de/zeynep-tufekci/7065-gefahren-von-big-data
20. Wrobel, S.: Big Data Analytics - Vom Maschinellen Lernen zur Data Science. In: Plödereder, E., Grunske, L., Schneider, E., Ull, D. (eds.) INFORMATIK 2014: Big Data - Komplexität meistern, p. 53. Lecture Notes in Informatics, Gesellschaft für Informatik (2014)

Sharing Data with Guaranteed Privacy

Arno Siebes[✉]

Algorithmic Data Analysis Group,
Universiteit Utrecht, Utrecht, The Netherlands
arno@cs.uu.nl

Abstract. Big Data is both a curse and a blessing. A blessing because the unprecedented amount of detailed data allows for research in, e.g., social sciences and health on scales that were until recently unimaginable. A curse, e.g., because of the risk that such – often very private – data leaks out though hacks or by other means causing almost unlimited harm to the individual.

To neutralize the risks while maintaining the benefits, we should be able to randomize the data in such a way that the data at the individual level is random, while statistical models induced from the randomized data are indistinguishable from the same models induced from the original data.

In this paper we first analyse the risks in sharing micro data – as statisticians tend to call it – even if it is anonymized, discretized, grouped, and perturbed. Next we quasi-formalize the kind of randomization we are after and argue why it is safe to share such data. Unfortunately, it is not clear that such randomizations of data sets exist. We briefly discuss why, if they exist at all, will be hard to find. Next I explain why I think they do exist and can be constructed by showing that the code tables computed by, e.g., KRIMP are already close to what we would like to achieve. Thus making privacy safe sharing of micro-data possible.

1 Introduction

The attitude of many people with respect to privacy may seem rather ambivalent. On the one hand they freely share rather personal – often very personal – data on social networks such as Facebook and Twitter. While on the other hand they baulk at the detailed information about them that is gathered both by these social networks as well as by websites such as search engines, companies, and also government agencies. To a great extent this seeming ambivalence is due to a lack of awareness. Most people only now start to understand how much detailed information on them is stored in countless databases that can, moreover, be easily linked with each other. To understand what the advantages and opportunities but also what the disadvantages and the risks of these vast amounts of data are.

S. Michaelis et al. (Eds.): Morik Festschrift, LNAI 9580, pp. 96–108, 2016.
DOI: 10.1007/978-3-319-41706-6_4

The reason why they become aware is "Big Data"[1]. Both the virtues – such as automatic translation – and the vices – such as hacks of large customer databases or seemingly private collections of photos – of a data driven society are presented in the media under this heading. And with awareness, the feelings vis a vis Big Data are changing. This can, e.g., be seen from more or less popular books on Big Data. Whereas [14] mostly extols the huge opportunities offered by Big Data, later books such as [11] – more from a technical point of view – and [17] – more from a political point of view – discuss the threats to privacy as deeply as – if not deeper than – the opportunities brought by Big Data.

Unfortunately, this growing awareness poses real threats to opportunities of Big Data. For example, under European regulations it is already hard – if not impossible – for hospitals to pool detailed information on rare diseases – just a few occurrences, if any, per hospital per year – with possible detrimental consequences for the recognition and treatment of patients. It is true that the involuntary release of healthcare information can have disastrous consequences for individual but the choice not to share seems about as bad, if not worse. With a growing distaste for vast data collections such situations will only occur more and more.

The data mining community was well aware of these threats, well before the general public started to worry. Under the umbrella term "privacy aware data mining" techniques such as k-anonymity [16] and differential privacy [8] – both originating in the Statistical databases community – have been embraced, studied and expanded upon. Yet, it is the question whether or not the deployment of such techniques will satisfy the population or the possibly much more severe privacy guidelines that might result from this unrest. After all, the data miner might not be able to learn individual data, but that data is still in the database. That is, it can potentially still leak.

In other words, privacy preserving data mining is not the answer to the problem of data sharing. What we want is to be able to share detailed data – micro data as it is called in Statistics – with an absolute guarantee for privacy. That is we would like to anonymize databases in such a way that no-mater what, none of the entries in the database can be related to a person. At the same time, of course, this anonymized database should be related to the original in the sense that any – sensible – statistical analysis on the one yields (mostly) similar results to that same analysis on the other database.

Note that such an anonymized database guarantees privacy. It is OK if someone has access to such a database, even if you are in the original database. The access doesn't give any information about you at all. As an aside, note that such anonymization techniques may also make companies more willing to share data.

---

[1] Surprisingly often terminology arising in marketing and/or journalism enters the scientific vernacular. While this is understandable from the point of view that funding agencies want to fund research that society needs and researchers need funding, but one can but wonder what would have happened if the term hypology – from the classical Greek $\upsilon\pi o\lambda o\gamma\iota\sigma\mu o$ (calculate) – once coined as an alternative for the term computer science [21], would have caught on.

Large data driven companies such as Facebook have data that is very valuable to, e.g., the social sciences. However, since that data is also very valuable to the company itself, it will certainly not share the data and be very careful to give access to individual researchers. This means that it is hard, if not impossible, to replicate research like in [2] and even more that there are problems that could be studied but never will. The anonymization we are discussing here removes most, if not all, commercial value of the data – after all, there are no real individuals one could target in that database – while its scientific value is mostly retained.

The second requirement – statistical indistinguishability – simply means that the anonymized database should exhibit, more or less, the same set of patterns as the original one does; at least that is what it seems to mean to a pattern miner like me.

The problem discussed in this paper is: is this possible? Can databases be transformed so that both privacy and statistical indistinguishability can be guaranteed? To be upfront, this question is not answered in this paper. What this paper presents is firstly a detailed discussion of the problem as well as a (quasi) formal problem statement. Secondly, I'll explain why I think that the problem can be solved in a positive way; note that it is far from clear that such transformed databases exist let alone be computed from an original database. For this second part, the data is assumed to be simply a table with categorical data. While I am convinced that everything discussed here is largely independent of the type of data under consideration, sticking to categorical data simplifies the discussion greatly.

2   Profiles

If you know exactly what a prospective customer wants, say X, you can make her a truly personalized offer: you can now order X for only $\$y$, this offer is valid for the next z days. But knowing exactly what a prospective client wants is not that easy. She may have examined, among other things, X at your website, but that doesn't mean that she is still interested in X. If not, adverts for X will be a waste of money and e-mail offers for X will most likely just annoy her.

Hence, personalized offers are often based on *profiles*, (simple) characterisations of clients based on their attributes. For example,

$$\text{Age} \in [18, 25] \wedge \text{Sex} = male,$$

with the offer being something very much liked by that client group and not yet owned, as far as you know(!), by this particular client.

In fact, profiles have a far greater use than just webvertising. Insurance companies create profiles to identify high-risk groups, health researchers profile to better understand who suffers from a certain disease or has a high(er) risk of doing so, and literature researchers profile authors to be able to attribute an anonymously published book.

In most, if not all, these cases, the profiles consist of simple selection statements as above. What varies is what queries are deemed to be interesting and how they are discovered. One way to formally define this type of profiling is through *theory mining* [13]:

Definition 1. *Given a query language $\mathcal{Q}$, a selection predicate ϕ, and a database db the theory $\mathcal{T}(\mathcal{Q}, \phi, db)$ is defined by*

$$\mathcal{T}(\mathcal{Q}, \phi, db) = \{q \in \mathcal{Q} \mid \phi(q, db) = true\}$$

In other words, the theory consists of all those queries – i.e., profiles – that are *interesting* in *db* according to ϕ.

The predicate ϕ could simply count the size of $q(db)$ and check whether or not it exceeds a threshold, then we have frequent pattern mining [1]. Or it could check whether or not the size of $q(db)$ increases by much for two consecutive time points, in which case we have emerging patterns [7]. Or it could compute an aggregate on some attribute for $q(db)$ and when that is higher than a threshold (which may the value of that aggregate on the complete database or on (some) complement) giving us subgroup mining [10]. And so on, and so on.

Whether or not we can compute the complete theory or have to be satisfied with a heuristically computed set of good results depends very much on both Q and ϕ. But there is a large collection of both *pattern languages* (as Q is often called) and interestingness predicates (as ϕ is usually called) together with algorithms to discover interesting patterns on a wide variety of data types.

That is, there are algorithms that allow us to discover profiles for a wide variety of tasks, making profiling easily one of the most important applications of Big Data. For the purposes of this paper we will mostly restrict ourselves to one specific type of profiles, viz., frequent patterns:

$$\mathcal{T}(\mathcal{Q}, \phi, db) = \{q \in \mathcal{Q} \mid\mid q(db) \mid \geq \theta\}$$

In which θ is some user defined threshold and Q consists of conjunctive select queries like

$$A_i = v_{i,l} \wedge A_j = v_{j,m} \wedge \cdots \wedge A_k = v_{k,n}$$

over the discrete domains of the attributes of the single table database.

The reason why profiles (patterns) are so popular is what social scientist call *homophily*, or as the English proverb has it: birds of a feather flock together. That is (in our case), the assumption that people who resemble each other in some (important)respects will also resemble each other in other respects.

Moreover, most predicates ϕ will ensure that $|q(db)|$ is large enough to ensure that it is likely that the profile q generalizes to unseen data. Which is in turn usually tested with out of sample techniques such as cross-fold validation or other techniques.

Another, slightly unrelated, reason why patterns – and especially frequent patterns on categorical data – are important is that they form the basis of many other data analysis techniques. Most, if not all, categorical data analysis is based on count data; count data on the complete table but even more count data on sub-tables like marginals and conditionals. These sub-tables are identified by patterns, the count is the support of that pattern.

Again for reasons of generalizability, also in this case one is mostly interested in patterns with a reasonably large support. Think, e.g., of decision trees [3] – which are also based on count data, i.e., the support of patterns. One may grow trees to small-support patterns, but these will quickly be pruned away by statistical hypothesis testing methods.

In other words, an important use of micro-data is to mine for patterns with reasonably large support. Clearly, that is not the only use as it would preclude the targeting of a single client, but for *data analysis* purposes it is arguably the most important use. And large-support patterns are reasonably safe with regard to privacy. After all, external knowledge will not help to identify you in a large peer-group. Or does it?

3   Re-identification

Anonymization, especially together with perturbation and/or generalization, may seem an excellent way to guarantee the privacy of those who are in the database. After all, anonymization means that sensitive attributes, such as names and social security numbers, are removed; perturbation means that random noise is added to the remaining attributes; and generalization means that individual attribute values are grouped, e.g., by discretization or by using hierarchies. What possible private information can remain after such a thorough sanitization of the data?

Unfortunately, it only seems an excellent way. Time and time again it has been proven that with outside data sources – background knowledge – it is possible to identify individuals in such anonymized data sets. Because this pertains directly to what we want to achieve, we discuss two such cases.

The first case is the famous Netflix prize [24]. To allow data miners outside of Netflix to build a recommender system which beats Netflix own system – and get a substantial reward for that – Netflix released an anonymized dataset to the public. The entries in that data set are of the form

$$< user,\ movie,\ date\ of\ grade,\ grade >$$

in which both *user* and *movie* are integer ID's. For each movie, title and year of release are given in a separate data set, for users there is, of course, no further information whatsoever. This may seem a completely safe data set to release. Unfortunately it isn't. The reason is that Netflix isn't the only site to which users upload ratings, IMDb is another one.

Very soon after the release of the data set, in 2007, two researchers, Arvind Narayanan and Vitaly Shmatikov, from the University of Texas at Austin were able to link users from Netflix to users from IMDb.

In [15] they showed how to do this. The main cause to make it possible is that the data is sparse in the sense of the following definition.

Definition 2. *A dataset D is (ϵ, δ)-sparse with regard to the similarity measure s if*

$$Pr(sim(r, r\prime) \geq \epsilon\ \forall r\prime \neq r) \leq \delta$$

That is, a data set is sparse if for the fast majority of records in the data set there are no other records that are remotely similar.

In our pattern language this can be expressed as: a data set is sparse if the vast majority of records satisfy (almost) unique patterns; i.e., patterns with a support of one or close to it. Note that the paper doesn't detail the length of these patterns, just their existence. This is different for the second paper we briefly discuss, on the re-identification of credit card data [6].

In [6], the data set D consists of 3 months of credit card transactions for 1.1 million users in 10,000 shops. It is anonymized by replacing the user details by an abstract user-id, the time of purchase is truncated to the date only and the transaction-amount is discretized in progressively larger bins. The tool of choice of the authors is *unicity*.

Definition 3. *For $p \in \mathbb{N}$, define for each user $I \in D$ $S(I_p)$ to be the set of traces in D that match the trace $t(I)$ of I on p randomly selected data points in $t(I)$. The unicity at p is then defined by:*

$$\epsilon_p = Pr(|S(I_p)| = 1)$$

That is, ϵ_p is the probability that knowing p transactions of a user – background knowledge – is enough to identify the complete trail (all purchases) of that user. For that particular data set, the authors show that $\epsilon_4 > 0.9$, i.e., over 90 % of the users are identifiable from only 4 known transactions. In other words, a little outside knowledge gives you a lot of new knowledge. With external knowledge you are not safe, even if one only knows that you are in the complete database.

4   The Problem

From the previous two sections we have learned that

1. for data analysis we need the support of patterns that have a reasonably large support,
2. but unfortunately individuals happen to satisfy very low support patterns as well.

Hence, guaranteeing privacy when sharing data is not easy if we want to allow (useful) pattern mining. The – very naive, a bit differential privacy like – approach of disclosing the support only if it is big enough obviously fails. If the support of p_1 is large, so will the support of $p_1 \vee p_2$ be, regardless of the support of p_2 and, thus, disclosing a lowerbound on the support of the latter.

The only way to guarantee that private information cannot be retrieved from a shared data set is by randomizing it. For, if the individual tuples in our table are completely random, no amount of external knowledge can link them to persons. The tuples are random, they do not correspond to any person.

The downside of such randomized data is, of course, that their use for data analysis is limited. Models derived from random data do not tell us very much about the real world. In fact, truly random data cannot be modelled. In the terminology of Algorithmic Information Theory [12], data is precisely random if it cannot be compressed. So, while truly random data guarantees privacy, it isn't the answer as it renders the shared data useless.

The randomized data would not be useless if the support of large(r) patterns would be more or less preserved; only more or less as models should be robust against small changes anyway. In other words, our problem is can we randomize the data such that

– small support patterns have a mostly random support
– while large(r) support patterns have mostly their original support?

Or as a pseudo[2] formal problem statement:

Problem. Given a database db, thresholds θ_1 and θ_2 and parameters ϵ and δ, create a database r such that

1. for all patterns p such that $supp_{db}(p) \leq \theta_1$:

$$Pr(supp_{db}(p) \neq supp_r(p)) \geq \delta$$

2. for all patterns p such that $supp_{db}(p) \geq \theta_2$:

$$Pr(|supp_{db}(p) - supp_r(p)| \leq \epsilon) \geq \delta$$

That sharing r instead of db mitigates privacy risks while allowing to draw statistical analysis as if one was given db should be clear. For complete (unique) tuples – or by extension high-unicity patterns – one doesn't know whether or not they correspond to a real world entity, whatever external information you can link it to. While requirement 2 guarantees that models derived from r will not be too far from models derived from db.

As already mentioned in the Introduction, I'm unfortunately not – yet(!) – able to give a proven algorithm that constructs such a randomized version. Instead I'll discuss why I believe that such algorithms exist. The task to design one is future work.

5   Do Solutions Exist?

It is not straightforward that our problem has a solution. In fact, it is obvious that for some settings no solution can exist. If one sets, e.g., $\epsilon_1 = 1$, $\epsilon_2 = 2$, $\epsilon = 0$ and $\delta = 1$, i.e. requiring that all support = 1 patterns in db get a different support in r, while all patterns with a support ≥ 2 in db get the same support. Clearly, this is an inconsistent set of requirements per the $p_1 \vee p_2$ argument above. Hence there are at least cases in which no solution exists.

It is easy to see that both criteria on their own are satisfiable, i.e., allow for solutions to exist. Let, e.g., db be a 0/1 database. If we only regard requirement 1, a simple bit-flip will suffice; if every 0 is set to 1 while simultaneously every 1

---

[2] If only because of the large number of parameters we use.

is set to 0 all patterns that happen to have a support that is different from half the database will get a new support. It is even easier to satisfy just the second requirement. Simply make r a copy of db. Unfortunately, these two solutions are about as different as it gets. Hence, we still have no reason to believe that solutions that satisfy both criteria simultaneously exist.

A different way to approach this is by counting. Again assume that db is a binary database with m columns and n rows. That is, $db \in D$, in which D is the set of all $ntimesm$ binary databases. Obviously

$$|D| = 2^{nm}$$

Clearly, r will also be in D. In other words, if we wouldn't care about the support of patterns we could replace db by any of the 2^{nm} elements of D. But we do care about the support of patterns and each pattern limits the set of elements of D we can use.

If a pattern π covers p columns and s rows in db an equality requirement for condition 2, diminishes D by a factor 2^{-ps} for the cover of the pattern and a factor $\left(\frac{2^p - 1}{2^p}\right)^{(n-s)}$ for the rows not covered by π. That is, $r \in D\prime$ with

$$|D\prime| = 2^{-ps} \left(\frac{2^p - 1}{2^p}\right)^{(n-s)} |D|,$$

which may be a lot smaller than D and there are many such patterns. Having ϵ leeway makes the reduction smaller, but again, there are many such patterns.

Something similar can be said for the condition 1 pattern constraints. While in general their reduction, given (obviously) by:

$$|D\prime| = \left(1 - 2^{-ps} \left(\frac{2^p - 1}{2^p}\right)^{(n-s)}\right) |D|,$$

will be smaller, there are usually even more small support patterns than there are larger support patterns.

Hence, also if one counts, it is not obvious that our problem admits solutions. However, the situation is not necessarily as bleak as it seems. For, the patterns one discovers in db are not independent, i.e., it is not true that each new pattern considered will decrease the number of possible candidates for r as drastically as the above calculation shows. This is, e.g., witnessed by the success of condensed representations such as non-derivable item sets [4].

Unfortunately, the complexity of non-derivable item set mining, based on the inclusion/exclusion principle, makes it hard to give a good estimate on how large the pool $D\prime$ of candidates for the randomized r actually is.

Hence, we still have no compelling reason to believe that such randomized versions exist. At most we have evidence that if they exist they'll be probably hard to find. So, why do i believe they exist? The reason is that given a database db I can generate databases that are close to what r should be.

6   Code Tables and Database Generation

The crux to the database generation is a code table [19,23]. A code table CT is a two column table in the first column there are patterns in the second column there are code words from some prefix code. To be a valid code table, CT has to contain all the singleton patterns, i.e., all patterns of the form

$$A_i = v_{i,j}$$

Moreover, both the patterns and the code words should be unique; i.e., occur at most once in CT. To encode a database we first cover it. To find the cover of a tuple $t \in db$, we go down the first column of CT and search for the first pattern p_i such that

$$p_i \subseteq t$$

If $t \backslash p_i = \emptyset$ we are done and $cov(t) = \{p_i\}$. If not, we recursively cover $t \backslash p_i$. To encode t we simply replace each pattern p_i in its cover by its related code word $c_i \in CT$.

Not all code tables are equally good for database generation, but the optimal one in the Minimum Description Length principle (MDL)-sense of the word is. The MDL principle [9] can be paraphrased as: *Induction by Compression*. Slightly more formal, it can be described as follows. Given a set of models $\mathcal{H}$, the best model $H \in \mathcal{H}$ for data set D is the one that minimises

$$L(H) + L(D|H)$$

in which

- $L(H)$ is the length, in bits, of the description of H
- $L(D|H)$ is the length, in bits, of the description of the data when encoded with H.

Given the left-hand side of CT – we fix the patterns – it is easy to determine the optimal code words. To compute this code length, we encode each transaction in the database db. The usage of an pattern $p \in CT$ is the number of tuples $t \in db$ which have c in their cover. The relative frequency of $c \in CT$ is the probability that c is used to encode an arbitrary $t \in db$. For optimal compression of db, the higher $P(c)$, the shorter its code should be. In fact, from information theory [5], we have the Shannon code length for c, which is optimal, as:

$$l_{CT}(c) = -\log(P(c|db)) = -\log\left(\frac{usage(c)}{\sum_{d \in CT} usage(d)}\right)$$

So, if we know which patterns to use – as well as their order in CT – it is easy to determine the optimal code table. Unfortunately, it is less easy, to use an understatement, to determine the optimal set of patterns as well as their order. Fortunately heuristic algorithms such as KRIMP [23] and SLIM [20] are known to produce good approximations.

In [22], we showed how code tables can be used to generate data. The crucial observation is that CT defines a probability distribution over D, the domain of db as follows. For $t \in D$,

$$l_{CT}(t) = \sum_{c \in cover(t)} l_{CT}(c) = \sum_{c \in cover(t)} -\log\left(P(c \mid db)\right)$$

$$= -\log\left(\prod_{c \in cover(t)} P(c \mid db)\right) = -\log\left(P(t \mid db)\right)$$

Clearly, the last equal sign is a bit optimistic if we want our distribution to be exactly the one db was sampled under. For, in that case it is only true if all the patterns in the code table are mutually independent and they are not. They are not independent for the simple reason that we use the order in CT to encode the data.

However, when

$$P(c_1 \mid db) \times P(c_2 \mid db) < P(c_1 \cup c_2 \mid db)$$

it becomes favourable to add $c_1 \cup c_2$ to the code table and if it is added, it will be *above* c_1 and c_2 (otherwise it wouldn't be used in encoding the database). Hence, problems may only occur for those cases where:

$$P(c_1 \mid db) \times P(c_2 \mid db) > P(c_1 \cup c_2 \mid db)$$

But this means that $c_1 \cup c_2$ doesn't occur very often in db. In other words, we expect that the distribution over db defined above is not too far off from the true distribution that gave us db.

To generate transactions from the distribution, we need the notion of a partial cover, which is defined as follows:

1. pc a partial cover, i.e., a subset of CT, such that:
   (a) $\forall p_1, p_2 \in pc : p_1 \neq p_2 \rightarrow p_1 \cap p_2 = \emptyset$; i.e., the patterns are defined on disjoint sets of attributes)
   (b) $\exists t \in D : \cup_{c \in pc} c \subseteq t$; i.e., there exists a tuple in D which is partially covered by pc – pc is consistent.
   note that $pc = \emptyset$ is, of course, fine
2. $pc^c = \{c \in CT \mid pc \cup \{c\}$ is a partial cover$\}$

Given a partial cover $pc \notin Dom(db)$, define for $p \in pc^c$ its *selection probability* as follows:

$$P_{sel}(p \mid pc) = \frac{P(p \mid db)}{\sum_{q \in pc^c} P(q \mid db)}$$

With these notions defined, we generate tuples as follows.

– start with $pc = \emptyset$,
– iterate

- choose an $p \in pc^c$ according to P_{sel}
- $pc := pc \cup \{p\}$
- until $pc \in D$
- return pc.

By iterating this simple procedure we can generate complete data sets. Since our probability distribution is, however, not necessarily the same as the one db was sampled from there is no guarantee that the generated database satisfies the requirements we stated for a randomized version.

Fortunately, in both [18,22]. We showed experimentally that databases generated in this way are remarkably close to what we want in this paper. More specifically we showed in [22] that the support of patterns in the original and in the generated data sets are close to each other. Moreover, in that same paper we defined an anonymity score by

$$AS(db_g, db_o) = \sum_{sup \in db_g} \frac{1}{sup} P(t \in db_g \mid t \in db_o)$$

$$NAS(db_o, db_o) = \frac{AS(db_g, db_o)}{AS(db_g, db_g)}$$

So a small NAS means that small support item sets in the original data set have a small chance that they end up in the generated data set. And for the experiments in [22] the NAS score was indeed small.

In [18], we showed that queries computed on the generated data give answers close to the same queries computed on the original data; especially for select queries, which are actually the patterns we are interested in here. Moreover we showed that one can even answer such queries by only generating tuples that satisfy the query.

Hence, both these papers indicate that we can already generate databases that are close to the kind of randomized databases we want. The biggest problem in that respect is that we cannot *prove* they satisfy the requirements. It is, however, the existence of these close to ideal database through which I believe randomized databases, fit for sharing, exist and can be constructed.

7   Conclusion

Big Data is both a blessing and a curse. A blessing because it allows us to study phenomena that were inaccessible before. From the behaviour of people, to health research, to the intricate workings of the cell. It is a curse because it means that details of the life of persons are recorded at an unprecedented scale opening up the risk by hacking or other means supposedly private data becomes public. What is worse is that the curse might outweigh the blessing. If privacy cannot be guaranteed, people will be unwilling to share information.

In this paper I advocate not to share the original data, but a randomized version. A version in which the details – the tuples in the database – are completely random, but in which the larger statistics (counts) are almost always close to the same statistics computed on the original database.

At the moment I cannot prove that such a randomization is possible. I have, however, shown that by using code tables such as computed by sc Krimp [23] we can already get close to our ideal; unfortunately not provably so, only experimentally. In the not to far future I do hope to publish an algorithm that will provably randomize a database in the sense advocated in this paper. Thus making it possible to share micro-data privacy-safe.

Acknowledgements. The author is supported by the Dutch national COMMIT project.

References

1. Agrawal, R., Mannila, H., Srikant, R., Toivonen, A.H., Verkamo, I.: Fast discovery of association rules. In: Usama, M.F., Piatetsky-Shapiro, G., Smyth, P., Uthurusamy, R. (eds.) Advances in Knowledge Discovery and Data Mining, pp. 307–328. AAAI/MIT Press, Menlo Park (1996)
2. Bakshy, E., Messing, S., Adami, L.: Exposure to ideologically diverse news and opinion on Facebook. Science **348**(6239), 1130–1132 (2015)
3. Breiman, L., Friedman, J.H., Olshen, R.A., Stone, C.J.: Classification and Regression Trees. Chapman & Hall, Wadsworth (1984)
4. Calders, T., Goethals, B.: Non-derivable itemset mining. Science **14**(1), 171–206 (2007)
5. Cover, T.M., Thomas, J.A.: Elements of Information Theory. Wiley, New York (2006)
6. de Montjoye, Y.-A., Radaelli, L., Singh, V.K., Sandy, A.: Pentland: unique in the shopping mall: on the reidentifiability of credit card metadata. Science **347**(6221), 536–539 (2015)
7. Dong, G., Li, J.: Efficient mining of emerging patterns: discovering trends and differences. In: Fayyad, S.C., Madigan, D., (eds.) Proceedings of the Fifth ACM SIGKDD International Conference on Knowledge Discovery and Data Mining, San Diego, CA, USA, 15–18 August 1999, pp. 43–52 (1999)
8. Dwork, C., Roth, A.: The algorithmic foundations of differential privacy. Data Min. Knowl. Disc. **9**(3–4), 211–407 (2014)
9. Grünwald, P.: The Minimum Description Length Principle. MIT Press, Cambridge (2007)
10. Klösgen, W.: Subgroup patterns. In: Klösgen, W., Zytkow, J.M. (eds.) Handbook of Data Mining and Knowledge Discovery, pp. 47–51. Oxford University Press, New York (2002)
11. Lane, J., Stodden, V., Bender, S., Nissenbaum, H. (eds.): Privacy, Big Data, the Public Good: Frameworks for Engagement. Cambridge University Press, New York (2015)
12. Li, M., Vitányi, P.: An Introduction to Kolmogorov Complexity and Its Applications. Springer, New York (1993)
13. Mannila, H., Toivonen, A.H., Verkamo, I.: Levelwise search and borders of theories in knowledge discovery. Data Min. Knowl. Disc. **1**(3), 241–258 (1997)

14. Mayer-Schönberger, V., Cukier, K.: Big Data, A Revolution That Will Transform How We Live, Work and Think. John Murray, London (2013)
15. Narayanan, A., Shmatikov, V.: Robust de-anonymization of large sparse datasets. In: 2008 IEEE Symposium on Security and Privacy (S&P 2008), 18–21 May 2008, Oakland, California, USA, pp. 111–125 (2008)
16. Samarati, P., Sweeney, L.: Protecting privacy when disclosing information: k-anonymity and its enforcement through generalization and suppression. In: Proceedings of the IEEE Symposium on Research in Security and Privacy (1998)
17. Schneier, B.: Data and Goliath: The Hidden Battles to Collect Your Data and Control Your World. W.W. Norton and Company, New York (2015)
18. Siebes, A., Puspitaningrum, D.: Mining databases to mine queries faster. In: Buntine, W., Grobelnik, M., Mladenić, D., Shawe-Taylor, J. (eds.) ECML PKDD 2009, Part II. LNCS, vol. 5782, pp. 382–397. Springer, Heidelberg (2009)
19. Siebes, A., Vreeken, J., van Leeuwen, M.: Item sets that compress. In: Ghosh, J., Lambert, D., Skillicorn, D.B., Srivastava, J. (eds.) SDM 2006 Proceedings, pp. 393–404. SIAM (2006)
20. Smets, K., Vreeken, J.: Slim: directly mining descriptive patterns. In: Ghosh, J., Liu, H., Davidson, I., Domeniconi, C., Kamath, C. (eds.) SDM 2012 Proceedings, pp. 236–247. SIAM (2012)
21. Tedre, M.: The Science of Computing: Shaping a Discipline. CRC Press/Taylor & Francis, Boca Raton (2014)
22. Vreeken, J., van Leeuwen, M., Siebes, A.: Preserving privacy through data generation. In: Ramakrishnan, N., Zaïane, O.R., Shi, Y., Clifton, C.W., Wu, X. (eds.) ICDM 2007 Proceedings, pp. 685–690. IEEE (2007)
23. Vreeken, J., van Leeuwen, M., Siebes, A.: Krimp: mining itemsets that compress. Science 23(1), 169–214 (2011)
24. Wikipedia: Netflix prize (2015). Accessed 31 July 2015

Distributed Support Vector Machines: An Overview

Marco Stolpe[1]([⊠]), Kanishka Bhaduri[2], and Kamalika Das[3]

[1] TU Dortmund, Computer Science, LS 8, 44221 Dortmund, Germany
marco.stolpe@tu-dortmund.de
[2] Netflix Inc., Los Gatos, CA 94032, USA
kanishka.bh@gmail.com
[3] UARC, NASA Ames, Mountain View, CA 94035, USA
kamalika.das@nasa.gov

Abstract. Support Vector Machines (SVM) have a strong theoretical
foundation and a wide variety of applications. However, the underlying
optimization problems can be highly demanding in terms of runtime and
memory consumption. With ever increasing usage of mobile and embedded systems, energy becomes another limiting factor. Distributed versions of the SVM solve at least parts of the original problem on different
networked nodes. Methods trying to reduce the overall running time and
memory consumption usually run in high performance compute clusters,
assuming high bandwidth connections and an unlimited amount of available energy. In contrast, pervasive systems consisting of battery-powered
devices, like wireless sensor networks, usually require algorithms whose
main focus is on the preservation of energy. This work elaborates on this
distinction and gives an overview of various existing distributed SVM
approaches developed in both kinds of scenarios.

Keywords: Distributed data mining · Support Vector Machines · High-
performance computing · Wireless sensor networks

1 Introduction

Every day, more and more data is getting stored on personal computers, electronic consumer devices, company servers, the world wide web or, more recently,
in the cloud. In the past, such data was mostly generated by humans. Due to
tremendous advances in hardware technology, however, data today is also automatically assessed by sensors which are deployed across devices as diverse as
mobile phones, embedded systems in cars, satellites or wireless sensor networks
that monitor, for instance, conditions in harsh environments. In many cases,
the individual machines and devices are connected to one or even several networks, which sometimes even include the internet. Such connectivity opens up
new opportunities for inferring information and answering questions that relate
to data not only assessed or stored by a single node, but across many nodes.
For example, data from spatially distributed sensors may be used to predict

© Springer International Publishing Switzerland 2016
S. Michaelis et al. (Eds.): Morik Festschrift, LNAI 9580, pp. 109–138, 2016.
DOI: 10.1007/978-3-319-41706-6_5

global events like catastrophes, e.g. tsunamis, earth-quakes or floods. Some other applications include predicting the traffic flow in smart cities, which improve the tracking and monitoring of objects with RFID sensors in production settings, or that help physicists to analyze huge amounts of data assessed by different telescopes.

Inferring information from raw data is a data analysis task. In the past, many methods for the analysis of data have been developed by such fields as diverse as signal processing, statistics, artificial intelligence, machine learning, data mining, information retrieval and research in databases. In their basic form, the developed methods usually expect all data to be available at a single node in main memory, or at least to be stored in a centralized database. For small amounts of data, this restriction can easily be accounted for by transferring all data first to a central node which then performs the analysis. However, the amounts of data generated today are more and more often too large to be transferred, stored and processed at a single node. Existing systems are not scalable. Either the size of the data is so big that it cannot even fit into the main memory of a supercomputer. This case will be referred to as *big data* scenario. Or the systems and devices recording and analysing data are so constrained that sending all data to a more powerful central node would be too costly in terms of bandwidth or energy consumption. This case will be denoted as the *small devices* scenario. Both cases require algorithms which are able to analyse the data in a distributed fashion, instead of working fully centrally.

However, the types of algorithms required for both types of scenarios may be different. For big data analysis, it usually can be assumed that all compute nodes have a continuous power supply. While the reduction of energy consumption may still be an important aim for economical or environmental reasons, energy isn't necessarily a limiting technical factor (though it can be). Similarly, the available communication bandwidth in cluster and cloud computing is also high. In contrast, for small battery-powered devices connected to wireless networks, energy and bandwith are usually the most scarce resources. Distributed data analysis algorithms for small devices therefore must be designed explicitly for taking the energy consumption of their actions and the remaining energy into account [2]. They also have to be more fault tolerant, e.g. by giving guarantees on the quality of their solution even when some devices fail.

This paper presents several distributed versions of the Support Vector Machine (SVM) [34], a popular method for data analysis with a strong foundation in statistical learning theory. Despite the many existing decomposition techniques for the underlying quadratic optimization problem, it will become apparent how challenging it is to distribute SVM computations over different nodes. This is especially true for non-linear problems in the vertically partitioned data scenario, where not observations, but their features are distributed over different nodes. Each method will be discussed in relation to its suitability for big data analysis and usage on small devices.

The next section briefly introduces regularized risk minimization and a variety of accompanying SVM problems in their non-distributed, original form.

Section 3 elaborates on the distinction between high-performance computing and pervasive systems and introduces two different types of data partitioning methods which have different implications for learning. Section 4 presents distributed SVMs for horizontally partitioned data, while Sect. 5 does the same for the vertically partitioned scenario. The summary and conclusions in Sect. 6 give a final overview over all methods, discuss their shortcomings and point to future research opportunities.

2   Support Vector Machines

The following subsections shortly describe the problem of supervised function learning in the context of structural risk minimization. They then introduce several variants of the Support Vector Machine (SVM), for which distributed versions are presented in Sects. 4 and 5.

2.1   The Problem of Supervised Function Learning

Let X be a space of *observations* and Y be a set of possible *labels*. The task of *supervised function learning* aims at deriving a function $f : X \to Y$ from a sample $S = \{(\mathbf{x}_1, y_1), \ldots, (\mathbf{x}_n, y_n)\}$ of n observation/label pairs $(\mathbf{x}_i, y_i) \in X \times Y$, drawn i.i.d. from an unknown joint probability distribution $P(X, Y)$, such that the expected risk

$$R_{\exp} = \int L(y, f(\mathbf{x})) dP(\mathbf{x}, y)$$

is minimized. Here, L is a convex *loss function* $L : Y \times Y \to \mathbb{R}_0^+$ which measures the cost of assigning the wrong label to individual observations. Sample S is also often called the *training data*.

The challenge of supervised function learning is that the expected risk cannot be calculated explicitly, since the underlying joint distribution of observations and labels is unknown. What can be directly estimated is the empirical risk

$$R_{\mathrm{emp}} = \frac{1}{n} \sum_{i=1}^{n} L(y_i, f(\mathbf{x}_i)), \ (\mathbf{x}_i, y_i) \in S$$

which measures the loss of function f on the training data. The empirical risk R_{emp} approximates $R_{\exp}$ as $n \to \infty$. However, it is well known that the minimization of R_{emp} alone on a small finite number of sample observations may yield functions f that perform arbitrarily bad on other samples. This problem is also known as *overfitting*.

2.2   Structural Risk Minimization

An empirical approach for obtaining a better estimate of R_{emp} is to train function f on a subset of sample S and to test it on a hold-out test set of labeled observations not used for training. Given a set of candidate functions $f_1, \ldots, f_c$, the best function f_{opt} is then the one with the lowest estimated test error. While such a black-box approach is able to work with arbitrary functions, it can be time consuming.

Whenever it is possible to measure the so called *structural complexity* of functions, *structural risk minimization* yields a more principled way of chosing f_{opt}. Intuitively, given two functions with the same empirical risk R_{emp}, the less complex function should generalize better and thus overfit less likely. For a function class $f(\mathbf{x}, \boldsymbol{\gamma})$ with parameter vector $\boldsymbol{\gamma}$, the *structural risk* (also called *regularized risk*) is therefore defined as

$$R_{\mathrm{reg}}(\boldsymbol{\gamma}) = R_{\mathrm{emp}}(\boldsymbol{\gamma}) + \lambda \Omega(\boldsymbol{\gamma}) \,,$$

where Ω is a strictly monotonic increasing function which measures the *capacity* of function class f depending on parameter vector $\boldsymbol{\gamma}$. The trade-off between the empirical training error and the capacity is managed by λ. The capacity can for example be measured with help of the *Vapnik-Chervonenkis dimension* (VC dimension), which yields a probabilistic bound for the regularized risk. The existence of guaranteed error bounds may be seen as one reason for the great success of methods following the structural risk minimization principle, like the large margin methods explained in the following.

2.3 Support Vector Classification

Large margin approaches for classification follow the regularized risk minimization principle by maximizing a margin between a linear function and the nearest data points.

Let observations be vectors consisting of p real-valued components, i.e. $\mathbf{x}_i \in \mathbb{R}^p$. Let further constrain Y to two values, -1 and +1, which is the problem of *binary classification*, and for simplicity assume that all observations are linearly separable. Then there must exist a hyperplane

$$H = \{h | \langle \mathbf{w}, h \rangle + b = 0\}$$

with $\mathbf{w}$ being normal to H, bias b with $|b|/||\mathbf{w}||$ being the perpendicular distance of H to the origin and $||\mathbf{w}||$ being the Euclidean norm of $\mathbf{w}$, such that

$$\forall_{i=1}^n y_i(\langle \mathbf{w}, \mathbf{x}_i \rangle + b) \geq 0,$$

i.e. that all observations of a particular class are lying in the same halfspace as given by H. Given $\mathbf{w}$ and b, observations $\mathbf{x}$ may be then be classified by function

$$f(\mathbf{x}, \mathbf{w}, b) = \mathrm{sgn}(\langle \mathbf{w}, \mathbf{x} \rangle + b) \,.$$

The parameters $\mathbf{w}$ and b define the position and orientation of H and can be seen as the parameter vector $\boldsymbol{\gamma}$ of function f.

There are infinitely many hyperplanes which correctly separate positive and negative training examples and thereby minimize the empirical risk R_{emp}. However, there is only one hyperplane which also minimizes the structural risk R_{reg}.

This hyperplane separates positive and negative observations with the largest possible *margin*, which is defined as the perpendicular distance of points closest to the hyperplane. For normalized $\mathbf{w}, b$ such that points $\mathbf{x}_i$ closest to the hyperplane satisfy $|\langle \mathbf{w}, \mathbf{x}_i \rangle + b| = 1$, the margin is given by $1/\|\mathbf{w}\|$. Instead of maximizing $1/\|\mathbf{w}\|$, we may as well minimize $\frac{1}{2}\|\mathbf{w}\|^2$. Furthermore, one can allow for non-separable data points that lie inside the margin or even in the wrong halfspace by introducing slack variables ξ_i and minimizing the sum of such errors.

Primal SVM Problem for Non-separable Data. The primal optimization problem for non-separable data then becomes

$$\min_{\mathbf{w}} \quad \frac{1}{2}\|\mathbf{w}\|^2 + C\sum_{i=1}^{n}\xi_i \tag{1}$$

$$\text{s.t.} \quad \forall_{i=1}^{n} : y_i(\langle \mathbf{w}, \mathbf{x}_i \rangle + b) \geq 1 - \xi_i \,.$$

In terms of structural risk minimization, parameter C in (1) trades off the empirical risk (the sum over all slack variables) against the structural risk (the size of the margin). This relationship can be even easier seen when replacing (1) by the *hinge function notation*

$$\min_{\mathbf{w}} \quad \sum_{i=1}^{n}[1 - y_i(\langle \mathbf{w}, \mathbf{x} \rangle + b)]_+ + \lambda\|\mathbf{w}\|^2 \,, \tag{2}$$

where $\xi_i = [1 - y_i(\langle \mathbf{w}, \mathbf{x}_i \rangle + b)]_+$ is the so called *hinge loss function*. It can be shown that this problem is a quadratic optimization problem with inequality constraints. Such problems are sometimes easier to solve by introducing Lagrange multipliers α_i, μ_i, $i = 1, \ldots, n$ for each inequality constraint, resulting in the Lagrangian

$$L_P(\mathbf{w}, b, \boldsymbol{\alpha}, \boldsymbol{\mu}) = \frac{1}{2}\|\mathbf{w}\|^2 - C\sum_{i=1}^{n}\xi_i - \sum_{i=1}^{n}\alpha_i(y_i(\langle \mathbf{w}, \mathbf{x}_i \rangle + b) - 1 + \xi_i) - \sum_{i=1}^{n}\mu_i\xi_i \tag{3}$$

Dual SVM Problem for Non-separable Data. By setting the partial derivatives of L_P for $\mathbf{w}$ and b to zero and inserting the solutions $\mathbf{w} = \sum_{i=1}^{n}\alpha_i y_i \mathbf{x}_i$ and $0 = \sum_{i=1}^{n}\alpha_i y_i$ into (3), one obtains the dual SVM problem for non-separable data

$$\max_{\boldsymbol{\alpha}} \quad \sum_{i=1}^{n}\alpha_i - \frac{1}{2}\sum_{i=1}^{n}\sum_{j=1}^{n}\alpha_i\alpha_j y_i y_j \langle \mathbf{x}_i, \mathbf{x}_j \rangle \tag{4}$$

$$\text{s.t.} \quad \forall_{i=1}^{n} : 0 \leq \alpha_i \leq C \quad \text{and} \quad \sum_{i=1}^{n}\alpha_i y_i = 0$$

The solution $\mathbf{w} = \sum_{i=1}^{n}\alpha_i y_i \mathbf{x}_i$ is a linear combination of data points for which $0 \leq \alpha_i \leq C$. Such data points are also called *support vectors* (SVs), as they sufficiently determine the computed hyperplane.

Kernel Functions. For better or non-linear separation of observations, it may help to map them to another space, called *feature space*, by a transformation function $\Phi : X \to H$. Space H often needs to have a higher dimension than X. Hence, an explicit calculation of the dot product in (4) on the mapped observations can become quite time-consuming. However, it can be shown that for certain mappings Φ, there exist kernel functions $k(\mathbf{x}, \mathbf{x}') = \langle \Phi(\mathbf{x}), \Phi(\mathbf{x}') \rangle$ which correspond to dot products in H. Often, replacing $\langle \Phi(\mathbf{x}), \Phi(\mathbf{x}') \rangle$ by $k(\mathbf{x}, \mathbf{x}')$ allows for a much more efficient computation of the dot product. Moreover, since solving the dual problem only depends on values of the dot product, but not the observations themselves, instance space X may not only consist of real-valued vectors, but arbitrary objects like strings, trees or graphs which have an associated similarity measure. Further, there exist kernel functions with H being infinite. Popular kernel functions are, for instance, the

$$\text{Polynomial Kernel} \quad k(\mathbf{x}, \mathbf{x}') = (\kappa \langle \mathbf{x}, \mathbf{x}' \rangle + \delta)^d \,, \tag{5}$$

$$\text{RBF Kernel} \quad k(\mathbf{x}, \mathbf{x}') = e^{-\frac{||\mathbf{x} - \mathbf{x}'||^2}{2\sigma^2}} \text{ and} \tag{6}$$

$$\text{Sigmoid Kernel} \quad k(\mathbf{x}, \mathbf{x}') = \tanh(\kappa \langle \mathbf{x}, \mathbf{x}' \rangle - \delta) \,. \tag{7}$$

As will become clear in the following sections, non-linear classification by the use of kernel functions is often difficult to achieve in distributed settings, especially in the vertically distributed scenario described in Sect. 5.

2.4   Solvers

There exist several methods that solve the centralized SVM problem. Interior point methods [4] replace the constraints with a barrier function. This results in a series of unconstraint problems which can be solved efficiently with Newton or Quasi-Newton methods. However, the general methods have a cubic run-time and quadratic memory requirements. More popular approaches are chunking and decomposition methods [18,23,25], which work on a subset of dual variables at a time. Finally, gradient methods like Pegasos and SVM-perf iteratively update the primal weights. Their convergence rate is usually $O(1/\varepsilon)$.

2.5   Support Vector Regression (SVR)

For real-valued outputs $y_i \in \mathbb{R}$, the primal SVM problem can be stated like

$$\min_{\mathbf{w}} \quad \frac{1}{2} ||\mathbf{w}||^2 + C \left(\sum_{i=1}^{n} \xi_i + \sum_{i=1}^{n} \xi_i' \right)$$
$$\text{s.t.} \quad \forall_{i=1}^{n} : \langle \mathbf{w}, \mathbf{x}_i \rangle + b \le y_i + \epsilon + \xi_i' \quad \text{and}$$
$$\langle \mathbf{w}, \mathbf{x}_i \rangle + b \ge y_i - \epsilon - \xi_i \,.$$

The dual formulation then contains two αs, one for each ξ_i and ξ_i'.

2.6 Support Vector Data Description (SVDD) and 1-Class SVM

Supervised classifiers are trained on two or more classes. Their accuracy may suffer if the distribution of observations over classes is highly imbalanced. For instance, this may happen in applications where unusual events and therefore data about them is scarce, like faults in machines, quality deviations in production processes, network intrusions or environmental catastrophes. In all such cases, many positive examples are available, but only few or even no examples of the negative class.

The task of data description, or 1-class classification [22], is to find a model that well describes the observations of a single class. The model can then be used to check whether new observations are similar or dissimilar to the previously seen data points and mark dissimilar points as anomalies or outliers. Support Vector Data Description (SVDD) [32] computes a spherical boundary around the data. The diameter of the enclosing ball and thereby the volume of the training data falling within the ball are user-chosen. Observations inside the ball are classified as normal whereas those outside the ball are treated as outliers or anomalies.

More formally, given a sample of training observations $S = \{\mathbf{x}_1, \ldots, \mathbf{x}_n\} \subseteq X$ that all belong to the same class, the primal SVDD problem is to find a minimum enclosing ball (MEB) with radius R and center $\mathbf{c}$ around all data points $\mathbf{x}_i \in S$:

$$\min_{R, \mathbf{c}} \quad R^2 : \|\mathbf{c} - \mathbf{x}_i\|^2 \leq R^2, \ i = 1, \ldots, n$$

Similar to the previously presented support vector methods, kernel functions may be applied whenever observations are arbitrary objects or the decision boundary in the original space is non-spherical. The dual problem after the kernel transformation then becomes

$$\max_{\alpha} \quad \sum_{i=1}^{n} \alpha_i k(\mathbf{x}_i, \mathbf{x}_i) - \sum_{i,j=1}^{n} \alpha_i \alpha_j k(\mathbf{x}_i, \mathbf{x}_j) \tag{8}$$

$$\text{s.t.} \quad \forall_{i=1}^{n} : \alpha_i \geq 0, \ \sum_{i=1}^{n} \alpha_i = 1.$$

The primal variables can be recovered using

$$\mathbf{c} = \sum_{i=1}^{n} \alpha_i \Phi(\mathbf{x}_i), \quad R = \sqrt{\alpha^T \text{diag}(\mathbf{K} - \alpha^T \mathbf{K} \alpha)}$$

where $\mathbf{K} = (k_{ij})$ with $k_{ij} = k(\mathbf{x}_i, \mathbf{x}_j)$ is the $n \times n$ kernel matrix. Support vectors are data points for which $\alpha_i > 0$. An observation $\mathbf{x}$ belongs to the training set distribution if its distance from the center $\mathbf{c}$ is smaller than radius R, where distance is expressed by the set of support vectors SV and the kernel function:

$$\|\mathbf{c} - \Phi(\mathbf{x})\|^2 = k(\mathbf{x}, \mathbf{x}) - 2 \sum_{i=1}^{|SV|} \alpha_i k(\mathbf{x}, \mathbf{x}_i) + \sum_{i,j=1}^{|SV|} \alpha_i \alpha_j k(\mathbf{x}, \mathbf{x}_j) \leq R^2$$

It can be shown [33] that for kernels $k(\mathbf{x}, \mathbf{x}) = \kappa$ (κ constant) that map all input patterns to a sphere in feature space, (8) can be simplified to the optimization problem (where $\mathbf{0} = (0, \ldots, 0)^T$ and $\mathbf{1} = (1, \ldots, 1)^T$)

$$\max_{\boldsymbol{\alpha}} \quad -\boldsymbol{\alpha}^T \mathbf{K} \boldsymbol{\alpha} : \boldsymbol{\alpha} \geq \mathbf{0}, \, \boldsymbol{\alpha}^T \mathbf{1} = 1 \tag{9}$$

Whenever the kernel satisfies $k(\mathbf{x}, \mathbf{x}') = \kappa$, any problem of the form (9) is an MEB problem. For example, Schölkopf [27] proposed the 1-class ν-SVM that, instead of minimizing an enclosing ball, separates the normal data by a hyperplane with maximum margin from the origin in feature space. If $k(\mathbf{x}, \mathbf{x}') = \kappa$, the optimization problems of the SVDD and the 1-class ν-SVM with $C = 1/(\nu n)$ are equivalent, and yield identical solutions.

3   Distributed Systems and Computation

The components of distributed SVM algorithms concurrently work on subproblems of the previously presented quadratic optimization problems. Although the underlying problems are quite similar, algorithms may specialize on certain *types* of distributed systems, *architectures* and network *topologies*. For our discussion, important distinctions are to be made between *parallel* and *distributed* computing, and between *high-performance* computing and *pervasive* distributed systems. This section elaborates on the differences, as far as they concern the distributed SVM algorithms presented in the following sections of this paper. Most of the following material is based on [31], which is recommended for further details. Finally, we present two different ways of how data can be partitioned in distributed settings and discuss their implications for learning.

3.1   Parallel vs. Distributed Computing

Although parallel and distributed systems share common properties and problems, they are also fundamentally different. Parallel algorithms are assumed to run on different processors, but on the same machine. The components of parallel algorithms usually have access to shared memory segments, which can be used for the coordination of different processes or threads of execution. Communication is only as costly as reading from or writing to main memory. Critical conditions, like program exceptions or hardware failures, will usually leave the system in a well-defined state, which may also include shutdowns of the whole system. Moreover, parallel processes share the same system clock. In comparison, distributed components run (or at least are expected to run) on different physical entities. By definition, they cannot share the same physical memory, but have a local memory and must coordinate themselves by exchanging messages over physical communication lines. The time for sending and receiving messages over such lines is usually several orders of magnitudes higher than the time to access main memory. On battery-powered embedded or mobile devices, communication usually also consumes lots of energy and thus doesn't come for free. Further, in distributed systems partial failures, i.e. failures of individual components, are common and sometimes hard to distinguish from valid system states. For instance, a sender may have difficulties to determine if other components have received a message or not. A message may have been dropped due to failure of communication lines that are not even directly linked to the sender, it may get lost due to a hardware failure at the receiving end, or the receiver got the message, but cannot acknowledge it in the required time frame, caused by a high computational load. It might also occur that the message has been delivered, but the acknowledgement is getting lost on its way back to the sender. In each case, it is hard for the sender to determine the state of the receiving end, leading to difficult synchronization problems. Similar problems may be caused by differing system clocks of different physical entities.

Considering the differences of parallel and distributed systems, distributed algorithms usually must be designed to be much more fault-tolerant and autonomous than parallel algorithms. Moreover, communication costs must be taken into account and possibly traded off for computation. With the advent of pervasive systems, another limiting factor that must be taken into account is the consumption of energy. Due to these differences, we have excluded from this paper parallel SVMs like [8] or [38] that rely on shared memory implementations instead of message passing.

3.2 High-Performance vs. Pervasive Computing

Despite the differences between parallel and distributed systems, algorithms designed for running in high-performance clusters often follow the paradigm of parallel computation. This partly can be justified by the control exerted over such systems, since machines in a cluster operate in a well-defined and closed environment. In particular, it may be assumed that communication lines are highly reliable and have a high bandwidth for fast communication. Also, the network topology stays the same during the run of a distributed algorithm. The computational resources per node usually may be assumed to be at least as high as that of contemporary hardware and furthermore, each node potentially may use an unlimited amount of energy. Frameworks and libraries for high-performance computing, like MPI, Storm or Hadoop, try to make the design of distributed algorithms as transparent as possible, giving the impression of working with a single system of parallel processors. Intricate details of fault handling, synchronization and coordination between distributed components are usually automated to a large degree and shielded from the developer as much as possible. Nevertheless, to benefit for example from automated fault handling procedures, developers must usually provide extra code. Another important property of high-performance clusters is that the data to work on oftentimes is transferred to such systems explicitly, with the ability to control how data is stored and partitioned across different networked nodes. This may allow for easier and more performant scheduling of distributed processes.

In comparison, pervasive distributed systems like wireless sensor networks (WSNs) consist of small devices that communicate with each other over sometimes highly unreliable communication lines. They may have a low bandwidth, prohibiting excessive data transmission. Data is usually directly assessed at each networked node and, due to communication restrictions, must be processed locally as much as possible. However, the computational units at the same time are tiny in terms of CPU power and main memory. In addition, battery-powered devices have severe energy constraints. Especially with mobile devices, the network topology might change continously. Hence, distributed algorithms that can work in such restricted environments must be differently designed than those following a parallel computation paradigm. The predominant concern often is not speed of computation, but to find a solution with satisfying accuracy at all, given the circumstances.

The quadratic optimization problems underlying most variants of support vector learning usually have high demands on resources, like CPU power, main memory and energy. Due to the iterative nature of most solvers, partial results like current support vectors, predictions or optimization variables often need to be transmitted several times to other nodes. With unlimited access to energy and a sufficiently high bandwidth, high performance clusters may solve moderately sized SVM problems in less time than their centralized counterpart, even if more data than the whole dataset is transmitted. In contrast, transmitting large amounts of data is prohibitive in big data or pervasive scenarios, where the most limiting factors are bandwidth and energy. In the following, an algorithm will be called *communication-efficient* if it transmits less data than the whole dataset. Though many distributed SVM algorithms have been designed for high performance distributed computation, there are more and more authors who claim that their algorithms might run in highly restricted environments like WSNs. Therefore, we will not only discuss the performance of distributed SVM algorithms in terms of accuracy and speed, but also with regard to their communication costs.

3.3   Types of Data Partitioning

Many data analysis methods expect observations and labels in a fixed-size $n \times p$ data matrix $\mathbf{D}$, whose rows store the feature values of observations in S. There exist two main scenarios for the partitioning of $\mathbf{D}$ across distributed nodes [6]. In the *horizontally partitioned data scenario*, each node stores a subset of observations (usually together with their labels). This means for $j = 1, \ldots, m$ nodes, we have datasets $S_j \subseteq S$ and $S = S_1 \cup \cdots \cup S_m$ with $S_u \cap S_v = \emptyset$ ($u \neq v$ for $u, v = 1, \ldots, n$) and corresponding data matrices $\mathbf{D}_1, \ldots, \mathbf{D}_m$. In the *vertically partitioned data scenario*, each node stores only partial information about observations, i.e. subsets of their features, but for all observations. Let $\mathbf{x}[j] \in \mathbb{R}^{p_j}$ denote a vector which contains p_j features of observation $\mathbf{x}$ available at node j. Columns of the data matrix $\mathbf{D}$ are then split over the nodes, i.e. each node j stores a $n \times p_j$ submatrix $\mathbf{D}[j]$ whose rows consist of vectors $\mathbf{x}_1[j], \ldots, \mathbf{x}_n[j]$.

From a learning point of view, both scenarios exhibit different properties. Local datasets in the horizontally partitioned scenario may be seen as potentially skewed subsamples of the whole training sample S. In principle, in a preprocessing step, one could use a distributed uniform sampling algorithm like the one introduced in [10] to adjust for the skewedness. Anyhow, each training example belongs to the same global instance space X. In comparison, partial information about observations in the vertically partitioned data is not just a skewed representation of the whole dataset, but the local vectors do not even share the same space. Hence, it is much harder to design communication-efficient algorithms in cases where features from different nodes are conditionally dependent, given the label. Modeling such dependencies necessarily requires to join features from different nodes, potentially resulting in a large communication overhead. Since both scenarios require different learning strategies, the rest of the paper is divided into two different sections accordingly.

4 Horizontally Distributed SVMs

For horizontally partitioned data, one can identify two main strands of distributed algorithms. The first type of algorithms is based on an exchange of summary information about the local datasets, like support vectors, with a central coordinator or neighboring peer nodes. The different variants are presented in Sect. 4.1. The second type of algorithms is discussed in Sect. 4.2. The different variants solve parts of the underlying optimization problem in a distributed fashion and only exchange partial results with other nodes. Finally, Sect. 4.3 presents a way to cast the original problem into a distributed least squares formulation.

4.1 Distributed Learning on Summarized Data

There exist several distributed approaches for horizontally partitioned data that are based on the exchange of summary information about observations. Many of such approaches are inspired by early incremental versions of the SVM which repeatedly keep only the support vectors of previous learning steps for training. More sophisticated versions have demonstrated that although support vectors are not sufficient representations of a dataset, correct results can be achieved by exchanging support vectors in multiple iterations or by keeping other relevant data points. The following subsections first describe the aforementioned incremental approaches and then discuss distributed methods that are also based on the exchange of summary information between nodes.

Incremental SVMs. Instead of learning on a single batch of data, the incremental SVM proposed in [30] assumes a training set S to be divided into disjunct subsets $S_1, \ldots, S_m$. The training procedure works incrementally. In the initial first step $t = 1$, the algorithm trains a SVM on set S_1, but only keeps the support vectors SV_1. At each following time step t, an SVM model f_t is trained on the union $S_t \cup SV_{t-1}$ of the current training set S_t and the support vectors SV_{t-1} found in the previous step. Empirical results on UCI datasets suggest the incremental SVM achieves a similar performance as the standard batch SVM trained on all data up to the corresponding time step t, i.e. also at the end of training. The performance drops significantly if the SVM is only trained on a subset (90 %) of the determined support vectors, from which the authors conclude that the incremental SVM finds a minimal set of support vectors. The authors further point out that the incremental SVM can be seen as a lossy approximation of the chunking method [23], with the incremental approach considering each support vector only once.

The incremental procedure proposed in [30] appears plausible as long as the distributions of training examples in each subset $S_1, \ldots, S_m$ are similar to the distribution of data points in the whole training set S. In cases where the training algorithm has full control over how S is split into subsets, the aforementioned condition could be achieved by a uniform sampling of examples from S. However, that type of control can neither be assumed in a real streaming case, nor in many distributed settings, where the data partitioning is already given.

The case where the statistical properties of each batch $S_1, \ldots, S_m$ may differ from those of S has been investigated, among others, by [26]. In contrast to detecting concept drift, the focus is on learning a single concept from all data. However, though training examples in each batch consistently represent that concept, their distributions differ. The author notes that while support vectors provide a condensed and sufficient description of the learned decision boundary, they do not represent the examples themselves. That is, in terms of empirical risk minimization, the support vectors provide an estimate of $P(X|Y)$, but not of $P(X)$. If the number of support vectors is small in comparison to the number of examples in the next batch, their influence on the decision boundary will be small. It is demonstrated how decision boundaries can differ between an SVM trained on all data and one trained on a subset of the data, with SVs from another subset added. The few support vectors are treated as mere outliers, which the SVM is known to be robust against. Therefore, [26] proposes to weight prediction errors on support vectors higher than errors on training examples in the new batch by replacing the original SVM objective (1) with

$$\min_{\mathbf{w}} \quad \frac{1}{2}||\mathbf{w}||^2 + C\left(\sum_{i \in S} \xi_i + L \sum_{i \in SV} \xi_i\right),$$

where S is the set of new training examples, SV is the set of old support vectors and $L = 2\frac{n}{|SV|}$. It is shown that the modified incremental algorithm empirically achieves a higher accuracy than the plain version proposed in [30].

Iterative Exchange of Support Vectors. While [26] gives a counter example which shows that local sets of support vectors may differ strongly from the global set of support vectors, [7] includes a formal proof. Like [1], the authors further show that points on the convex hull provide a sufficient statistic for SVM learning from distributed data sources. However, for higher dimensions, computing the convex hull is exponential and thus not efficient.

In [5], the same authors propose a scheme that is, according to their reasoning, efficient and exact. The idea consists of exchanging support vectors iteratively with a central node. At iteration t, each local site j determines its current set of support vectors SV_j^t, based on the dataset $S_j \cup GSV^{t-1}$, where S_j is the local data stored at node j and GSV^{t-1} is the global set of support vectors j has received from the central node at the previous iteration $t-1$. Each node sends its local set of support vectors to the central node, which merges them to the global set GSV^t and communicates it to all local nodes. The authors sketch a proof which shows that after a finite number of iterations, the local sets of support vectors converge to the globally optimal set of support vectors.

Cascade SVM. The Cascade SVM introduced in [16] is based on a similar idea as the previously presented incremental SVMs, which is to identify and eliminate non-support vectors as early as possible and only communicate support vectors between distributed nodes. Proposed is a hierarchical binary tree topology of cascading SVMs. At the beginning, disjunct subsets $S_1, \ldots, S_m$ of S are distributed over the leaves of the tree. An SVM is trained at each leave and the resulting support vectors are communicated to the parent node in the next layer of the hierarchy. At the parent nodes, SVMs are trained on unions of support vectors of the previous layer. The root node communicates the finally determined support vectors to each leave, and each leave decides if any of its input vectors would become new support vectors. When the set of support vectors has stabilized over all leaves, the algorithm stops, otherwise the hierachical cascade is traversed again. The algorithm thus includes a similar feedback loop as the approach proposed in [5], but due its hierarchical design, it may earlier filter data points that are not in the global set of support vectors.

It is proven that the Cascade SVM converges to a set of support vectors that is globally optimal in finite time. However, no bound is given for the total number of iterations. For the standard datasets tested, like the MNIST dataset, the authors report a low number of iterations between 3 and 5. Moreover, with 16 machines in a cluster, the final number of support vectors and the size of subsets at each leave is about 16 times smaller than the total number of 1 M data points. Training time was reduced from about one day on a single machine to one hour with the distributed approach.

In practice, run-time and communication costs will largely depend on the ratio of training examples to support vectors. Often, bad choices of hyperparameters, like C or σ for the RBF kernel, result in an unnecessarily large number of support vectors. Unfortunately, optimal hyperparameters are hard to determine in advance, but must be found experimentally. Similarly, a high number of support vectors can be expected for complicated non-linear decision boundaries.

[20] prove and demonstrate that the iterative exchange of support vectors converges to the global optimum also in case of other network topologies. Particularly, they train local SVMs and exchange support vectors with their ancestors and descendants in a strongly connected network. It is shown that the binary cascade proposed in [16] is a special case of a strongly connected network and that a random strongly connected network topology may lead to faster convergence. A ring topology has the slowest convergence. Further tested are synchronous

and asynchronous versions of the algorithm. The synchronous implementation dominates in terms of training time, while the asynchronous version leads to less data accumulation (number of exchanged support vectors) in sparser networks.

Energy-Efficient Distributed SVMs. In [12], the incremental procedure proposed in [26] is brought into the context of distributed wireless sensor networks. Subsets of S are assumed to be stored at clusterheads. Such clusterheads may be determined by already existing energy-efficient clustering network protocols (for an overview, see e.g. [2]). Similar to the original incremental algorithm, models are consecutively trained on the data S_i at clusterhead i and support vectors received from the previous clusterhead in a chain of clusterheads. The authors regard varying distributions of observations by a different weighting of examples and support vectors, as already proposed in [26]. Empirically the algorithm is shown to be similarly accurate, but more energy-efficient than transmitting all data to a central node and training a single SVM on all data. However, in comparison to [5,16], the algorithm is not guaranteed to find a globally optimal solution. Moreover, it was only evaluated on a single synthetic two-dimensional dataset consisting of two Gaussian distributions. Like the Cascade SVM, the communication costs will very much depend on the number of support vectors found. Here, it may happen that clusterheads at the end of the chain always receive more support vectors than those at the beginning. Balancing the network's total energy consumption would thus require a technique for changing the order of communication dynamically.

For solving the last problem, in [13] the same authors propose two gossip algorithms that exchange summary information between one-hop neighboring nodes. A single iteration of the minimum selective gossip algorithm (MSG-SVM) at time step t consists of training SVMs at each node, based on the currently available local information. Each node then communicates its current set of support vectors to all one-hop neighbors and all nodes update their current model at time step $t + 1$. Although the authors give no explicit stopping criterion for their algorithm, they argue that over time, all nodes will converge to the same SVM model. However, they also argue that their algorithm is sub-optimal and will not converge to the same solution as a centralized SVM trained on all data. The idea of this proof is based on the same argument as already given in [7]. It remains unclear if between iterations, nodes only keep the determined support vectors or if exchanged support vectors are added to the already available local data points. In the first case, data points that might later become support vectors could be thrown away and would thus be missed. In the second case, however, the iterative exchange of support vectors closely resembles the filtering mechanism and feedback loops of the approaches introduced in [5] and [16], which both converge to the global optimum. The second proposed strategy, the sufficient selective gossip algorithm (SSG-SVM), ensures convergence to the global optimum by exchanging points that lie on the convex hull of each class. While this algorithm might work efficiently on the synthetic two-dimensional datasets used for evaluation, it is inefficient for higher dimensions (see [5]).

In [14], the authors propose to trade-off communication costs for accuracy by exchanging only a pre-determined percentage of observations between neighboring nodes. The observations to be transmitted by each node are ranked by their distance from the current determined local hyperplane. For a single type of synthetic data, consisting of two 2-dimensional Gaussians, the authors demonstrate that accuracy can be increased by transmitting slightly more observations than only the support vectors. While the algorithm allows for trading off communication costs for accuracy, it remains unclear how much accuracy decreases if much less is sent than the set of support vectors. Since the number of support vectors may be high, it appears somewhat questionable that any of the aforementioned approaches could truly work in highly energy-constraint systems like WSNs.

4.2   Distributed Optimization and Consensus

As discussed in the previous sections, communication costs for the exchange of summary information largely depend on the structure of the data. When both classes are well separated and the decision border is simple, the SVM model will be simple, consisting only of a small number of support vectors. However, the number of support vectors necessarily grows with the complexity of the decision border in non-linear cases and may also be high for bad choices of hyperparameters. In the worst case, where all data points become support vectors, the whole dataset needs either to be exchanged between local nodes and a fusion center or between neighboring peer nodes. According to our definition from Sect. 3, such algorithms are not communication-efficient.

An important observation is that all previously presented approaches are globally solving the SVM problem in its dual formulation (see Sect. 2.3), where each constraint on training example $\mathbf{x}_i$ in the primal formulation introduces an optimization variable α_i. The total number of variables is thus n. In comparison, in the primal problem, the number of optimization variables equals the dimension p of weight vector $\mathbf{w}$ plus one variable for the bias. In many practical applications, the number of features, and therefore the dimension of $\mathbf{w}$, is much smaller than the number of training examples. Further, it can be noted that the constraints on subsets of the training examples, as given in the horizontally partitioned data scenario, can be checked independently from each other (i.e. locally). The question therefore arises if instead of exchanging special data points, like support vectors, an iterative exchange of $\mathbf{w}$ and b among nodes could be feasible as well and more communication efficient at the same time.

Alternating Direction Method of Multipliers (ADMM). In [15], the centralized SVM problem is divided into a set of decentralized convex optimization problems which are coupled by consensus constraints on the weight vector $\mathbf{w}$. The approach followed is the Alternating Direction Method of Multipliers (ADMM), which in general solves the problem

$$\min_{\mathbf{v}} \quad f_1(\mathbf{v}) + f_2(\mathbf{Av})$$

$$\text{s.t.} \quad \mathbf{v} \in \mathcal{P}_1, \mathbf{Av} \in \mathcal{P}_2,$$

where $f_1 : \mathbb{R}^{p_1} \to \mathbb{R}$ and $f_2 : \mathbb{R}^{p_2} \to \mathbb{R}$ are convex functions, A is a $p_2 \times p_1$ matrix, while $\mathcal{P}_1 \subset \mathbb{R}^{p_1}$ and $\mathcal{P}_2 \subset \mathbb{R}^{p_2}$ denote non-empty polyhedral sets. The problem is made separable by introducing an auxiliary variable $\boldsymbol{\omega} \in \mathbb{R}^{p_2}$:

$$\min_{\mathbf{v},\boldsymbol{\omega}} \quad f_1(\mathbf{v}) + f_2(\boldsymbol{\omega})$$

$$\text{s.t.} \quad \mathbf{Av} = \boldsymbol{\omega}, \mathbf{v} \in \mathcal{P}_1, \boldsymbol{\omega} \in \mathcal{P}_2$$

Let $\boldsymbol{\alpha} \in \mathbb{R}^{p_2}$ denote the Lagrange multipliers corresponding to the constraint $\mathbf{Av} = \boldsymbol{\omega}$. The augmented Lagrangian is

$$\mathcal{L}(\mathbf{v}, \boldsymbol{\omega}, \alpha) = f_1(\mathbf{v}) + f_2(\boldsymbol{\omega}) + \alpha^T(\mathbf{Av} - \boldsymbol{\omega}) + \frac{\eta}{2}\|\mathbf{Av} - \boldsymbol{\omega}\|^2 \,,$$

where $\eta > 0$ controls how much equality constraints may be violated. ADMM minimizes $\mathcal{L}$ in an alternating fashion, i.e. first for the primal variable $\mathbf{v}$ and then for the auxiliary variable $\boldsymbol{\omega}$. After each iteration, it updates the multiplier vector $\boldsymbol{\alpha}$. With t denoting the current iteration, the ADMM iterates at $t + 1$ are given by

$$\mathbf{v}^{t+1} = \underset{\mathbf{v} \in \mathcal{P}_1}{\operatorname{argmin}} \ \mathcal{L}(\mathbf{v}, \boldsymbol{\omega}(t), \boldsymbol{\alpha}^t),$$

$$\boldsymbol{\omega}^{t+1} = \underset{\boldsymbol{\omega} \in \mathcal{P}_2}{\operatorname{argmin}} \ \mathcal{L}(\mathbf{v}^{t+1}, \boldsymbol{\omega}, \boldsymbol{\alpha}^t),$$

$$\boldsymbol{\alpha}^{t+1} = \boldsymbol{\alpha}^t + \eta(\mathbf{Av}^{t+1} - \boldsymbol{\omega}^{t+1})$$

The first two optimization problems may be solved on different processors or machines. In a distributed setting, their results must be communicated over the network, such that each node can update its multiplier vector $\boldsymbol{\alpha}$. It can be proven that after a finite amount of iterations, the iterates $\boldsymbol{\alpha}^t$ will converge to the globally optimal solution $\boldsymbol{\alpha}^*$ of the dual problem.

In [15], the centralized SVM problem is cast into an ADMM formulation. The network is modeled by an undirected graph $G(P, E)$, where vertices $P = \{1, \ldots, m\}$ represent nodes and the set of edges E describes communication links between them. The graph is assumed to be connected and each node j only communicates with nodes in a one-hop neighborhood $\mathcal{N}_j \subseteq P$. Each node $j \in P$ stores a sample $S_j = \{(\mathbf{x}_{j1}, y_{j1}), \ldots, (\mathbf{x}_{jn_j}, y_{jn_j})\}$ of labeled observations, where $\mathbf{x}_{ji}$ is a $p \times 1$ vector from $\mathbb{R}^p$ and $y_{ji} \in \{-1, +1\}$ is a binary class label. The original primal SVM problem (see Sect. 2.3) is cast into the distributed ADMM framework by putting consensus constraints on the weight vectors $\mathbf{w}_j, \mathbf{w}_l$ and bias variables b_j, b_l of each node j and its one-hop neighboring nodes $l \in \mathcal{N}_j$:

$$\min_{\{\mathbf{w}_j, b_j, \xi_{ji}\}} \quad \frac{1}{2} \sum_{j=1}^{m} ||\mathbf{w}_j||^2 + mC \sum_{j=1}^{m} \sum_{i=1}^{n_j} \xi_{ji}$$

$$\text{s.t.} \quad y_{ji}(\langle \mathbf{w}_j, \mathbf{x}_{ji} \rangle + b_j) \geq 1 - \xi_{ji} \qquad \forall j \in P, i = 1, \ldots, n_j$$

$$\xi_{ji} \geq 0 \qquad \forall j \in P, i = 1, \ldots, n_j$$

$$\mathbf{w}_j = \mathbf{w}_l, b_j = b_l \qquad \forall j \in P, l \in \mathcal{N}_j.$$

For ease of notation, the authors define the augmented vector $\mathbf{v}_j :=$ $[\mathbf{w}_j^T, b_j]^T$, the augmented matrix $\mathbf{X}_j := [[\mathbf{x}_{j1}, \ldots, \mathbf{x}_{jn_j}]^T, \mathbf{1}_j]$, the diagonal label matrix $\mathbf{Y}_j := \text{diag}([y_{j1}, \ldots, y_{jn_j}])$, and the vector of slack variables $\boldsymbol{\xi}_j := [\xi_{j1}, \ldots, \xi_{jn_j}]^T$. $\boldsymbol{\Pi}_{p+1}$ is a $(p+1) \times (p+1)$ matrix with zeros everywhere except for the $(p+1), (p+1)$-st entry. It follows that $\mathbf{w}_j = (\mathbf{I}_{p+1} - \boldsymbol{\Pi}_{p+1})\mathbf{v}_j$ for $[\boldsymbol{\Pi}_{p+1}]_{(p+1)(p+1)} = 1$. With these vector and matrix notations, the problem can be rewritten as

$$\min_{\{\mathbf{v}_j, \boldsymbol{\xi}_j, \boldsymbol{\omega}_{ji}\}} \quad \frac{1}{2} \sum_{j=1}^{m} \mathbf{v}_j^T (\mathbf{I}_{p+1} - \boldsymbol{\Pi}_{p+1})\mathbf{v}_j + mC \sum_{j=1}^{m} \mathbf{1}_j^T \boldsymbol{\xi}_j$$

$$\text{s.t.} \quad \mathbf{Y}_j \mathbf{X}_j \mathbf{v}_j \succeq \mathbf{1}_j - \boldsymbol{\xi}_j \qquad \forall j \in P$$

$$\boldsymbol{\xi}_j \succeq \mathbf{0}_j \qquad \forall j \in P$$

$$\mathbf{v}_j = \boldsymbol{\omega}_{jl}, \boldsymbol{\omega}_{jl} = \mathbf{v}_l \qquad \forall j \in P, \forall l \in \mathcal{N}_j$$

where the auxiliary variables $\{\boldsymbol{\omega}_{jl}\}$ decouple parameters $\mathbf{v}_j$ at node j from those of its neighbors $l \in \mathcal{N}_j$. The augmented Lagrangian for the problem is

$$\mathcal{L}(\{\mathbf{v}_j\}, \{\boldsymbol{\xi}_j\}, \{\boldsymbol{\omega}_{jl}\}, \{\boldsymbol{\alpha}_{jlk}\}) = \frac{1}{2} \sum_{j=1}^{m} \mathbf{v}_j^T (\mathbf{I}_{p+1} - \boldsymbol{\Pi}_{p+1})\mathbf{v}_j + mC \sum_{j=1}^{m} \mathbf{1}_j^T \boldsymbol{\xi}_j$$

$$+ \sum_{j=1}^{m} \sum_{l \in \mathcal{N}_j} \boldsymbol{\alpha}_{jl1}^T (\mathbf{v}_j - \boldsymbol{\omega}_{jl}) + \sum_{j=1}^{m} \sum_{l \in \mathcal{N}_j} \boldsymbol{\alpha}_{jl2}^T (\boldsymbol{\omega}_{jl} - \mathbf{v}_l)$$

$$+ \frac{\eta}{2} \sum_{j=1}^{m} \sum_{l \in \mathcal{N}_j} ||\mathbf{v}_j - \boldsymbol{\omega}_{jl}||^2 + \frac{\eta}{2} \sum_{j=1}^{m} \sum_{l \in \mathcal{N}_j} ||\boldsymbol{\omega}_{jl} - \mathbf{v}_l||^2 \quad (10)$$

where the Lagrange multipliers $\boldsymbol{\alpha}_{jl1}$ and $\boldsymbol{\alpha}_{jl2}$ correspond to the constraints $\mathbf{v}_j = \boldsymbol{\omega}_{jl}$ and $\boldsymbol{\omega}_{jl} = \mathbf{v}_l$. The quadratic terms $||\mathbf{v}_j - \boldsymbol{\omega}_{jl}||^2$ and $||\boldsymbol{\omega}_{jl} - \mathbf{v}_l||^2$ ensure strict convexity, while parameter η allows for trading off speed of convergence against approximation error.

The distributed iterations that solve (10) are

$$\{\mathbf{v}_j^{t+1}, \boldsymbol{\xi}_j^{t+1}\} = \underset{\{\mathbf{v}_j, \boldsymbol{\xi}_j\}}{\text{argmin}} \, \mathcal{L}(\{\mathbf{v}_j\}, \{\boldsymbol{\xi}_j\}, \{\boldsymbol{\omega}_{jl}^t\}, \{\boldsymbol{\alpha}_{jlk}^t\})$$

$$\{\boldsymbol{\omega}_{jl}^{t+1}\} = \underset{\{\boldsymbol{\omega}_{jl}\}}{\text{argmin}} \, \mathcal{L}(\{\mathbf{v}_j^{t+1}\}, \{\boldsymbol{\xi}_j^{t+1}\}, \{\boldsymbol{\omega}_{jl}\}, \{\boldsymbol{\alpha}_{jlk}^t\})$$

$$\alpha_{jl1}^{t+1} = \alpha_{jl1}^t + \eta(\mathbf{v}_j^{t+1} - \omega_{jl}^{t+1}) \quad \forall j \in P, \forall l \in \mathcal{N}_j$$
$$\alpha_{jl2}^{t+1} = \alpha_{jl2}^t + \eta(\omega_{jl}^{t+1} - \mathbf{v}_l^{t+1}) \quad \forall j \in P, \forall l \in \mathcal{N}_j.$$

For details of how to simplify these iterations further and how to formulate the corresponding dual, see [15]. In each iteration, each node $j \in P$ must solve a quadratic optimization problem that is similar to the original SVM problem, but the local datasets S_j on which it needs to be solved can be considerably smaller than the whole dataset $S = S_1 \cup \ldots \cup S_m$. After each iteration, nodes must communicate their local solutions v_j to each one-hop neighbor and then update their local multiplier vectors. Casting the SVM problem into the ADMM framework guarantees that after a finite number of iterations, the local solutions v_j at each node $j \in P$ equal the global solution $\mathbf{w}, b$ that would have been found by training a centralized SVM classifier on all data S.

As long as the number of iterations is smaller than the total number of training examples n, the algorithm communicates less than the whole dataset. On the MNIST dataset used for evaluation in [15], the algorithm only needs about 200 iterations for reaching a similar error as a centralized SVM. Since in each iteration, only $p + 1$ scalars are transmitted, the total communication costs are considerably smaller than, for instance, those of the Cascade SVM that transmits about 60 k p-dimensional support vectors.

Even although each local optimization problem is solved in its dual formulation, only primal weight vectors are exchanged between nodes. The non-linear case is thus much harder to solve, since direct application of the Φ transformation may lead to high-dimensional weight vectors and therefore to high communication costs. Direct application of the kernel trick is not possible. The authors of [15] therefore propose to enforce consensus of the local discriminants on a subspace of reduced rank. This however requires preselected vectors common to all nodes, which introduce a subset of basis functions common to all local functional spaces. The choice of such vectors isn't necessarily straightforward and potentially requires the algorithm to be run again for each new classification query. Due to its complexity, the non-linear version is not discussed here. For further details, see [15].

Distributed Dual Ascend. [17] presents a method for distributed SVM learning based on distributed dual ascend. Let $P_j(\mathbf{d})$ with $\mathbf{d} = \mathbf{0}$ be the solution of a standard linear SVM with zero bias trained on the data S_j at node j:

$$P_j(\mathbf{d}) = \operatorname*{argmin}_{\mathbf{w}[j]} \frac{\lambda}{2m}||\mathbf{w}_j||^2 + \mathbf{w}_j^T \mathbf{d} + \frac{1}{m} \sum_{(\mathbf{x}_i, y_i) \in S_j} \max\{0, 1 - y_i\langle \mathbf{w}_j, \mathbf{x}_i \rangle\}.$$

The proposed distributed scheme uses $\mathbf{d} \neq \mathbf{0}$ for tying together the local results $\mathbf{w}_j$ while iterating over the local solutions $P_j(\cdot)$. At the beginning, the algorithm sets $\boldsymbol{\lambda}_j^{(0)} = \mathbf{0}$ and $\boldsymbol{\mu}_j^{(0)} = \mathbf{0}$. Then, in each iteration, each node j computes updates $\boldsymbol{\lambda}_j^{(t)} \leftarrow -\boldsymbol{\mu}_j^{(t-1)} - \frac{\lambda}{m} P_j(\boldsymbol{\mu}_j^{(t-1)})$ and passes its solution to a central node, which calculates $\boldsymbol{\mu}_j^{(t)} \leftarrow -\boldsymbol{\lambda}_j^{(t)} + \frac{1}{m}\sum_{j=1}^m \boldsymbol{\lambda}_j^{(t)}$ and communicates the solution back to each local node. The final output at iteration T is $\mathbf{w}^* = -\frac{1}{\lambda}\sum_{j=1}^m \boldsymbol{\lambda}_j^{(T)}$.

The algorithm is based on principles of Fenchel Duality (for further details, see [17]) and thus has a linear convergence rate, i.e. it takes $O(\log(1/\varepsilon))$ iterations to get ε-close to the optimal solution. With a linear kernel, only p scalars need to be transmitted in each iteration. The authors have also extended the algorithm to non-linear kernels. There, in the worst case, all components of vector $\boldsymbol{\alpha}$ need to be exchanged, meaning that each node j must transmit n/m scalars and receive all remaining αs in each iteration. Therefore, if $T > p$, the algorithm is not communication-efficient.

Distributed Block Minimization. For the linear SVM, [24] rewrite the dual SVM problem (see Sect. 2.3) as

$$\min_{\boldsymbol{\alpha}} \quad \boldsymbol{\alpha}^T \mathbf{Q}\boldsymbol{\alpha}/2 - \mathbf{1}^T\boldsymbol{\alpha}$$
$$\text{s.t.} \quad \forall_{i=1}^n : 0 \leq \alpha_i \leq C$$

where $\mathbf{Q}_{uv} = y_u y_v \mathbf{x}_u \mathbf{x}_v$. [35] have shown that this problem can be used with sequential block minimization (SBM), i.e. that at each iteration t, only a single block S_j of matrix $\mathbf{Q}$ is considered. The authors show that when solving for the variables in this block, the variables from other blocks don't need to be kept in memory. Suppose that $\boldsymbol{\alpha}^t$ is a solution after t iterations and that at $t + 1$, the focus is on block S_j, with $\mathbf{d}^j = \boldsymbol{\alpha}^{t+1}[j] - \boldsymbol{\alpha}^t[j]$ being the direction for components of the $\boldsymbol{\alpha}$ vector that are associated with block S_j. Then, according to [24], $\mathbf{d}_j$ may be obtained by solving the optimization problem

$$\min_{\mathbf{d}_j} \quad \mathbf{d}_j^T \mathbf{Q}[j,j]\mathbf{d}_j/2 + (\mathbf{w}^t)^T \mathbf{U}_j\mathbf{d}_j - \mathbf{1}^T\mathbf{d}_j \tag{11}$$
$$\text{s.t.} \quad 0 \preceq \boldsymbol{\alpha}^t[j] + \mathbf{d}_j \preceq C,$$

where $\mathbf{Q}[j,j]$ is a submatrix of $\mathbf{Q}$ consisting only of entries associated with the training examples in S_j and $\mathbf{U}_j$ is a $p \times |S_j|$ matrix where the i-th column is the i-th example in S_j, multiplied by its label y_i. For solving the problem, all that needs to be kept in memory are the training examples in S_j and the p-dimensional vector $\mathbf{w}^t$. After solving (11), $\mathbf{w}^t$ is updated as $\mathbf{w}^{t+1} = \mathbf{w}^t + \sum_{\mathbf{x}_i \in S_j} \mathbf{d}_j[i]y_i\mathbf{x}_i$.

The distributed block minimization (DBM) with averaging scheme proposed in [24] is then straightforward: Instead of processing each block S_j sequentially, they are all optimized in parallel. That is, given a central coordinator and j local nodes, per iteration t each node j solves (11) for S_j, sends $\Delta\mathbf{w}_j^t = \sum_{\mathbf{x}_i \in S_j} \mathbf{d}_j[i]y_i\mathbf{x}_i$ to the central coordinator and sets $\boldsymbol{\alpha}^{t+1}[j] = \boldsymbol{\alpha}^t[j] + 1/m \cdot \mathbf{d}_j$. The central coordinator then computes $\mathbf{w}^{t+1} = \mathbf{w}^t + 1/m\sum_{j=1}^m \Delta\mathbf{w}_j^t$ from the deltas received by each local node. The new vector $\mathbf{w}$ must then be transmitted to each local node, before a new iteration starts. The authors also discuss another variant than averaging, using line search for updating $\mathbf{w}$.

In each iteration, the algorithm communicates $O(mp)$ values. The authors argue that for a constant number of iterations, the communication complexity becomes independent from the number of training examples n. However, they haven't provided a proof of global convergence or for the rate of convergence. Empirically, the number of iterations needed to achieve sufficient accuracy on two different datasets was only 20. On both tested datasets, one proprietary from Akamai with 79 M training examples and the other a public learning to rank dataset from Microsoft with 37 M training examples, the algorithm achieves a higher accuracy than LIBLINEAR-CDBLOCK (see [35]) in a much shorter time.

4.3   Incremental Least Squares SVM

[11] introduces a distributed SVM based on a slightly different formulation of the SVM, the so called Least Squares SVM [29]. In the least squares formulation, the inequality constraints of the original SVM problem (see Sect. 2.3) are replaced by equality constraints and a 2-norm error, leading to the unconstraint optimization problem

$$\min_{\mathbf{w},b} \quad \frac{1}{2}||\mathbf{w}||^2 + \frac{\lambda}{2}||\mathbf{1} - \mathbf{Y}(\mathbf{D}\mathbf{w} - \mathbf{1}b)||^2 \,,$$

where $\mathbf{Y}$ is a diagonal matrix with $\mathbf{Y}_{ii} = y_i$. Setting the gradient w.t.r. $\mathbf{w}$ and b to zero, instead of a quadratic optimization problem one obtains a system of $(p+1)$ linear equations

$$[\mathbf{w}\,b]^T = \left(\frac{1}{\lambda}\mathbf{I}^\circ + \mathbf{E}^T\mathbf{E}\right)^{-1} \mathbf{E}^T\mathbf{Y}\mathbf{1} \,,$$

where $\mathbf{E} = [\mathbf{D}\,-\mathbf{1}]$ and $\mathbf{I}^\circ$ denotes a $(p+1) \times (p+1)$ diagonal matrix whose $(n+1)$-th diagonal entry is zero and the other diagonal entries are 1.

As the authors show, it is possible to solve this system of linear equations incrementally:

$$[\mathbf{w}\,b]^T = \left(\frac{1}{\lambda}\mathbf{I}^\circ + \sum_{j=1}^{m}\mathbf{E}_j^T\mathbf{E}_j\right)^{-1} \sum_{j=1}^{m}\mathbf{E}_j^T\mathbf{Y}_j\mathbf{1} \,. \tag{12}$$

In the distributed version of their algorithm, each node j computes the local sums $\mathbf{E}_j^T\mathbf{E}$ and $\mathbf{E}_j^T\mathbf{Y}_j\mathbf{1}$ independently from each other and communicates them to a central coordinator. The coordinator can then sum up these matrices and globally solve the linear system of Eq. (12).

The algorithm obviously can speed up computations, because the sums involved in solving the linear system of equations can be computed in parallel over different nodes j. With a linear kernel, the algorithm is communication-efficient if $n > p^2$. For cases where $p^2 > n$, the authors applied the Sherman-Morrison-Woodbury formula to the linear system of equations, resulting in a $n \times n$ instead of a $(p+1) \times (p+1)$ matrix. For non-linear kernels, the algorithm usually is not communication-efficient, since the original data matrix $\mathbf{D}$ is replaced by the kernel matrix $\mathbf{K}$, resulting in an n-dimensional weight vector $\mathbf{w}$ and thus a system of n linear equations.

5   Vertically Distributed SVMs

The vertically partitioned data scenario has been especially examined in the context of privacy-preserving data mining. However, as will be shown in Sect. 5.1, the approaches followed are generally not communication-efficient. Hybrid solutions that combine local and global models are presented in Sect. 5.2. They may reduce communication costs by several orders of magnitude, but in some cases at the expense of a decreased prediction performance. Finally, Sect. 5.3 discusses consensus algorithms, which guarantee convergence to a global optimum, but might transmit more data than the original dataset.

5.1   Privacy-Preserving SVMs

[21,36,37] present privacy-preserving SVMs that are mainly based on the communication of kernel matrices. A central observation in each work is that entries of the $n \times n$ kernel matrix $\mathbf{K}$ are separable in the sense that

$$k([\mathbf{E}\,\mathbf{F}], [\mathbf{G}\,\mathbf{H}]^T) = k(\mathbf{E}, \mathbf{G}^T) + k(\mathbf{F}, \mathbf{H}^T) \text{ or} \tag{13}$$

$$k([\mathbf{E}\,\mathbf{F}], [\mathbf{G}\,\mathbf{H}]^T) = k(\mathbf{E}, \mathbf{G}^T) \odot k(\mathbf{F}, \mathbf{H}^T) \tag{14}$$

where $k : \mathbb{R}^{n \times p} \times \mathbb{R}^{p \times n} \to \mathbb{R}^{n \times n}$ denotes the kernel function for whole matrices, $+$ denotes standard addition and $\odot$ denotes the Hadamard componentwise product of two matrices with same dimensions. In [21] it is shown that the linear dot product kernel $k(\mathbf{x}, \mathbf{x}') = \langle \mathbf{x}, \mathbf{x}' \rangle$ satisfies (13), while the RBF kernel (6) satisfies (14). Moreover, separability can be extended to polynomial kernels (5).

In a distributed setting, $\mathbf{D}[j]$ is the $n \times p_j$ data matrix whose rows consist of the (partial) training examples S_j at each local node j and $\mathbf{D}$ the $n \times p$ data matrix for the whole dataset S. Given kernel matrices $\mathbf{K}_1, \ldots, \mathbf{K}_m$ with entries of the linear kernel for data matrices $\mathbf{D}[1], \ldots, \mathbf{D}[m]$ at m different nodes, the global kernel matrix $\mathbf{K}$ for $\mathbf{D}$ can be calculated as

$$\mathbf{K} = \mathbf{K}_1 + \cdots + \mathbf{K}_m = \mathbf{D}[1]\mathbf{D}[1]^T + \cdots + \mathbf{D}[m]\mathbf{D}[m]^T.$$

In [36], it is proposed that each local node j first calculates its local kernel matrix $\mathbf{K}_j$. Each node might then send $\mathbf{K}_j$ to a central coordinator, which builds $\mathbf{K}$ and trains a centralized SVM on the full kernel matrix as usual. The scheme preserves the privacy of each local data matrix $\mathbf{D}[j]$, since it doesn't reveal the original attribute values. For added privacy, i.e. not even revealing the entries of the local kernel matrices, the authors propose an extended scheme with a secure addition mechanism (for details, see [36]). There, m nodes communicate in a ring topology where each node sends an $n \times n$ matrix to the next node and then back to the first node.

A slightly different approach is followed in [21]. There, it is proposed to replace the standard kernel by a reduced kernel $k(\mathbf{D}, \mathbf{B}^T) : \mathbb{R}^{n \times p} \times \mathbb{R}^{p \times \tilde{n}} \to \mathbb{R}^{n \times \tilde{n}}$, where $\tilde{n} < n$ and $\mathbf{B}$ is a random matrix. The $\tilde{n}$ columns of the random matrix are privately generated in m blocks corresponding to the m nodes which hold the corresponding feature values in their local data matrices $\mathbf{D}[1], \ldots, \mathbf{D}[m]$. Each node communicates its reduced local kernel matrix to a central coordinator, which reconstructs the global (reduced) kernel matrix $\mathbf{K}^r$ according to (13) or (14) and then trains a centralized SVM based on $\mathbf{K}^r$ as usual. It is empirically shown for several standard datasets that learning with the reduced kernel matrix achieves a similar error rate as a centralized SVM trained on the full kernel matrix.

The authors of [37] propose a similar scheme as [21,36], but argue that the secure addition procedure proposed in [36] or the reduced kernel are not necessary for the preservation of privacy. Instead, local kernel matrices could be sent directly to a central coordinator.

While all of the aforementioned approaches preserve the privacy of each local dataset, only [21] may improve the total run-time, due to a reduced kernel matrix. None of the approaches is communication-efficient for most practical purposes. The data matrix $\mathbf{D}[j]$ at node j consists of $n \times p_j$ real values and each kernel matrix $\mathbf{K}_j$ of $n \times n$ (or $n \times \tilde{n}$) values. Only if $p_j > n$ (or $p_j > \tilde{n}$), less data is sent than transmitting the original data matrices $\mathbf{D}[j]$ to a central node. However, usually $p_j \ll n$, especially since the total number of features p is split among m different nodes.

5.2 Local vs. Global Optimization

As already discussed in Sect. 3.3, distributed learning in the vertically partitioned data scenario is particularly challenging if the label depends on features from different nodes. Since each local dataset S_j is based on an entirely different feature space than the whole dataset S, separate SVM models trained on each local dataset will have different $\boldsymbol{\alpha}$ values and thus cannot easily be merged. However, in cases where features from different nodes are conditionally independent, given the label, the predictions of locally trained models or kernel values may be centrally combined by different means. The following subsections present approaches which combine local and global models in hybrid learning schemes.

Separable SVM. [19] solves the primal SVM problem locally at each node with stochastic gradient descent (SGD). The global optimization problem consists of learning a weighting for the combination of local predictions. While [19] addresses the tasks of 1-class learning, binary classification and regression, the following discussion is restricted to binary classification. The primal optimization problem to solve is denoted in hinge function notation (see (2)) as

$$\min_{\mathbf{w} \in H} \quad \frac{\lambda}{2} \|\mathbf{w}\|^2 + \frac{1}{n} \sum_{u=1}^{n} \max\{0, 1 - y\langle \mathbf{w}, \Phi(\mathbf{x}_u) \rangle\}, \tag{15}$$

where feature mapping $\Phi : \mathbb{R}^p \to H$ induces a positive semidefinite kernel $k(\mathbf{x}, \mathbf{x}') = \langle \Phi(\mathbf{x}), \Phi(\mathbf{x}') \rangle$. (Here, without loss of generality, the intercept b is ignored). The kernel function is split across nodes by definition of a *composite* kernel k, which is a conic combination of local kernels $k_j : \mathbb{R}^{p_j} \times \mathbb{R}^{p_j} \to \mathbb{R}$ defined on the partial feature vectors $\mathbf{x}_i[j]$, $i = 1, \ldots, n$ stored at node j:

$$k(\mathbf{x}, \mathbf{x}') = \sum_{j=1}^{m} \mu_j^2 k_j(\mathbf{x}[j], \mathbf{x}'[j]).$$

With Lagrange multipliers $\boldsymbol{\alpha}$, the optimization problem becomes

$$\min_{\boldsymbol{\alpha} \in \mathbb{R}^n} \quad \frac{\lambda}{2} \sum_{j=1}^{m} \mu_i^2 \sum_{u=1}^{n} \sum_{v=1}^{n} \alpha_u \alpha_v k_j(\mathbf{x}_u[j], \mathbf{x}_v[j])$$

$$+ \frac{1}{n} \sum_{u=1}^{n} \max \left\{ 0, 1 - y_u \sum_{j=1}^{m} \mu_j^2 \sum_{v=1}^{n} \alpha_v k_j(\mathbf{x}_u[j], \mathbf{x}_v[j]) \right\}. \quad (16)$$

Problem (16) points to the fundamental difficulty of distributing SVMs in the vertically partitioned data scenario: All optimization variables $\alpha_1, \ldots, \alpha_n$ are coupled with each node $j = 1, \ldots, m$, and therefore cannot be split over the nodes. In [19], separability is achieved by two different means. The first observation is that the terms $||\mathbf{w}||^2$ and $\langle \mathbf{w}, \Phi(\mathbf{x}_u) \rangle$ in the primal problem (15) become separable over the components of $\mathbf{w}$ if $\mathbf{w}$ and $\Phi(\mathbf{x}_u)$ are in a finite dimensional space. The authors therefore propose to replace local feature mappings Φ_j with approximate mappings φ_j which can be directly constructed using the technique of random projections (for details, see [19]). The second observation is that the hinge loss can be upper bounded as follows:

$$\max \left\{ 0, \sum_{j=1}^{m} \mu_j(1 - y\langle \mathbf{w}[j], \varphi_j(\mathbf{x}[j]) \rangle) \right\} \leq \sum_{j=1}^{m} \mu_j(1 - y\langle \mathbf{w}[j], \varphi_j(\mathbf{x}[j]) \rangle). \quad (17)$$

Summing up the inequalities (17) over training examples $u = 1, \ldots, n$, the local objective solved by each node $j = 1, \ldots, m$ becomes

$$\min_{\mathbf{w}[j]} \quad \frac{\lambda}{2} ||\mathbf{w}[j]||^2 + \frac{1}{n} \sum_{u=1}^{n} \mu_j(1 - y\langle \mathbf{w}[j], \varphi_j(\mathbf{x}_u[j]) \rangle).$$

A global classifier may then be constructed by combining local predictions, i.e. $\langle \mathbf{w}, \varphi(\mathbf{x}) \rangle = \sum_{j=1}^{m} \mu_j \langle \mathbf{w}[j], \varphi(\mathbf{x}[j]) \rangle$. Hence, for each test point, m scalars $\langle \mathbf{w}[j], \varphi(\mathbf{x}[j]) \rangle$ need to be transmitted.

Approximating the non-separable hinge loss by separable upper bounds necessarily leads to a gap in accuracy. In addition to the local objectives, which are solved by SGD, [19] therefore poses a central quadratic optimization problem for finding optimal weights $\mu_1, \ldots, \mu_m$. The problem is solved iteratively in an alternating fashion. Per iteration, each node j solves its local objective and transmits predictions for all n observations to the central node. The central node finds optimal weights μ_j and transmits them back to the corresponding local nodes. The loop stops after a user-specified number of iterations or if the central objective cannot be further improved. The algorithm has been evaluated on synthetic data and five standard datasets. While on one dataset, prediction accuracy could be improved by six percentage points with the central optimization, improvement on the other datasets was marginal. The use of random projections and a composite kernel reduces accuracy in the range of 5.5–27.8 % points. However, the method is highly communication-efficient, since without the central optimization, *no* data needs to be transmitted during training. It is also communication-efficient during application, since each node only transmits a single scalar value per test point, instead of p_j feature values.

Distributed 1-Class SVM. [9] introduces a synchronized distributed anomaly detection algorithm based on the 1-class ν-SVM (see also Sect. 2.6). A local 1-class model is trained at each node and points identified as local outliers are sent to the central node P_0, together with a small sample of all observations. A global model trained on the sample at node P_0 is used to decide if the outlier candidates sent from the data nodes are true global outliers or not. The method cannot detect outliers which are global due to a combination of attributes. However, the algorithm shows good performance if global outliers are also local outliers. Moreover, in the application phase, the algorithm is highly communication-efficient, since the number of outlier candidates is often only a small fraction of the data. A drawback is that the fixed-size sampling approach gives no guarantees or bounds on the correctness of the global model. Moreover, during training, no other strategies than sampling are used for a reduction of communication costs.

Vertically Distributed Core Vector Machine. The Vertically Distributed Core Vector Machine (VDCVM) [28] replaces the global 1-class ν-SVM from [9] by the Core Vector Machine (CVM) [33]. The CVM algorithm is guaranteed to find a $(1 + \epsilon)$ approximation of the MEB with high probability. It starts with a so called *core set S* consisting of two data points far away from each other and their center c_0. In each iteration t, the CVM samples a fixed-size batch V_t of data points. For this batch, the algorithm determines a point $z_t \in V_t$ that is furthest away from the current center c_t and adds it to the current core set, i.e. $S_{t+1} = S_t \cup \{z_t\}$. The CVM stops if all core set points lie inside a $(1 + \varepsilon)$-ball, i.e. $\|c_t - \Phi(x_i)\| \leq R_t(1+\varepsilon)$ for all $x_i \in S_{t+1}$. If not, it calculates a new MEB_{t+1} around all core set points, resulting in a new center c_{t+1} and radius R_{t+1}, and goes on with the next iteration. It can be shown that the number of iterations, and thus the core set size, is constant (see [33]) and does neither depend on the number of observations n, nor the number of features p.

The distributed CVM algorithm proposed in [28] is based on the idea that with a composite kernel like the one in [19], the furthest point calculation

$$z_t = \operatorname*{argmax}_{x_\ell \in V_t} \left[-\sum_{x_i \in S_t} \alpha_i^{(t)} \tilde{k}(x_i, x_\ell) \right]$$

for positive kernels $\tilde{k}(\mathbf{x}_u, \mathbf{x}_v) = k(\mathbf{x}_u, \mathbf{x}_v) + \delta_{uv}/C$ with $\delta_{uv} = 1$ for $u = v$ and $\delta_{uv} = 0$ for $u \neq v$ may be distributed across local nodes $j = 1, \ldots, m$ in each iteration. Kernel k is defined as a combination of RBF kernels on features of each local node, i.e.

$$k(\mathbf{x}_u, \mathbf{x}_v) = \sum_{j=1}^{m} e^{-\gamma_j \|\mathbf{x}_u[j] - \mathbf{x}_v[j]\|^2}.$$

With this kernel, the furthest point calculation becomes separable. In each iteration t, a central coordinator sends random indices of observations which would be contained in V_t to each local node. Each local node $j = 1, \ldots, m$ calculates a partial sum

$$v_\ell^{(j)} = \sum_{\mathbf{x}_i[j] \in \mathcal{S}_t} \alpha_i^{(t)} e^{-\gamma_j \|\mathbf{x}_i[j] - \mathbf{x}_\ell[j]\|^2}.$$

for each point $\mathbf{x}_\ell \in V_t$ and sends it back to the central coordinator. These partial sums are centrally aggregated (added) and used to determine the index of the furthest point. The coordinator asks each local node for the furthest point's feature values, adds $\mathbf{z}_t$ to $\mathcal{S}_t$ and then either stops or goes on with the MEB calculation. Since the partial sum computations are based on the current $\boldsymbol{\alpha}^{(t)}$ vector, in each iteration the central coordinator must further transmit $\boldsymbol{\alpha}^{(t)}$ to each local node.

On many of the tested synthetic and standard datasets, the algorithm reaches a similar prediction performance as a centrally trained standard 1-class SVM and the distributed 1-class SVM introduced in [9]. The algorithm's communication costs have been compared theoretically as well as empirically to sending all data to a central node. It is demonstrated that up to a fixed number of iterations T, which depends on the number of features p stored at each node, the algorithm is highly communication-efficient and may reduce communication by several orders of magnitude. The reduction in early iterations is mainly achieved by sending feature values only for core set points, while the partial sums sent for the remaining points in each subsample are just single scalar values. However, in later iterations the transmission of vector $\boldsymbol{\alpha}$ dominates communication costs, which are asymptotically quadratic in the number of iterations (i.e. the core set size).

5.3 Distributed Optimization and Consensus

[3] casts the vertically distributed SVM problem into the ADMM framework. The general problem of model fitting on vertically partitioned data is posed in terms of regularized risk minimization as

$$\min_{\{\mathbf{w}[j]\}} \quad l\left(\sum_{j=1}^{m} \mathbf{D}[j]\mathbf{w}[j] - \mathbf{y}\right) + \sum_{j=1}^{m} r_j(\mathbf{w}[j]),$$

where $\mathbf{y} = y_1, \ldots, y_n$ is the vector of all labels, l measures the loss and $\mathbf{w}[j]$ is a partial weight vector whose dimension corresponds to the number of features

p_j stored at node j. Multiplication of the local $n \times p_j$ data matrix $\mathbf{D}[j]$ with the partial weight vector $\mathbf{w}[j]$ results in a vector of dimension n, which consists of the local predictions at node j for the partial observations stored at j. The loss over all nodes should be minimized. The regularization function $r(x)$ is assumed to be separable. For solving with ADMM, the authors introduce an auxiliary variable $\mathbf{z}_j$, resulting in the optimization problem

$$\min_{\{\mathbf{w}[j]\}} \quad l\left(\sum_{j=1}^{m} \mathbf{z}_j - \mathbf{y}\right) + \sum_{j=1}^{m} r_j(\mathbf{w}[j])$$
$$\text{s.t.} \quad \mathbf{D}[j]\mathbf{w}[j] - \mathbf{z}_j = 0, \ j = 1, \ldots, m.$$

For the SVM problem in particular, the distributed ADMM iterates are

$$\mathbf{w}^{t+1}[j] = \operatorname*{argmin}_{\mathbf{w}[j]} \left(\frac{\eta}{2}||\mathbf{D}[j]\mathbf{w}[j] - \mathbf{D}[j]\mathbf{w}^t[j] - \overline{\mathbf{z}}^t + \overline{\mathbf{Dw}}^t + \mathbf{u}^t||^2 + \lambda||\mathbf{w}[j]||^2\right)$$

$$\overline{\mathbf{z}}^{t+1} = \operatorname*{argmin}_{\mathbf{z}} \left(\mathbf{1}^T(m\overline{\mathbf{z}} + 1)_+ + \frac{\eta}{2}||\overline{\mathbf{z}} - \overline{\mathbf{D}[j]} - \overline{\mathbf{Dw}}^{t+1} - \mathbf{u}^t||^2\right)$$

$$\mathbf{u}^{t+1} = \mathbf{u}^t + \overline{\mathbf{Dw}}^{t+1} - \overline{\mathbf{z}}^{t+1},$$

where the bar denotes averaging, η allows for trading off speed of convergence against approximation error and λ controls the structural risk. Updates of the weight vector $\mathbf{w}$ require solving local ridge regression problems at each node. The $\overline{\mathbf{z}}$ updates can be shown to split to the component level, i.e. they can be run on each node independently from each other (for details, see [3]). What needs to be communicated to other nodes in each iteration is vector $\overline{\mathbf{Dw}}$, which is the average of predictions over all nodes.

The communication costs will depend on the total number of iterations T. In comparison to the horizontally distributed consensus SVM (see [15]), consensus is not to be reached on p components of the weight vector $\mathbf{w}$, but on n predictions after applying the partial weight vectors $\mathbf{w}[j]$ to local data. Therefore, in each iteration n scalar values need to be communicated by each node. If $\exists j : T > p_j$, already more data would be transmitted by j than the local data, which is not communication-efficient. Unfortunately, the authors provide no empirical evaluation of how many iterations the algorithm needs to reach a sufficient accuracy on different datasets. In general, however, ADMM is known to have a slow convergence rate.

6   Summary and Conclusions

There exist several solutions for the distributed computation of SVMs in the horizontally partitioned data scenario which are communication-efficient, at least for the linear kernel. Algorithms exchanging summary information, like support vectors, speed up computation and can be communication-efficient if the number of support vectors is small. However, their number depends on the structure of the data, chosen hyperparameters, and on the type of used kernel. Running such types of algorithms in pervasive distributed systems with energy-constraints thus comes at a risk. They seem to be better suited for moderately sized problem instances solved on a high-performance cluster. Their communication complexity is $O(T|SV|)$, where T is the total number of iterations.

In contrast, the communication complexity of distributed optimization algorithms that iteratively update the primal weight vector $\mathbf{w}$ is $O(Tp)$. In practice, p is often much smaller than the number of observations n, leading to small communication costs per iteration. The total communication costs, however, depend on the convergence rate. ADMM is known to have slow convergence, while the dual ascend approach introduced in [17] converges in $O(\log 1/\varepsilon)$ steps to a global optimum. Distributed block minimization has no proof for the convergence rate. All algorithms show good performance in practice, with regard to prediction performance as well as the number of iterations. Moreover, the Consensus SVM is also highly fault tolerant, as it only communicates with neighboring peer nodes and has no single point of failure. However, it is very difficult to incorporate nonlinear kernels. The same is true for the distributed Least Squares SVM. However, for linear kernels, it is among the most communication-efficient methods, sending only $O(p^2)$ values per node.

Distributed SVM computing in the vertically partitioned data scenario is much more difficult to achieve and usually requires a central coordinator. The privacy-preserving methods have a communication complexity of $O(n^2)$, and thus communicate more than the entire data if $n > p$. Hybrid methods which combine local and global models can be highly communication-efficient, but accuracy may suffer in cases where features are conditionally dependent, given the label. In comparison to all other methods presented in this paper, the VDCVM is the only method which makes only a single pass over the data, due to the use of core sets. However, communication costs are quadratic in the size of the core set (i.e. $O(T^2)$). Nevertheless, on many of the datasets tested, the method could reduce communication costs by at least an order of magnitude, maintaining a similar accuracy. As an incremental sampling method, it is also fault-tolerant against transmission errors: If samples get lost, new data points can be easily re-transmitted. The ADMM approach for vertically partitioned data has a communication complexity of $O(Tn)$, which is not efficient if $T > p$. Since it has not been tested on any data, it remains unknown how it performs empirically.

Obviously, an implementation of the SVM algorithm on small embedded devices involves much more than the reduction of communication costs, which may be the reason why none of the above methods have yet been evaluated in a really constraint environment like WSNs. The distributed optimization approaches that work on horizontally partitioned data look most promising here, while the hybrid methods might work for vertically partitioned sensor data. With big horizontally partitioned data, distributed block minimization needs about six hours for 80 million data points on a high performance cluster. For big vertically partitioned data, the VDCVM looks most promising, since it only samples as many data points as needed to reach a sufficient accuracy with high probability.

An unsolved problem regarding all of the presented methods, however, is how to tune the SVM's hyperparameters in a communication-efficient way, i.e. without running the method several times with different parameter settings.

Acknowledgements. This work has been supported by the DFG, Collaborative Research Center SFB 876 (http://sfb876.tu-dortmund.de/), project B3.

References

1. Bennett, K.P., Campbell, C.: Support vector machines: hype or hallelujah? SIGKDD Explor. Newsl. **2**(2), 1–13 (2000)
2. Bhaduri, K., Stolpe, M.: Distributed data mining in sensor networks. In: Aggarwal, C.C. (ed.) Managing and Mining Sensor Data, pp. 211–236. Springer, Heidelberg (2013)
3. Boyd, S., Parikh, N., Chu, E., Peleato, B., Eckstein, J.: Distributed optimization and statistical learning via the alternating direction method of multipliers. Found. Trends Mach. Learn. **3**(1), 1–122 (2011)
4. Boyd, S., Vandenberghe, L.: Convex Optimization. Cambridge University Press, New York (2004)
5. Caragea, C., Caragea, D., Honavar, V.: Learning support vector machines from distributed data sources. In: Proceedings of the 20th National Conference on Artificial Intelligence (AAAI), vol. 4, pp. 1602–1603. AAAI Press (2005)
6. Caragea, D., Silvescu, A., Honavar, V.: Towards a theoretical framework for analysisand synthesis of agents that learn from distributed dynamic data sources. In: Proceedings of the Workshop on Distributed and Parallel Knowledge Discovery. ACM SIGKDD Int. Conf. on Knowledge Discovery and Data Mining (KDD) (2000)
7. Caragea, D., Silvescu, A., Honavar, V.: Agents that learn from distributed dynamic data sources. In: Proceedings of the Workshop on Learning Agents (2000)
8. Collobert, R., Bengio, S., Bengio, Y.: A parallel mixture of SVMs for very large scale problems. Neural Comput. **14**(5), 1105–1114 (2002)
9. Das, K., Bhaduri, K., Votava, P.: Distributed anomaly detection using 1-class SVM for vertically partitioned data. Stat. Anal. Data Min. **4**(4), 393–406 (2011)
10. Datta, S., Kargupta, H.: Uniform data sampling from a peer-to-peer network. In: Proceedings of the 27th International Conference on Distributed Computing Systems (ICDCS), pp. 1–8 (June 2007)
11. Do, T.N., Poulet, F.: Classifying one billion data with a new distributed SVM algorithm. In: International Conference on Research, Innovation and Vision for the Future, pp. 59–66 (February 2006)
12. Flouri, K., Beferull-Lozano, B., Tsakalides, P.: Training a SVM-based classifier in distributed sensor networks. In: EUSIPCO 2006 (2006)
13. Flouri, K., Beferull-Lozano, B., Tsakalides, P.: Distributed consensus algorithms for SVM training in wireless sensor networks. In: EUSIPCO (2008)
14. Flouri, K., Beferull-Lozano, B., Tsakalides, P.: Optimal gossip algorithm for distributed consensus SVM training in wireless sensor networks. In: 16th International Conference on Digital Signal Processing, pp. 1–6 (July 2009)
15. Forero, P.A., Cano, A., Giannakis, G.B.: Consensus-based distributed support vector machines. J. Mach. Learn. Res. **11**, 1663–1707 (2010)

16. Graf, H.P., Cosatto, E., Bottou, L., Dourdanovic, I., Vapnik, V.: Parallel support vector machines: the cascade SVM. In: Saul, L., Weiss, Y., Bottou, L. (eds.) Advances in Neural Information Processing Systems, vol. 17, pp. 521–528. MIT Press, Cambridge (2005)
17. Hazan, T., Man, A., Shashua, A.: A parallel decomposition solver for SVM: distributed dual ascend using fenchel duality. In: IEEE Conference on Computer Vision and Pattern Recognition (CVPR), pp. 1–8 (June 2008)
18. Joachims, T.: Making large-scale support vector machine learning practical. In: Advances in Kernel Methods, pp. 169–184. MIT Press, Cambridge (1999)
19. Lee, S., Stolpe, M., Morik, K.: Separable approximate optimization of support vector machines for distributed sensing. In: Flach, P.A., De Bie, T., Cristianini, N. (eds.) ECML PKDD 2012, Part II. LNCS, vol. 7524, pp. 387–402. Springer, Heidelberg (2012)
20. Lu, Y., Roychowdhury, V.P.: Parallel randomized support vector machine. In: Ng, W.-K., Kitsuregawa, M., Li, J., Chang, K. (eds.) PAKDD 2006. LNCS (LNAI), vol. 3918, pp. 205–214. Springer, Heidelberg (2006)
21. Mangasarian, O.L., Wild, E.W., Fung, G.M.: Privacy-preserving classification of vertically partitioned data via random kernels. ACM Trans. Knowl. Discov. Data 2(3), 12:1–12:16 (2008)
22. Moya, M., Koch, M., Hostetler, L.: One-class classifier networks for target recognition applications. In: Proceedings of World Congress on Neural Networks, pp. 797–801. International Neural Network Society (1993)
23. Osuna, E., Freund, R., Girosi, F.: An improved training algorithm for support vector machines, pp. 276–285. IEEE (1997)
24. Pechyony, D., Shen, L., Jones, R.: Solving large scale linear SVM with distributed block minimization. In: NIPS 2011 Workshop on Big Learning: Algorithms, Systems and Tools for Learning at Scale (2011)
25. Platt, J.C.: Fast training of support vector machines using sequential minimal optimization. In: Advances in kernel methods, pp. 185–208. MIT Press, Cambridge (1999)
26. Rüping, S.: Incremental learning with support vector machines. In: Proceedings of the 2001 IEEE International Conference on Data Mining (ICDM), pp. 641–642 (2001)
27. Schölkopf, B., Platt, J.C., Shawe-Taylor, J.C., Smola, A.J., Williamson, R.C.: Estimating the support of a high-dimensional distribution. Neural Comp. 13(7), 1443–1471 (2001)
28. Stolpe, M., Bhaduri, K., Das, K., Morik, K.: Anomaly detection in vertically partitioned data by distributed core vector machines. In: Blockeel, H., Kersting, K., Nijssen, S., Železný, F. (eds.) ECML PKDD 2013, Part III. LNCS, vol. 8190, pp. 321–336. Springer, Heidelberg (2013)
29. Suykens, J., Vandewalle, J.: Least squares support vector machine classifiers. Neural Process. Lett. 9(3), 293–300 (1999)
30. Syed, N.A., Huan, S., Kah, L., Sung, K.: Incremental learning with support vector machines. In: International Joint Conference on Artificial Intelligence (IJCAI) (1999)
31. Tanenbaum, A., van Steen, M.: Distributed Systems: Principles and Paradigms, 2nd edn. Prentice Hall, Upper Saddle River (2006)
32. Tax, D.M.J., Duin, R.P.W.: Support vector data description. Mach. Learn. 54, 45–66 (2004)
33. Tsang, I., Kwok, J., Cheung, P.: Core vector machines: fast svm training on very large data sets. J. Mach. Learn. Res. 6, 363–392 (2005)

34. Vapnik, V.N.: The Nature of Statistical Learning Theory. Springer, New York (1995)

35. Yu, H.F., Hsieh, C.J., Chang, K.W., Lin, C.J.: Large linear classification when data cannot fit in memory. In: Proceedings of the 16th ACM SIGKDD International Conference on Knowledge Discovery and Data Mining, pp. 833–842. ACM, New York (2010)

36. Yu, H., Vaidya, J., Jiang, X.: Privacy-preserving svm classification on vertically partitioned data. In: Ng, W.-K., Kitsuregawa, M., Li, J., Chang, K. (eds.) PAKDD 2006. LNCS (LNAI), vol. 3918, pp. 647–656. Springer, Heidelberg (2006)

37. Yunhong, H., Liang, F., Guoping, H.: Privacy-preserving SVM classification on vertically partitioned data without secure multi-party computation. In: 5th International Conference on Natural Computation (ICNC), vol. 1, pp. 543–546 (August 2009)

38. Zanghirati, G., Zanni, L.: A parallel solver for large quadratic programs in training support vector machines. Parallel Comput. **29**(4), 535–551 (2003)

Big Data Classification – Aspects on Many Features

Claus Weihs[(✉)]

Department of Statistics, TU Dortmund University, Dortmund, Germany
claus.weihs@tu-dortmund.de

Abstract. In this paper we discuss the performance of classical classification methods on Big Data. We concentrate on the case with many more features than observations and discuss the dependence of classification methods on the distance of the classes and their behavior for many noise features. The examples in this paper show that standard classification methods should be rechecked for Big Data.

1 Introduction

This paper is on Big Data Analytics (BDA). But what is Big Data? Unfortunately, the answer depends on whom you have asked when. In Machine Learning (ML) benchmarks in the 1990s (e.g. in the UCI repository) maximum 100 s to 1000 s of data points were available. In modern benchmarks we often have more than 10^6 data points. When you ask, e.g., Google, the answer might be 'Big Data means that data are much too big for your computer storage, only streaming is possible from a cloud, only distributed analytics, ...' Another possibility is to define a 'Big Data problem' by the impossibility to exactly solve the learning problem by computation time complexity.[1] Therefore, information in the data is not optimally utilizable.

In this paper we will discuss typical classification methods in the context of Big Data Analytics. The message of this paper is that for BDA not all classical methods are adequate in all Big Data situations and that Big Data might even long for special methods. We concentrate here on one extreme when there are many more features than observations[2]. With the advent of high throughput biotechnology data acquisition platforms such as micro arrays, SNP chips and mass spectrometers, data sets with many more features than observations are now routinely being collected (see, e.g., [4]). Most often, however, only a small part of these p features or a small number of directions in p-space are important for classification. Therefore, one might be tempted to thoughtlessly apply standard methods which are known to be only adequate for $p < n$ (not too big), but problematic in high dimensions (curse of dimensionality) and for very large n. In this paper, we will discuss some of the many available classification methods in this context.

---

[1] Thanks to T. Glasmachers for suggesting this definition.

[2] We base on the corresponding part of an earlier version of this paper [7].

© Springer International Publishing Switzerland 2016
S. Michaelis et al. (Eds.): Morik Festschrift, LNAI 9580, pp. 139–147, 2016.
DOI: 10.1007/978-3-319-41706-6_6

The paper is structured as follows. In Sect. 2 some theoretical results from the literature will be considered. In Sect. 3 related simulations will be reported. The paper is concluded with a summary and ideas for extensions.

2 Some Theoretical Results

One of the best known and most used classification methods in statistics is Fisher discrimination. The performance of this method in the case of more features than observations is discussed by [1] showing the following property.

Property 1: Consider 2 classes with Gaussian distributions $\mathcal{N}(\mu_1, \Sigma)$, $\mathcal{N}(\mu_2, \Sigma)$. Let the corresponding a priori probabilities be equal, i.e., $\pi_1 = \pi_2 = 0.5$. Then, for Fisher discrimination the classification function has the form $\delta_F(x) = (x - \mu)^T \Sigma^{-1}(\mu_1 - \mu_2)$ with $\mu = (\mu_1 + \mu_2)/2$.

Let the corresponding samples be observed with equal sample sizes, i.e. $n_1 = n_2$. Then, the sample version of the classification rule is: Assign class 1 iff $\hat{\delta}_F(x) = (x - \bar{x})^T S^{-1}(\bar{x}_1 - \bar{x}_2) > \log(\pi_2/\pi_1) = 0$. If $p > n$, then the inverse of the estimated pooled covariance matrix S does not exist and the Moore-Penrose generalized inverse is used instead.

For this situation, the following result is true under some regularity conditions, which particularly state that the norm of the mean vectors should be limited and the true covariance matrix Σ is not ill-conditioned for each p. If $p \to \infty, n \to \infty$, and $p/n \to \infty$, then in the worst case $\text{error}(\hat{\delta}_F(x)) \to 0.5$, i.e. the class assignment is no better than random guessing.

This result states a strong warning concerning the application of Fisher discrimination in the case of many more features than observations. As have been motivated by [1], the bad performance of Fisher discriminant analysis is due to the fact that the condition number of the estimated covariance matrix goes to infinity as dimensionality diverges even though the true covariance matrix is not ill-conditioned.

Noise accumulation might be reduced by ignoring the full covariance structure, e.g. by using a diagonal matrix as an estimate of the covariance matrix. In this context, [1] derived the following asymptotic result for the so-called *independence rule (ir)*, i.e. linear discriminant analysis with diagonal covariance matrix.

Property 2: Let Γ be a 'regular' space of possible means and covariance matrices of the two classes, Σ the full covariance matrix in the two classes, Σ_0 the corresponding correlation matrix, $\lambda(\Sigma_0)$ an eigenvalue of Σ_0, and Φ the distribution function of the standard normal. Then, the following result is true:

If $\log(p)/n \to 0$, then $\limsup_{n \to \infty}(\text{maximal error in } \Gamma) = 1 - \Phi\left(\frac{\sqrt{K_0}}{1+K_0}c\right) \leq 0.5$, where $K_0 = \max_\Gamma \left(\frac{\lambda_{\max}(\Sigma_0)}{\lambda_{\min}(\Sigma_0)}\right)$ and $c^2 = \min_\Gamma \left((\mu_2 - \mu_1)^T \Sigma^{-1}(\mu_2 - \mu_1)\right)$.

Therefore, if p is going slower to infinity than e^n, then for big data sets there is a bound for the maximal error in the space of possible data situations. In practice, this property may lead to a superiority of *ir* over the full *lda*.

Note that for normal distributions the independence rule is equivalent to the Naive Bayes method. In practice, however, the Naive Bayes method (*NB*) is typically implemented in a non-parametric way and not by assuming a certain type of distribution like the normal distribution. This generally leads to implementations different from the independence rule. For normal distributions as in our examples, *NB* is thus expected to be inferior to *ir*.

Let us now take a closer look at the distance dependency of classification quality for a general class of distance-based classifiers. For a plausible *distance-based classifier g* we only assume the following two properties:

(a) g assigns x to class 1 if it is closer to each of the x_i in class 1 than it is to any of the x_j in class 2.

(b) If g assigns x to class 1, then x is closer to at least one of the x_i in class 1 than to the most distant x_j in class 2.

Property 3: For such a method the following property is true [3]: Consider the model $x_{ij} = \mu_{kj} + \epsilon_{ij}$, $i \in \{g_k, k = 1, 2\}$, where x_{ij} is the jth component of x_i, μ_{kj} the jth component of the mean vector μ_k, and the ϵ_{ij} are independently identically distributed with mean 0 and finite 4th moment. Then, the probability that a distance based classifier of the above kind classifies a new observation correctly converges to 1 iff $p = o(||\mu_2 - \mu_1||^4)$ for $p \to \infty$.

This property shows that with distance based classifiers perfect class prediction is possible, but only if the distance of class means grows with the number of influential features so that $p^{1/4}/||\mu_2 - \mu_1|| \to 0$, i.e. that $||\mu_2 - \mu_1||$ grows faster than $p^{1/4}$. Note that this result is independent of sample size n.

Also note that the above definition of general distance-based classifiers includes the nearest neighbor classifiers *kNN* and the linear support vector machine (*svm*) which is also looking for linear separations. The latter method will be discussed here as a representative of methods which can be adapted to the actual data by tuning hyperparameters, e.g. the cost parameter. Note that we expect a linear separation in our examples so that nonlinear *svm*s are not expected to be sensible. Another type of classifiers we include in our comparison is the decision tree generated by CART (*tree*). Such trees have the advantage of automatically selecting the important features and thus ignoring features which mainly represent noise for the classification problem. Finally, we include one representative of ensemble techniques. Here, we restrict ourselves to bagged CART-trees (*baggedtree*), in order to again benefit from automatic feature selection.

3   Simulations

Let us now discuss the above theoretical results by means of simulations. We start with Generic Data Generation.

3.1   Generic Data Generation (GDG)

In this paper, we will always consider the ideal situation for linear discriminant analysis (*lda*), i.e. two classes where the influential features are multivariate normally distributed with different mean vectors and the same covariance matrix. When pr features influence class separation we choose the class means $m_1(i) = -md/2$, $m_2(i) = md/2$, where md = difference between the two class means in dimension i, $i = 1, \ldots, pr$. The covariance matrices are built so that $\Sigma = \Sigma_R + d \cdot I$, where Σ_R is built of independent uniform random numbers between 0.1 and 1 and the multiple d of the identity is added in order to generate positive definiteness. Note that if d is large, then Σ is nearly diagonal, making our above discussion on error rates for diagonal covariance matrices relevant. By choosing different distances md between the mean vectors or different d the difficulty of the classification problem can be varied.

Sometimes we add noise by means of features which do not have any influence on class separation by adding $(p - pr)$ normally distributed features with mean 0 and variance d. Overall, we assume that we have p features. Note that possibly $p = pr$.

We typically use $n = 2 \cdot nel << p$ observations, nel observations for each class. Thus, p tends to be much bigger than n, the case we discuss in this paper. The generation of n data points from the above normal distributions in p dimensions is repeated $rp = 200$ times using different random covariance matrices Σ. For the estimation of error rates, corresponding test samples with $nelt = 1000$ observations per class are generated from training distributions.

3.2   Properties 1 and 2: Error Rate Convergence

Let us now discuss properties 1 and 2 by means of simulations. First, let us state that in the case of two normal distributions with identical invertible covariance matrices Σ and identical a-priori probabilities like in properties 1 and 2 it is known that the Bayes error is given by
$err_{Bayes} = \Phi(-0.5((\mu_2 - \mu_1)^T \Sigma^{-1} (\mu_2 - \mu_1))^{0.5})$ (see [5]). Note that the Bayes error is mainly influenced by the Mahalanobis distance of the mean vectors of the classes. If Σ is diagonal with identical diagonal elements d and if the mean class distance is equal to md for all pr individual dimensions, then
$err_{Bayes} = \Phi(-0.5\sqrt{pr} \cdot md/d^{0.5})$.
We will consider two distinct cases:

1. The mean vectors drift away from each other the higher the no. of dimensions pr by setting the distance in the pr individual coordinates equal to $md = 2.5$ for all i. Then, if the covariance matrix is diagonal, namely $\Sigma = 25 \cdot I$, the Bayes error is
   $err_{Bayes} = \Phi(-0.5\sqrt{pr} \cdot md/d^{0.5}) = \Phi(-\sqrt{pr}/4) = 0.19, 0.003, \ldots, 0$ for $pr = 12, 120, \ldots, 2040$.
2. The distance $||\mu_1 - \mu_2||$ of the mean vectors of the classes stay the same for different pr by means of shrinking the distance in the individual coordinates as $md = 20/\sqrt{pr}$. Then, if the covariance matrix is diagonal, namely $\Sigma = 25 \cdot I$, the Bayes error is
   $err_{Bayes} = \Phi(-0.5\sqrt{pr} \cdot md/d^{0.5}) = \Phi(-2) = 0.023$.

Note that in both cases all pr features influence classification equally. Obviously, if the class distance is increasing with increasing pr, the classification problem

Table 1. Comparison of mean error rates (%): (a) all features influence, (b) only $p/6$ features influence, no selection, (c) only $p/6$ features influence, $p/6$ features selected

p	12	120	240	360	480	600	1080	2040	12	120	240	360	480	600	1080	2040	sec
(a) all	md = 2.5								md = 20 / $\sqrt{p}$								
lda	41	23	16	14	12	11	8	7	23	32	35	37	38	39	41	43	1.0
ir	32	14	10	9	8	7	7	6	7	24	31	34	37	38	41	45	0.1
NB	37	23	19	16	14	13	10	8	13	33	39	41	42	43	46	47	1.5
1NN	38	23	19	16	15	13	12	10	12	33	39	41	43	43	46	47	0.0
svm	35	16	11	9	8	8	7	6	9	27	32	35	38	39	42	44	134
tree	41	41	41	41	41	41	41	41	29	44	46	47	48	48	49	50	0.6
baggedtree	41	35	30	27	24	22	18	14	25	41	44	45	46	47	48		223
(b) p/6	md = 2.5								md = 20 / $\sqrt{p/6}$								
lda	48	45	45	43	43	41	38	32	22	37	42	43	44	44	46	46	1.0
ir	47	38	34	31	29	27	22	16	6	20	27	30	32	34	38	41	0.05
NB	47	44	42	40	39	38	34	31	12	32	37	39	41	42	44	46	1.5
1NN	48	43	40	37	36	35	31	27	10	29	34	37	39	40	43	45	0.0
svm	47	39	35	32	29	27	22	16	15	22	27	31	33	34	38	41	138
tree	47	45	44	44	44	44	43	41	11	36	42	44	45	46	47	48	0.6
baggedtree	48	46	46	44	43	43	41	37	13	37	42	44	45	45	47	48	228
(c) p/6, fs/6	md = 2.5								md = 20 / $\sqrt{p/6}$								
lda	44	40	36	34	32	31	25	19	5	25	30	33	34	37	38	40	10
ir	43	35	29	26	24	22	17	12	4	16	21	25	28	29	34	37	10
NB	44	39	36	33	32	31	27	23	5	22	29	32	35	37	40	43	10
1NN	47	38	34	30	28	26	23	19	5	19	26	29	32	33	38	42	10
tree	46	45	44	44	43	44	43	42	10	37	41	44	45	46	47	48	10

Note: p = no. of dimensions, sec = mean training time over both md in seconds for p = 2040,
md = mean difference of classes in each dimension,
lda = linear discriminant analysis (lda, package MASS, software R [6]),
fs/6 = feature selection (best p/6 features, mutual information (symmetrical.uncertainty)
criterion), package FSelector in R),
ir = independence rule = lda with diagonal covariance matrix
(sda, package sda in R, no shrinkage, diagonal = TRUE)
NB = naive Bayes rule (naiveBayes, package e1071 in R),
1NN = 1 nearest neighbor rule (knn, package class in R),
svm = linear support vector machine
(svm in R, package e1071, cost parameter tuned on grid $2^{-4}, \ldots, 2^4$ by leave-one-out).
tree = CART decision tree (rpart, minsplit=4, minbucket=2 in R), baggedtree = bagged
decision tree (bagging.rpart =
makeBaggingWrapper(base.rpart,bw.iters=500,bw.replace=FALSE,bw.size=1,bw.feats=0.75),
makeDownsampleWrapper(bagging.rpart,dw.perc=0.75,dw.stratify=TRUE) in R)

gets simpler. On the other hand, if the overall class distance $||\mu_1 - \mu_2||$ stays the same for different pr, the Bayes error, representing the difficulty of the classification problems, stays the same for different numbers of dimensions pr.

For method ir, taking $\Sigma = d \cdot I$ leads to $K_0 = 1$ and to a limit for the maximal error of $1 - \Phi(c/2) = \Phi(-0.5c)$ when the sample size $n \to \infty$. This is again equal to the Bayes error above.

Let us start our simulations with assuming that all involved features in fact influence the class choice, i.e. $p = pr = 12, \ldots, 2040$, and let $d = 25$,

$nel = 6, md \in \{2.5, 20/\sqrt{p}\}$ representing the above first and second case of class distance choice. By means of this variation of p with constant $n = 2 \cdot 6 = 12$ we vary the ratio p/n from 1 to 170. For $md = 2.5$ the classification problem tends to become easier for increasing p than for the problem with $md = 20/\sqrt{p}$. On the accordingly generated data (see GDG) the classification methods introduced in Sect. 2 are compared. Let us start with the discussion of the mean error rates in Table 1(a).[3] Obviously, all methods but *tree* benefit from higher dimensions in the case $md = 2.5$ as expected. For *tree* the error rate stays constant when p is increasing. This is because CART-trees are splitting in individual dimensions and the distance in these dimensions is staying the same for all numbers of dimensions p. Obviously, bagging helps, probably by offering different observations for splitting. Method *ir* appears to be more adequate than full *lda*. In the case $md = 20/\sqrt{p}$ all methods are suffering from higher dimensions. This was expected for *lda*, but appears also to be true for the other methods. Here, method *ir* does not show lower error rates than full *lda* for higher numbers of dimensions p.

Notice that *svm* and *baggedtree* need by far the most training time (cp. column sec) and are not distinctly better than the other methods. Therefore, the choice of these methods cannot be justified for the studied problems. Also note that runtime is near zero for 1NN because the training data set only consists of $n = 12$ observations.

3.3   Property 3: Distance Dependency

Let us now illustrate the distance dependence of classifier quality.

Reconsider the situation in Sect. 3.2 and let the p-dimensional mean distance *pmd* between the two classes increase with dimension p so that
$pmd = p^{1/ip-0.5} \cdot 2.5/12^{1/ip-0.5}$, $ip = 1, 1.5, 2, 2.5, 3$,
guaranteeing a start distance of 2.5. Note that the case $ip = 2$ is identical with the left part in Table 1(a) since the norm of the mean vector automatically increases with $p^{0.5}$ if the distance between the classes is the same in every dimension. Also note that the distances in the individual dimensions are decreased for $ip > 2$. Moreover, note that we do not assume sampling from independent normal distributions but only from approximately independent distributions for $d = 25$ (cp. GDG and Property 3). Let us see whether the theoretical properties are approximately valid also. Table 2 shows that the start distance of 2.5 leads to a high mean test error rate between 32 % and 42 %. However, the error rate benefits from more features if $ip < 3$, confirming the theoretical result. Only *tree* leads to increasing error rates already for $ip = 2.5$. Also, *tree* benefits the least for $ip < 2$. For $ip = 3$ all methods except *lda* only benefit until $p = 120$. Overall, the theory for independent coordinates (Property 3) also mainly applies for our example.

---

[3] This simulation was carried out using the R-packages *BatchJobs* [2] and *mlr* on the *SLURM* cluster of the Statistics Department of TU Dortmund University.

Table 2. Comparison of mean error rates (%): p-dimensional mean distance increasing with $p^{0.5}, p^{0.167}, p^{-0.1}, p^{-0.133}$ corresponding to $ip = 1, 1.5, 2.5, 3$

p	12	120	240	360	480	600	1080	2040	12	120	240	360	480	600	1080	2040	sec
$ip < 2$	$pmd = 2.5 \cdot p^{0.5}/12^{0.5}, ip = 1$								$pmd = 2.5 \cdot p^{0.167}/12^{0.167}, ip = 1.5$								
lda	41	1	0	0	0	0	0	0	41	12	4	2	1	1	0	0	1.0
ir	32	0	0	0	0	0	0	0	32	4	1	0	0	0	0	0	0.1
NB	37	0	0	0	0	0	0	0	37	10	4	2	1	1	0	0	1.5
1NN	38	0	0	0	0	0	0	0	38	9	3	1	1	1	0	0	0.0
svm	35	0	0	0	0	0	0	0	34	4	1	1	0	0	0	0	137
tree	42	23	15	11	7	5	2	0	42	37	35	34	33	33	31	29	0.6
baggedtree	41	0	0	0	0	0	0	0	41	22	11	6	3	2	0	0	227
$ip > 2$	$pmd = 2.5 \cdot p^{-0.1}/12^{-0.1}, ip = 2.5$								$pmd = 2.5 \cdot p^{-0.133}/12^{-0.133}, ip = 3$								
lda	41	30	25	23	22	21	20	19	41	34	31	30	29	28	28	28	1.0
ir	32	22	19	18	18	19	18	19	32	27	26	26	26	27	27	29	0.1
NB	37	31	29	28	28	27	26	24	37	35	34	35	35	35	35	34	1.5
1NN	38	31	29	29	29	28	28	27	38	35	35	35	36	36	37	37	0.0
svm	34	23	21	20	19	19	19	19	34	28	27	28	27	27	28	29	144
tree	42	43	44	44	45	44	45	46	42	44	45	46	46	46	47	47	0.6
baggedtree	41	39	38	37	35	35	32	31	41	42	41	41	40	40	39	39	225

3.4 Feature Selection

Let us now reconsider the example in Sect. 3.2 in the case of noise factors. We use only $pr = p/6$ features influencing the classes. Looking at the results in Table 1(b), the benefit for higher dimensions is much slower in case $md = 2.5$ because there is a much smaller class distance increase since only $pr = p/6$ factors contribute to the distance. Notice, however, that the methods ir and svm distinctly benefit the most, ir with much less training time than svm. In the case $md = 20/\sqrt{pr}$ the behavior is similar as for $pr = p$.

In order to possibly eliminate noise factors, let us now have a look on feature selection methods in high dimensions. Simple filters are the fastest feature selection methods. In filter methods, numerical scores s_i are constructed for the characterization of the influence of feature i on the dependent class variable. Filters are generally independent of classification models. Easy example filters are the χ^2-statistic for the evaluation of independence between (discretized) feature i and the class variable, the p-value of a t-test indicating whether the mean of feature i is different for the two classes, the correlation between feature i and the class variable, and the mutual information in feature i and the class variable.

Filters can be easily combined with a classification method. First calculate filter values (scores). Then sort features according to scores and choose the best k features. Finally, train the classification method on these k features.

Let us demonstrate the possible effect of a filter by reconsidering the example in Sect. 3.2 (see Table 1(c)). When only $p/6$ features influence the classes, we apply filtering by mutual information so that the correct number of features is selected. The corresponding error rates are then much lower than without feature selection but at the price of higher computation times (see column "sec") caused by the usage of a mutual information criterion for feature selection. Note that

Table 3. Comparison of mean error rates (%): $md = 2.5, pr = p/6$, best $m \in \{p/12, 3p/12, p/3, p/2\}$ features selected

p	12	120	240	360	480	600	1080	2040	12	120	240	360	480	600	1080	2040	sec
	$m = p/12$								$m = 3p/12$								
lda	44	45	41	40	38	37	34	30	44	39	37	35	32	31	25	19	10
ir	46	39	37	33	32	30	25	20	43	34	29	25	22	21	16	12	10
NB	46	42	40	39	37	36	34	30	45	39	36	35	32	31	28	23	10
1NN	47	42	38	35	34	31	28	22	46	38	33	31	29	27	23	19	10
tree	47	44	44	44	44	44	43	42	46	44	44	44	44	44	43	42	10
	$m = p/3$								$m = p/2$								
lda	45	41	38	37	34	33	28	22	46	42	41	39	37	36	31	25	10
ir	44	34	30	26	23	22	17	13	45	36	32	28	25	23	18	14	10
NB	45	40	37	36	34	32	29	24	46	42	39	38	36	35	31	26	10
1NN	47	39	35	32	30	28	24	20	47	41	37	34	32	31	26	22	10
tree	46	44	44	44	44	44	43	42	47	44	44	44	44	44	43	42	10

the slowest methods *svm* and *baggedtree* are not tested because there were better methods which were much faster when feature selection was not used.

The most important problem with feature selection is the adequate choice of k. Let us discuss whether a correct finding of the real number of influential dimensions is helpful for the considered classification methods.

Reconsider the example in Sect. 3.2 in the case of noise factors with $pr = p/6$. We identify the best $m = p/12, 3p/12, p/3, p/2$ factors by filtering via mutual information and compare the corresponding error rates with those for $m = p/6$ already given in Table 1(c)(*fs/6*). From the results in Tables 3 and 1(b), (c)(left) it should be clear that choosing the correct number of features is best corresponding to error rates in the case of noise. However, a too small number of features (here $m = p/12$) appears to be more severe than a too big number (here $m = 3p/12, p/3$). In the former case, the error rates are similar to the case of no feature selection, in the latter case the error rates are only slightly too high.

4   Summary and Conclusion

In this paper we discussed the performance of standard classification methods on Big Data. We concentrated on the case with many more features than observations. For this case we studied class distance dependency and feature selection. If the class distance sufficiently increases for higher dimensions, then error rates are decreasing, whereas for constant Bayes error the estimated errors are increasing up to nearly 0.5 for higher dimensions. Also, feature selection might help to find better models in high dimensions in the case of noise. In our example, a too small number of features appears to be more severe than a too big number. *ir* and *svm* performed best in high dimensions, *ir* in much less time than *svm*.

We used special example data being normally distributed with nearly diagonal invertible covariance matrix and identical contributions of all features to classification except noise features. This might be extended in different ways.

A more general invertible covariance matrix might be more adequate. Other data distributions might be worthwhile studying. Also, different contributions to class separation should be considered for the non-noise features. Last but not least, for that many features as studied here it appears to be probable that there is much more complicated structure than assumed in GDG. This might lead to much more involved classification models not being covered by standard classification methods. For example, *deep learning* methods utilizing neural nets with more than one layer or complex Bayesian networks might be more adequate and such structures should be studied also.

References

1. Bickel, P.J., Levina, E.: Some theory for Fisher's linear discriminant function, naive bayes, and some alternatives when there are many more variables than observations. Bernoulli **10**, 989–1010 (2004)
2. Bischl, B., Lang, M., Mersmann, O., Rahnenfuehrer, J., Weihs, C.: BatchJobs and BatchExperiments: abstraction mechanisms for using R in batch environments. J. Stat. Softw. **64**, 1–25 (2015). doi:10.18637/jss.v064.i11. http://www.jstatsoft.org/index.php/jss/article/view/v064i11
3. Fan, J., Fan, Y., Wu, Y.: High-dimensional classification. In: Cai, T.T., Shen, X. (eds.) High-dimensional Data Analysis, pp. 3–37. World Scientific, New Jersey (2011)
4. Kiiveri, H.T.: A general approach to simultaneous model fitting and variable elimination in response models for biological data with many more variables than observations. BMC Bioinform. **9**, 195 (2008). doi:10.1186/1471-2105-9-195
5. McLachlan, G.J.: Discriminant analysis and statistical pattern recognition. Wiley, New York (1992)
6. Core Team, R.: R: A Language and Environment for Statistical Computing. R Foundation for Statistical Computing, Vienna, Austria (2014). http://www.R-project.org/
7. Weihs, C., Horn, D., Bischl, B.: Big data classification: aspects on many features and many observations. In: Wilhelm, A.F.X., Kestler, H.A. (eds.) Analysis of Large and Complex Data, pp. 113–122. Springer (2016)

Knowledge Discovery from Complex High Dimensional Data

Sangkyun Lee[1](✉) and Andreas Holzinger[2,3]

[1] Artificial Intelligence Unit LS8, Computer Science Department,
Technische Universität Dortmund, Dortmund, Germany
sangkyun.lee@tu-dortmund.de
[2] Research Unit HCI-KDD, Institute for Medical Informatics,
Statistics and Documentation, Medical University Graz, Graz, Austria
a.holzinger@hci-kdd.org
[3] Institute for Information Systems and Computer Media,
Graz University of Technology, Graz, Austria

Abstract. Modern data analysis is confronted by increasing dimensionality of problems, mainly contributed by higher resolutions available for data acquisition and by our use of larger models with more degrees of freedom to investigate complex systems deeper. High dimensionality constitutes one aspect of "big data", which brings us not only computational but also statistical and perceptional challenges. Most data analysis problems are solved using techniques of optimization, where large-scale optimization requires faster algorithms and implementations. Computed solutions must be evaluated for statistical quality, since otherwise false discoveries can be made. Recent papers suggest to control and modify algorithms themselves for better statistical properties. Finally, human perception puts an inherent limit on our understanding to three dimensional spaces, making it almost impossible to grasp complex phenomena. For aid, we use dimensionality reduction or other techniques, but these usually do not capture relations between interesting objects. Here graph-based knowledge representation has lots of potential, for instance to create perceivable and interactive representations and to perform new types of analysis based on graph theory and network topology. In this article, we show glimpses of new developments in these aspects.

1 Introduction

Thanks to modern sensing technology, we witness rapid increase in data dimensions in numerous domains, for example high-resolution images, large-scale social networks, high-throughput genetic profiles, just to name a few. In most cases, the number of measured entities (features) grows in a much faster rate than the number of observations: pictures taken with smart phones have few million pixels, whereas we may have only few hundreds or thousands of photos.

Our main interest is such "high dimensional" data: to be more specific, a data set is high dimensional when the number of features (p) is larger than

© Springer International Publishing Switzerland 2016
S. Michaelis et al. (Eds.): Morik Festschrift, LNAI 9580, pp. 148–167, 2016.
DOI: 10.1007/978-3-319-41706-6_7

the number of observations (n) by a few magnitude. A good example is gene expression study data.

Figure 1 shows a part of breast cancer data from the Gene Expression Omnibus[1], which contains expression values of $p = 22k$ transcripts measured by the Affymetrix GeneChip Human Genome U133A microarrays. Typically, the number of observations is much smaller in this type of data, due to the cost involved to handle human subjects in a limited time. In the figure, the color represents high (green/bright) or low (red/dark) values of expression, and a primary task using the color intensity values is to identify genes that have different expression patterns in different groups of subjects. Genes with differential expression are then further investigated by wet experiments to identify their roles in biochemical pathways, their relations to other genes, and so forth.

Fig. 1. Gene expression measurement samples of 100 genes (rows) from 50 breast cancer subjects (columns). GEO accession no. GSE11121. (Color figure online)

A surprising misconception about high dimensionality is that data analysis would produce better outcome with higher dimensional data, because of increased amount of available information. In a way, this makes sense, for instance we can see objects more clearly in high-resolution digital photographs. In data science, an increased number of input features may allow for building more accurate predictors. However, realizing such predictors comes with extra cost in several aspects.

First, high dimensionality brings computational challenges to data analysis. Obviously, extra memory space will be needed, but also efficient computation algorithms will be required to obtain the best hypothesis for explaining data. The task of finding such a hypothesis is typically described as an optimization problem, where a parametrized function is fitted to data minimizing the mismatch between predictions and observed responses of interest (e.g., categories of objects, severity levels of a disease, etc.)

Secondly, an important task of identifying a (possibly small) subset of features contributing to prediction becomes statistically more challenging as dimension grows. Simply speaking, the reason is that performing multiple hypothesis tests to distinguish important features takes more statistical power, in other words, requires larger sample sizes. There have been quite a few literature on the conditions when we can identify relevant features: later we will discuss some of the recent results on lasso-type regression.

Third, due to limitations in human perception, understanding structures in high dimensional spaces is inherently difficult for us. In particular for interdisciplinary research, the outcome of data analysis would have to be shaped in a form easily perceivable by domain experts who may not be computer scientists. Graph-based representations of data space and analysis outcomes have lots of potential for this purpose: we will demonstrate some examples in biomedical data analysis.

---

[1] Gene Expression Omnibus http://www.ncbi.nlm.nih.gov/geo/.

2   Sparse Variable Selection and Estimation

There have been a lot of improvements in convex optimization, in particular for dealing with composite objective functions which are interesting for extracting understandable structures from high-dimensional data.

We consider a standard setting for data analysis: a set of m training data points $\{(\mathbf{x}_i, y_i)\}_{i=1}^m$ are given, where $\mathbf{x}_i \in \mathcal{X}$ is an input point and $y_i \in \mathcal{Y}$ is a response of interest. Typically, $\mathbf{x}_i$ is a vector and $y_i \in \{-1, +1\}$ for binary classification and $y_i \in \mathbb{R}$ for regression tasks, but both $\mathbf{x}_i$ and y_i can be more structured objects such as strings [51] or trees [38]. A goal of data analysis is to find a function $h_\mathbf{w}(\mathbf{x})$ parametrized by a vector $\mathbf{w} \in \mathbb{R}^n$, which best reflects the data in terms of a certain error measure between responses and predictions made by $h_\mathbf{w}(\mathbf{x})$. Finding the best parameters vector $\mathbf{w}$ can be formulated as follows,

$$\mathbf{w}^* = \arg\min_{w \in \mathbb{R}^n} \frac{1}{m} \sum_{i=1}^m \ell(y_i, h_\mathbf{w}(\mathbf{x}_i)) + \Psi(\mathbf{w}) = f(\mathbf{w}) + \Psi(\mathbf{w}). \tag{1}$$

Here, $\ell(y_i, h_\mathbf{w}(\mathbf{x}_i)) : \mathbb{R}^n \to \mathbb{R}$ is a *loss* function between a prediction $h_\mathbf{w}(\mathbf{x}_i)$ and an observed response y_i, which is convex in terms of $\mathbf{w}$. A function $f(\mathbf{w}) : \mathbb{R}^n \to$ dom f is convex if for all $\mathbf{w}, \mathbf{v} \in$ dom f, the following holds for some $\alpha \geq 0$,

$$f((1 - \lambda)\mathbf{w} + \lambda\mathbf{v}) \leq (1 - \lambda)f(\mathbf{w}) + \lambda f(\mathbf{v}) - \frac{\alpha}{2}\lambda(1 - \lambda)\|\mathbf{w} - \mathbf{v}\|^2.$$

If there exists $\alpha > 0$, f is called α-strongly convex. The second part $\Psi(\mathbf{w}) : \mathbb{R}^n \to \overline{\mathbb{R}} := \mathbb{R} \cup \{+\infty\}$ in the objective is a *regularizer*, which is a proper ($\Psi(\mathbf{w}) \equiv +\infty$ is not true) convex function used to control certain statistical properties of the estimation process. Ψ also can be the indicator function of a convex set $\mathcal{W}$, i.e., $\Psi(\mathbf{w}) = 0$ if $\mathbf{w} \in \mathcal{W}$ and $\Psi(\mathbf{w}) = +\infty$ otherwise.

2.1   Sparsity-Inducing Regularization

An intriguing use of the convex minimization in (1) is to extract the most relevant features in data vectors $\mathbf{x}$ that contribute to minimizing the averaged loss. In particular, when a generalized linear model is considered so that $h_\mathbf{w}(\mathbf{x}) = f(\langle \mathbf{w}, \mathbf{x} \rangle)$ for a convex function f, where $\langle \cdot, \cdot \rangle$ is an inner product, we can set unimportant components of $\mathbf{w}$ to zero to turn-off their contribution to prediction. Such componentwise switching-off can be achieved by minimizing $\Psi(\mathbf{w}) = \lambda\|\mathbf{w}\|_1$ at the same time, where $\lambda > 0$ is a tuning parameter. With least squares loss function, i.e., $\ell(y_i, h_\mathbf{w}(\mathbf{x}_i)) = (y_i - h_\mathbf{w}(\mathbf{x}_i))^2$, the problem (1) is called as the lasso problem [66].

Variants. The idea can be extended to incorporate a combination of ℓ_2 and ℓ_1 regularization, i.e., $\Psi(\mathbf{w}) = \lambda\{(1 - \alpha)\|\mathbf{w}\|_2^2 + \alpha\|\mathbf{w}\|_1\}$ for some given $\lambda > 0$ and $\alpha \in [0, 1]$. This regularization is called the elastic net [80], which tends to select all correlated features together compared to the selection by lasso where some correlated features may not be selected. In addition, for $\alpha < 1$ the regularizer $\Psi(\mathbf{w})$ makes the objective strongly convex in $\mathbf{w}$, which can lead to better convergence rate e.g. in gradient descent algorithms.

When certain grouping of features is known a priori, then we can use $\Psi(\mathbf{w}) = \sum_{g \in G} \|\mathbf{w}_g\|_2$ for subvectors $\mathbf{w}_g$ of $\mathbf{w} \in \mathbb{R}^n$ corresponding to groups $g \subset \{1, 2, \ldots, n\}$. This particular setting is useful when it is preferable to select groups rather than individual components. For instance, a group of binary variables may encode a single multinomial variable of interest. This setting within (1) is known as group-lasso [74]. When groups may overlap, a modified version in [36] is recommended to avoid turning-off all groups sharing a demoted variable. Interested readers can find more details in an introductory article [48].

2.2   Accelerated Proximal Gradient Descent Algorithm

When the convex functions ℓ is smooth (continuously differentiable) and Ψ is possibly nonsmooth but "simple" (the meaning will be clarified later), one of the best algorithm for solving the optimization problem (1) is the accelerated proximal gradient descent algorithm, also known as FISTA [7].

Similarly to the gradient descent, the proximal gradient descent algorithm considers a simple quadratic approximation of the smooth part ℓ in the objective, augmented with Ψ, that is,

$$f(\mathbf{w}) + \Psi(\mathbf{w}) \approx f(\mathbf{w}_k) + \langle \nabla f(\mathbf{w}_k), \mathbf{w} - \mathbf{w}_k \rangle + \frac{L}{2}\|\mathbf{w} - \mathbf{w}_k\|_2^2 + \Psi(\mathbf{w}), \qquad (2)$$

where $L > 0$ is the Lipschitz constant of the gradients ∇f,

$$\|\nabla f(\mathbf{w}) - \nabla f(\mathbf{v})\| \leq L\|\mathbf{w} - \mathbf{v}\|_2^2, \;\; \forall \mathbf{w}, \mathbf{v} \in \mathrm{dom}\, f.$$

Given these, the proximal gradient method chooses the next iterate as the minimizer of the right-hand side expression of (2),

$$\mathbf{w}_{k+1} = \arg\min_{\mathbf{w}} \; \langle \nabla f(\mathbf{w}_k), \mathbf{w} - \mathbf{w}_k \rangle + \frac{L}{2}\|\mathbf{w} - \mathbf{w}_k\|^2 + \Psi(\mathbf{w})$$

$$= \arg\min_{\mathbf{w}} \; \frac{1}{2}\|\mathbf{w} - (\mathbf{w}_k - (1/L)\nabla f(\mathbf{w}_k))\|^2 + (1/L)\Psi(\mathbf{v})$$

$$= \mathrm{prox}_{(1/L)\Psi}(\mathbf{w}_k - (1/L)\nabla f(\mathbf{w}_k)). \qquad (3)$$

Here, we have defined the *proximal operator* associated with a function $h : \mathbb{R}^n \to \overline{\mathbb{R}}$ of a given point $\mathbf{z} \in \mathbb{R}^n$ as

$$\mathrm{prox}_h(\mathbf{z}) := \arg\min_{\mathbf{w} \in \mathbb{R}^n} \; \frac{1}{2}\|\mathbf{w} - \mathbf{z}\|^2 + h(\mathbf{w}).$$

From this definition, we can interpret that the update in (3) computes the next iterate $\mathbf{w}_{k+1}$ as a point which is close to the given gradient descent point $\mathbf{z} = \mathbf{w}_k - (1/L)\nabla f(\mathbf{w}_k)$ and minimizes $h = (1/L)\Psi$ at the same time. We call h (or Ψ) is "simple" if the proximal operator can be computed efficiently.

This procedure can be *accelerated* using an ingenious technique due to Nesterov [59]. The modified version uses another sequence of vectors $\mathbf{v}_k$ composed as a particular linear combination of the two past iterates,

$$\mathbf{v}_{k+1} = \mathbf{w}_k + \left(\frac{t_k - 1}{t_{k+1}}\right)(\mathbf{w}_k - \mathbf{w}_{k-1}), \quad t_{k+1} = \frac{1}{2}(1 + \sqrt{1 + 4t_k^2}).$$

Then, the next iterate $\mathbf{w}_{k+1}$ is computed based on $\mathbf{v}_k$, not $\mathbf{w}_k$,

$$\mathbf{w}_{k+1} = \text{prox}_{(1/L)\Psi}(\mathbf{v}_k - (1/L)\nabla f(\mathbf{v}_k))$$

This method generate iterates $\{\mathbf{w}_k\}$ converging to an optimal solution $\mathbf{w}^*$ with the a sublinear rate $\mathcal{O}(1/k^2)$ [7], that is,

$$[f(\mathbf{w}_k) + \Psi(\mathbf{w}_k)] - [f(\mathbf{w}^*) + \Psi(\mathbf{w}^*)] \leq \frac{2L\|\mathbf{w}_0 - \mathbf{w}^*\|_2^2}{(k+1)^2}.$$

This achieves the best convergence rate as a first-order optimization method [59], and it becomes slower only by a constant factor if line-search is involved.

2.3 Consistency in Variable Selection

One of the important questions regarding the solution $\mathbf{w}^*$ of (1) with ℓ_1 regularization is that if the "true" set of important variables (often called as the support) will be identified. This type of discussion is based on a data generation model that an $m \times n$ training data matrix $\mathbf{X} = (\mathbf{x}_1^T, \ldots, \mathbf{x}_m^T)$ and responses $\mathbf{y} \in \mathbb{R}^m$ are related by

$$\mathbf{y} = \mathbf{X}\mathbf{w}^\circ + \epsilon$$

where ϵ is a vector of m i.i.d. random variables with mean 0 and variance σ^2. Here, $\mathbf{w}^\circ$ defining the relation is the true weight vector we try to estimate, by a solution $\mathbf{w}^*$ of (1) with $\Psi(\mathbf{w}) = \lambda\|\mathbf{w}\|_1$.

Consistency results has been established first by Knight and Fu [41], for the cases where n and $\mathbf{w}^\circ$ are independent of m and some regularity conditions hold. In estimation consistency, they showed that $\mathbf{w}^* \to \mathbf{w}^\circ$ in probability as $m \to \infty$, and $\mathbf{w}^*$ is asymptotically normal when $\lambda = o(m)$. In variable selection consistency, they also showed that when $\lambda \propto \sqrt{m}$, the true set of important variables are identified in probability, that is,

$$\mathbb{P}(\{i : \mathbf{w}_i^* \neq 0\} = \{i : \mathbf{w}_i^\circ \neq 0\}) \to 1, \quad \text{as } m \to \infty.$$

In high dimensions, the growth of dimensions n is restricted in a way that $s\log(n) = o(m)$, where s is the sparsity of the true signal $\mathbf{w}^\circ$ [56,76]. In addition, other conditions are required for the design matrix $\mathbf{X}$, namely the *neighborhood stability conditions* [56] or the equivalent *irrepresentable conditions* [76,79] that are almost necessary and sufficient for lasso to identify the true support for the cases where n is fixed or n grows with m. Roughly speaking, these conditions state that the irrelevant covariates are orthogonal to relevant ones.

The conditions however may not be satisfied in practice, and finding weaker conditions is in active research, e.g. [37]. Also, more general notions of variable selection consistency have been discussed in other context, e.g. in stochastic online learning [49].

3 Sparse Graph Learning

From a sparse solution $\mathbf{w}^*$ of (1), we can find a set of relevant features, and also can prioritize them by the magnitude of the coefficient vector $\mathbf{w}^*$ for further investigation, e.g. bio-chemical studies of chosen genes to clarify their roles in a complex system. However, its outcome is essentially a ranked list of features which does not tell much about the relations of covariates: the latter type of information would be more helpful to understand the underlying system. In this view, we consider another learning model which produces a graph of features, where connections between nodes (features) represents a certain statistical dependency.

3.1 Gaussian Markov Random Field

The Gaussian Markov Random Field (GMRF) is a collection of n jointly Gaussian random variables represented as nodes in a graph $G = (V, E)$, with a set of n vertices V and a set of undirected edges E. In this model we consider random vectors $\mathbf{x} \sim \mathcal{N}(\mu, \Sigma)$ with a mean vector μ and a covariance matrix Σ, whose probability density is given as

$$p(\mathbf{x}) = (2\pi)^{-n/2} \det(\Sigma)^{-1/2} \exp\left(-\frac{1}{2}(\mathbf{x} - \mu)^T \Sigma^{-1}(\mathbf{x} - \mu)\right).$$

The edges represent conditional dependency structure: in GRMFs, the variables $\mathbf{x}_i$ and $\mathbf{x}_j$ associated with the nodes i and j are *conditionally independent* given all the other nodes [45] when there is no edge connecting the two nodes, or equivalently the corresponding entry in the precision matrix satisfies $\Sigma_{ij}^{-1} = 0$. That is,

$$\Sigma_{ij}^{-1} = 0 \quad \Leftrightarrow \quad \begin{aligned} &P(\mathbf{x}_i, \mathbf{x}_j | \{\mathbf{x}_k\}_{k \in \{1,2,\ldots,n\} \setminus \{i,j\}}) \\ &= P(\mathbf{x}_i | \{\mathbf{x}_k\}_{k \in \{1,2,\ldots,n\} \setminus \{i,j\}}) P(\mathbf{x}_j | \{\mathbf{x}_k\}_{k \in \{1,2,\ldots,n\} \setminus \{i,j\}}). \end{aligned}$$

This also implies that we can consider the precision matrix Σ^{-1} as a weighted adjacency matrix for an undirected graph representing a GMRF.

3.2 Sparse Precision Matrix Estimation

Assuming that $\mu = \mathbf{0}$ without loss of generality (i.e. subtract the mean from data points), the likelihood function to describe the chance to observe a collection of m i.i.d. samples $\mathcal{D} = \{\mathbf{x}_1, \mathbf{x}_2, \ldots, \mathbf{x}_m\}$ from $\mathcal{N}(\mathbf{0}, \Sigma^{-1})$ becomes

$$L(\Sigma^{-1}, \mathcal{D}) = \prod_{i=1}^{m} p(\mathbf{x}_i) \sim \prod_{i=1}^{m} \det(\Sigma)^{-1/2} \exp\left(-\frac{1}{2}\mathbf{x}_i^T \Sigma^{-1} \mathbf{x}_i\right).$$

Therefore the log likelihood function (omitting constant terms and scaling by $2/m$) becomes,

$$LL(\Sigma^{-1}, \mathcal{D}) = \log \det(\Sigma^{-1}) - \text{tr}\,(S\Sigma^{-1}).$$

Here we have defined $S := \frac{1}{m} \sum_{i=1}^{m} \mathbf{x}_i \mathbf{x}_i^T$ as the sample covariance matrix.

Minimizing the negative log likelihood plus a sparsity inducing norm on the prediction matrix $\Theta = \Sigma^{-1}$ can be stated as

$$\min_{\Theta \in \mathbb{R}^{n \times n}} -LL(\Theta, \mathcal{D}) + \lambda \|\Theta\|_1, \quad \text{subject to}\ \ \Theta \succ 0,\ \Theta^T = \Theta. \tag{4}$$

The ℓ_1 norm of Θ here is defined elementwise, that is, $\|\Theta\|_1 := \sum_{i=1}^{n} \sum_{j=1}^{n} |\Theta_{ij}|$.

The sparse precision matrix estimation in (4) is a convex optimization problem proposed by Yuan and Lin [75]. Due to its special structure maximizing the determinant of a matrix, they applied an interior point algorithm [68], which may not be suitable for high dimensions n due to the complexity $\mathcal{O}(n^6 \log(1/\epsilon))$ to obtain an ϵ-suboptimal solution. A more efficient block coordinate descent algorithm has been suggested by Banerjee et al. [3], to solve the dual problem of (4). Each subproblem of this block coordinate descent formulation can be cast as a lasso problem in forms of (1), and this fact has been used by Friedman, Hastie, and Tibshirani to build the graphical lasso algorithm [24]. However, each subproblem of these solvers still involves quite large $(n-1) \times (n-1)$ matrices, resulting in $\mathcal{O}(sn^4)$ complexity for s sweeps of all variables. Many research articles have contributed more efficient optimization algorithms (for a brief survey, see [47]).

3.3   Graph Selection Consistency

Regarding the statistical quality of the solution Θ^* of (4), we can ask similar questions to those in Sect. 2.3, that if the solution identifies the true graphical structure, or equivalently the true set of edges or the nonzero patterns in the true model Θ°. In other word, we check if following property holds:

$$P\left(\{(i,j) : \Theta_{ij}^* \neq 0\} = \{(i,j) : \Theta_{ij}^\circ \neq 0\}\right) \to 1\ \text{ as }\ m \to \infty.$$

The sparse graph learning problem (4) has a very similar structure to the sparse variable selection problem (1), and they share very similar consistency results, e.g. [75]. Algorithms using random sampling have been recently proposed, such as bolasso [2] and stability selection [57], which require weaker conditions to achieve variable selection consistency.

3.4   Breast Cancer Gene Dependency Graphs

To demonstrate graph extraction using the Gaussian MRF, we used a genomic data set consisting of gene expression profiles of $n = 20492$ features (genes, more specifically, transcripts) from $m = 362$ breast cancer patients. The data set was created combining three gene expression data sets available from the Gene Expression Omnibus, with the accession IDs GSE1456, GSE7390, and GSE11121.[2]

Figure 2 shows the graph learned separately on subgroups of patients determined by their "grade" of cancer progression: 1 (almost normal), 2 (faster growth) and 3 (much faster growth). The parameter $\lambda = 1.6$ was chosen for all cases which produced small numbers of connected components. Only the connected components with at least two nodes are shown for compact visualization. The color of node represents the p-values of the likelihood ratio test, for the case of using each node (gene) as an univariate predictor for overall survival time under the Cox proportional hazard model [16]. Colors are assigned to five p-value intervals in $[10^{-5}, 1)$, equally sized in logarithmic scale, where darker colors indicate smaller p-values.

The visualization in Fig. 2 looks quite easy to comprehend even for no biology expert. For example, genes with many neighbors in the graphs (so called hub nodes) turned out to have important roles in breast cancer development, including ASPN [11], SFRP1 [40], and ADII1B [50], even though some (e.g. ADH1B) may not be interesting as univariate predictors considering their p-value.

4 Graph-Based Discovery in Medical Research

An ongoing trend in many scientific areas is the application of network analysis for knowledge discovery. The underlying methodology is the representation of the data by a graph representing a relational structure. Benefits can be created in a blend of different approaches and methods and a combination of disciplines including graph theory, machine learning, and statistical data analysis. This is particularly applicable in the biomedical domain: large-scale generation of various data sources (e.g. from genomics, proteomics, metabolomics, lipidomics, transcriptomics, epigenetics, microbiomics, fluxomics, phenomics, cytomics, connectomics, environomics, exposomics, exonomics, foodomics, toponomics, etc.) allows us to build networks that provide a new framework for understanding the molecular basis of physiological and pathological health states. Many widespread diseases, for example diabetes mellitus, [20], involve enormous interactions between thousands of genes. Although, modern high-throughput techniques allows the identification of such genes amongst the resulting omics data, a functional understanding is still the grand challenge. A major goal is to find diagnostic biomarkers or therapeutic candidate genes.

Network-based methods have been used for quite a while in biology to characterize genomic and genetic mechanisms. Diseases can be considered as abnormal

---

[2] The CEL files from the GEO were normalized and summarized for transcripts using the frozen RMA algorithm [55]. Then only the verified (grade A) genes were chosen for further analysis according to the NetAffx probeset annotation v33.1 of Affymetrix ($n = 20492$ afterwards). Also, microarrays with low quality according to the GNUSE [54] error scores > 1 were discarded ($m = 392$ afterwards).

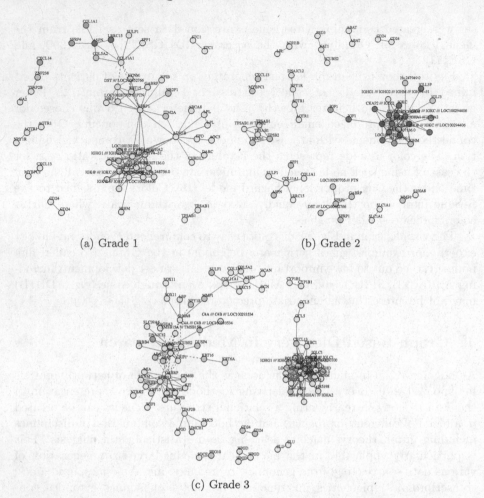

(a) Grade 1   (b) Grade 2

(c) Grade 3

Fig. 2. Graphical representation of transcript relations corresponding to breast cancer subgroups. Node color represents the p-value of each node (genes) as univariate predictor of overall survival times (darker color = smaller p-value). Edge types represent correlation: solid = positive and dashed = negative. Node labels show the corresponding gene symbols. (Color figure online)

perturbations of critical cellular networks. The progress and intervention in complex diseases can be analyzed today using network theory. Once the system is represented by a network, methods of network analysis can be applied, not only to extract useful information regarding important system properties, but also to investigate its structure and function. Various statistical and machine learning methods have been developed for this purpose and have already been applied to networks [19]. The underlying structure of such networks are graphs. Graph theory [25] provides tools to map data structures and to find unknown connections between single data objects [21, 65]. The inferred graphs can be further analyzed by using graph-theoretical, statistical and machine learning techniques [18].

A mapping of already existing and in medical practice approved *knowledge spaces* as a conceptual graph and the subsequent visual and graph-theoretical analysis may bring novel insights on hidden patterns in the data, which exactly is the goal of knowledge discovery [28]. Another benefit of the graph-based data structure is in the applicability of methods from network topology and network analysis and data mining, e.g. small-world phenomenon [4, 39], and cluster analysis [42, 72].

However, the biomedical domain is significantly different from other real world domains. Mostly, the processes are data-driven trial-and-error processes, used as help to extract patterns from large data sets by way of predefined models through an fully automated tool without human involvement [77]. Many machine learning researchers pay much attention to find algorithms, models and tools to support such fully automated approaches. The Google car is currently a best practice example [64], at the same time little attention is paid to include the human into this loop.

The reason for this huge difference is the high complexity of the biomedical research domain itself [14]. It is inevitable for the future biomedical domain expert to switch from the classical consumer-like role [44] to an active part in the knowledge discovery process [27, 30]. However, this is not so easy, because it is well known that many biomedical research projects fail due to the technical barriers that arise to the domain experts in data integration, data handling, data processing, data visualization and analysis [1, 34, 43]. A survey from 2012 among hospitals from Germany, Switzerland, South Africa, Lithuania, and Albania [60] showed that only 29 % of the medical professionals were familiar with any practical application of data mining methods and tools. Although this survey might not be representative globally, it clearly shows the trend that medical research is still widely based on standard statistical methods.

To turn the life sciences into data intensive sciences [28], consequently, there is urgent need for usable and useful data exploration systems - which are in the direct work flow of the biomedical domain expert [81]. A possible solution to solve such problems is in a hybrid approach to put the human into the machine learning loop [22, 63].

4.1   Medical Knowledge Space

This example shows the advantage of representing large data sets of medical information using graph-based data structures. Here, the graph is derived from a standard quick reference guide for emergency doctors and paramedics in the German speaking area; tested in the field, and constantly improved for 20 years: The handbook "Medikamente und Richtwerte in der Notfallmedizin" [58] (German for Drugs and Guideline Values in Emergency Medicine, currently available in the 11th edition accompanies every German-speaking emergency doctor as well as many paramedics and nurses. It has been sold 58,000 times in the

German-speaking area. The 92-pages handbook (size: 8×13 cm) contains a comprehensive list of emergency drugs and proper dosage information. Additionally, important information for many emergency situations is included.

The data includes more than 100 essential drugs for emergency medicine, together with instructions on application and dosage depending on the patient condition, complemented by additional guidelines, algorithms, calculations of medical scores, and unit conversion tables of common values. However, due to the traditional list-based interaction style, the interaction is limited to a certain extent. Collecting all relevant information may require multiple switches between pages and chapters, and knowledge about the entire content of the booklet. In consequence to the alphabetical listing of drugs by active agents, certain tasks, like finding all drugs with common indications, proved to be inefficient and time consuming.

Modeling relationships between drugs, patient conditions, guidelines, scores and medical algorithms as a graph (cf. Fig. 3) gives valuable insight into the structure of the data set. Each drug is associated with details about its active agent and pharmacological group; brand name, strengths, doses and routes of administration of different products; indications and contraindication, as well as additional remarks on application. Consequently, a single drug itself can be represented as connected concepts. Shared concepts create links between multiple drugs with medical relevance, and provide a basis for content-aware navigation.

The interconnection of two drugs, namely adrenaline and dobutamine, is shown in Fig. 4. The left-hand side illustrates the main three types of relations inducing medical relevance; shared indications, shared contra-indications and shared pharmacological groups. Different node colors are used to distinguish between types of nodes such as active agents, pharmacological groups, applications, dosages, indications and contra-indications. The right-hand side highlights the connection of adrenaline and dobutamine by a shared indication.

Links to and between clinical guidelines, tables and calculations of medical scores, algorithms and other medical documents, follow the same principle. On the contrast to a list-based interaction style, these connections can be used for identification and visualization of relevant medical documents, to reorganize the presentation of the medical content and to provide a fast and reliable contextual navigation.

The explosive growth of complexity of networks have overwhelmed conventional visualization methods and future research should focus on developing more robust and efficient temporally aware clustering algorithms for dynamic graphs, i.e. good clustering will produce layouts that meet general criteria, such as cluster colocation and short average edge length, as well as minimize node motion between time steps [52]. The use of multi-touch interfaces for graph visualization [32] extends graph manipulation capabilities of users and thereby can be used to solve some of the visualization challenges.

4.2 DrugBank

DrugBank is an comprehensive, open, online database that combines detailed drug data with drug target information, first released in 2006 [71]. The current

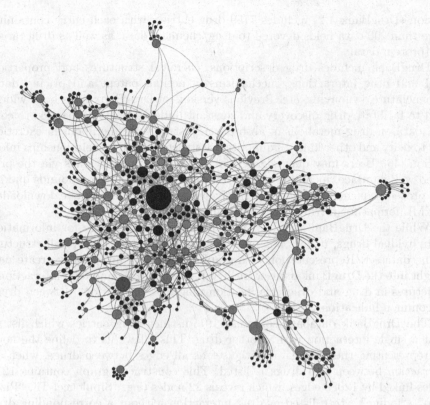

Fig. 3. Graph of the medical data set showing the relationship between drugs, guidelines, medical scores and algorithms.

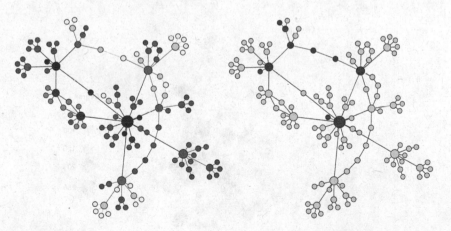

Fig. 4. Interconnection between two drugs, "Adrenaline" and "Dobutamine"; connections to and between clinical guidelines, tables and calculations of medical scores, algorithms and other medical documents, follow the same principle. (Color figure online)

version (DrugBank 4.2) includes 7759 drug entries, with each entry containing more than 200 data fields devoted to drug/chemical data, as well as drug target and protein data.

DrugBank includes drug descriptions, chemical structures and properties, food and drug interactions, mechanisms of action, patent and pricing data, nomenclature, synonyms, etc. Previous versions of DrugBank have been widely used to facilitate drug discovery and constant updates have it expanded to contain data on drug metabolism, absorption, distribution, metabolism, excretion and toxicity and other kinds of quantitative structure activity relationships information [46]. Users may query DrugBank in several different ways via the provided web interface, including simple text queries, chemical compounds queries and protein sequence searches. Alternatively the full database can be downloaded in XML format for further data processing and exploration.

While the DrugBank database is a comprehensive resource for information on individual drugs, it does not provide an illustration of the overall structure of the data set. Representation as a graph can quickly and clearly create new insight into the DrugBank dataset, such as pattern in drug and food interactions, structures in drug and drug classification relations, or relations between drugs by common indications.

The DrugBank database contains 1191 distinct drug entries which list at least a single interaction with another drug. This allows us to define the node set representing these drugs, and the set as all edges between drugs, when an interaction between two drugs is listed. This construced graph contains 1213 nodes linked by 12088 edges, which reveals 22 nodes (e.g. "Sipuleucel-T", "Pizotifen", "Iodine", etc.), listed as drug interaction without a corresponding drug entry in the DrugBank database. Figure 5 shows the visualization of the drug interaction graph, with the drug node size weighted by degree.

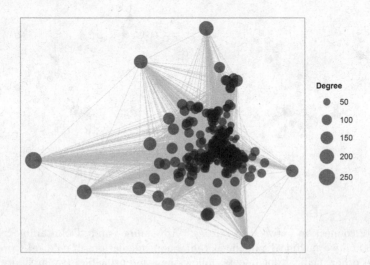

Fig. 5. Graph of drug interactions in the DrugBank database.

4.3   Biological Networks

Functions of life on a sub-cellular level rely on various complex interactions between different entities. Proteins, genes and metabolites interact to produce either healthy or diseased cellular processes. Our understanding of this network of interactions, and the interacting objects themselves, is continuously changing; and the graph structure itself is constantly changing and evolving as we age or as disease progresses.

Our methods for discovering new relationships and pathways change as well. A tool from Jurisica Group in Toronto may be of help here: NAViGaTOR 3 addresses such realities by having a very basic core rooted in graph theory, with the flexibility of a modular plugin architecture that provides data input and output, analysis, layout and visualization capabilities. NAViGaTOR 3 implements this architecture by following the OSGi standard[3]. Available API enables developers to expand standard distribution by integrating new features and extending the functionality of the program to suit their specific needs [61].

NAViGaTOR 3 was designed with the knowledge that a researcher may need to combine heterogeneous and distributed data sources. The standard distribution supports the loading, manipulation, and storage of multiple XML formats and tabular data. XML data is handled using a suite of file loaders, including XGMML, PSI-MI, SBML, KGML, and BioPAX, which store richly-annotated data and provide links to corresponding objects in the graph. Tabular data is stored using DEX (Martinez-Bazan, Gomez-Villamor and Escale-Claveras, 2011), a dedicated graph database from Sparsity Technologies[4].

Figure 6 shows an integrated graph by combining metabolic pathways, protein-protein interactions, and drug-target data. This metabolic data was collected in the Jurisca Lab, combining several steroid hormone metabolism pathways: androgen, glutathione, N-nitrosamine and benzo(a)pyrene pathway, the ornithine-spermine biosynthesis pathway, the retinol metabolism pathway and the TCA cycle aerobic respiration pathway. The figure highlights different pathways with different edge colors. The edge directionality highlights reactions and flow between the individual pathways. The data set was centred on steroid hormone metabolism and included data from hormone-related cancers [26]. The list of FDA-approved drugs used for breast, ovarian and prostate cancer was retrieved from the National Cancer Institute[5]. Afterwards the DrugBank[6] was searched for targets for each drug and those integrated in the graph structure.

5   Challenges and Future Research

A grand challenge is to discover relevant *structural* patterns and/or *temporal* patterns ("knowledge") in high dimensional data, which are often hidden and

---

[3] OSGi Standard http://www.osgi.org/Main/HomePage.

[4] DEX Graph Database http://www.sparsity-technologies.com/dex.

[5] National Cancer Institute, http://www.cancer.gov.

[6] DrugBank, http://www.drugbank.ca.

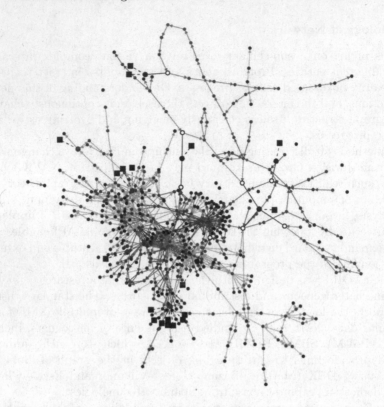

Fig. 6. Partially explored network: connecting drugs and metabolism. A network comprising metabolites, enzymes, and drugs of multiple pathways in the early stages of exploration. (Color figure online)

not accessible to the human expert but would be urgently needed for better decision support or for deeper investigation. Also, the fact that most data sets in the biomedical domain are weakly-structured or non-standardized add extra difficulties [28].

In medical research, these challenges are closely connected to the search for personalized medicine, which is a trend resulting in an explosion in data size (especially dimensionality): for instance "-omics" data, including data of genomics, proteomics, metabolomics, etc [35]. Whilst personalized medicine is the ultimate goal, stratified medicine has been the current approach, which aims to select the best therapy for groups of patients who share common biological characteristics. Here, machine learning approaches and optimization of knowledge discovery tools become imperative [53,61].

Optimization algorithms and techniques are now at the core of many data analysis problems. In high dimensional settings, statistical understanding of these algorithms is crucial not only to obtain quality solutions but also to invent new types of algorithms, as witnessed in recent literature [2,8,49,57]. Efficient and distributed algorithm implementations also become critical due to high computational demands. There are lots of active research in this regard based on optimization algorithms e.g. the ADMM [9] and block-coordinate descent methods [6,67].

Graph-based approaches introduced above are closely related to the graph-based data mining and topological data mining, which are amongst the most challenging topics [31–33,62]. Graph-based data mining was pioneerined about two decades ago [15,17,73], and based upon active research subjects including subgraph categories, isomorphism, invariance, measures, and solution methods [70]. It also can involve content-rich information, e.g. relationship among biological concepts, genes, proteins and drugs, such as in [13] or network medicine [5].

A closely related method is topological data mining, which focuses more on topological spaces (or manifolds) equipped with measures defined for data elements. The two most popular topological techniques in the study of data are *homology* and *persistence*. The connectivity of a space is determined by its cycles of different dimensions. These cycles are organized into groups, called homology groups. Given a reasonably explicit description of a space, the homology groups can be computed with linear algebra. Homology groups have a relatively strong discriminative power and a clear meaning, while having low computational cost. In the study of persistent homology the invariants are in the form of persistence diagrams or barcodes [23]. For interested readers, we suggest papers about point cloud from vector space models [69], and persistent homology [10,12,78].

The grand vision for the future is to effectively support human learning with machine learning. The HCI-KDD network of excellence[7] is an initiative proactively supporting this vision, bringing together people with diverse background but with a shared goal of finding solutions for dealing with big and complex data sets. We believe such an endeavor is necessary to deal with the complex and interdisciplinary nature of the problem. A recent outcome of the network can be found here [29]. This shows that diverse techniques and new ideas need to be integrated for successful knowledge discovery with big and complex real data. Still, there are many emergent challenges and open problems, which we believe deserve further research.

References

1. Anderson, N.R., Lee, E.S., Brockenbrough, J.S., Minie, M.E., Fuller, S., Brinkley, J., Tarczy-Hornoch, P.: Issues in biomedical research data management and analysis: needs and barriers. J. Am. Med. Inform. Assoc. **14**(4), 478–488 (2007)
2. Bach, F.R.: Bolasso: Model consistent Lasso estimation through the bootstrap. In: 25th International Conference on Machine Learning, pp. 33–40 (2008)
3. Banerjee, O., Ghaoui, L.E., d'Aspremont, A.: Model selection through sparse maximum likelihood estimation for multivariate Gaussian or binary data. J. Am. Med. Inform. Assoc. **9**, 485–516 (2008)
4. Barabasi, A.L., Albert, R.: Emergence of scaling in random networks. Science **286**(5439), 509–512 (1999)

---

[7] HCI-KDD Network: www.hci-kdd.org.

5. Barabási, A., Gulbahce, N., Loscalzo, J.: Network medicine: a network-based approach to human disease. Science **12**(1), 56–68 (2011)
6. Beck, A., Tetruashvili, L.: On the convergence of block coordinate descent type methods. Science **23**(4), 2037–2060 (2013)
7. Beck, A., Teboulle, M.: A fast iterative shrinkage-thresholding algorithm for linear inverse problems. Science **2**(1), 183–202 (2009)
8. Bogdan, M., van den Berg, E., Sabatti, C., Su, W., Candes, E.J.: SLOPE - adaptive variable selection via convex optimization. (2014). arXiv:1407.3824
9. Boyd, S., Parikh, N., Chu, E., Peleato, B., Eckstein, J.: Distributed optimization and statistical learning via the alternating direction method of multipliers. Science **3**(1), 1–122 (2011)
10. Bubenik, P., Kim, P.T.: A statistical approach to persistent homology. Science **9**(2), 337–362 (2007)
11. Castellana, B., Escuin, D., Peiró, G., Garcia-Valdecasas, B., Vázquez, T., Pons, C., Pérez-Olabarria, M., Barnadas, A., Lerma, E.: ASPN and GJB2 are implicated in the mechanisms of invasion of ductal breast carcinomas. Science **3**, 175–183 (2012)
12. Cerri, A., Fabio, B.D., Ferri, M., Frosini, P., Landi, C.: Betti numbers in multidimensional persistent homology are stable functions. Science **36**(12), 1543–1557 (2013)
13. Chen, H., Sharp, B.M.: Content-rich biological network constructed by mining PubMed abstracts. BMC Bioinformatics **5**(1), 147 (2004)
14. Cios, K.J., Moore, G.W.: Uniqueness of medical data mining. BMC Bioinformatics **26**(1), 1–24 (2002)
15. Cook, D.J., Holder, L.B.: Graph-based data mining. BMC Bioinformatics **15**(2), 32–41 (2000)
16. Cox, D.R., Oakes, D.: Analysis of Survival Data. Monographs on Statistics & Applied Probability. Chapman & Hall/CRC, London (1984)
17. Dehaspe, L., Toivonen, H.: Discovery of frequent DATALOG patterns. BMC Bioinformatics **3**(1), 7–36 (1999)
18. Iordache, O.: Methods. In: Iordache, O. (ed.) Polystochastic Models for Complexity. UCS, vol. 4, pp. 17–61. Springer, Heidelberg (2010)
19. Dehmer, M., Basak, S.C.: Statistical and Machine Learning Approaches for Network Analysis. Wiley, Hoboken (2012)
20. Donsa, K., Spat, S., Beck, P., Pieber, T.R., Holzinger, A.: Towards personalization of diabetes therapy using computerized decision support and machine learning: some open problems and challenges. In: Holzinger, A., Röcker, C., Ziefle, M. (eds.) Smart Health. LNCS, vol. 8700, pp. 237–260. Springer, Heidelberg (2015)
21. Dorogovtsev, S., Mendes, J.: Evolution of Networks: From Biological Nets to the Internet and WWW. Oxford University Press, Oxford (2003)
22. Duerr-Specht, M., Goebel, R., Holzinger, A.: Medicine and health care as a data problem: will computers become better medical doctors? In: Holzinger, A., Röcker, C., Ziefle, M. (eds.) Smart Health. LNCS, vol. 8700, pp. 21–39. Springer, Heidelberg (2015)
23. Epstein, C., Carlsson, G., Edelsbrunner, H.: Topological data analysis. BMC Bioinformatics **27**(12), 120201 (2011)
24. Friedman, J., Hastie, T., Tibshirani, R.: Sparse inverse covariance estimation with the graphical Lasso. BMC Bioinformatics **9**(3), 432–441 (2008)
25. Golumbic, M.C.: Algorithmic Graph Theory and Perfect Graphs. Elsevier, Amsterdam (2004)
26. Henderson, B.E., Feigelson, H.S.: Hormonal carcinogenesis. Carcinogenesis **21**(3), 427–433 (2000)

27. Holzinger, A.: Human-Computer Interaction and Knowledge Discovery (HCI-KDD): what is the benefit of bringing those two fields to work together? In: Cuzzocrea, A., Kittl, C., Simos, D.E., Weippl, E., Xu, L. (eds.) CD-ARES 2013. LNCS, vol. 8127, pp. 319–328. Springer, Heidelberg (2013)
28. Holzinger, A., Dehmer, M., Jurisica, I.: Knowledge discovery and interactive data mining in bioinformatics - state-of-the-art, future challenges and research directions. BMC Bioinformatics 15(Suppl 6), I1 (2014)
29. Holzinger, A., Jurisica, I. (eds.): Interactive Knowledge Discovery and Data Mining in Biomedical Informatics: State-of-the-Art and Future Challenges, vol. 8401. Springer, Heidelberg (2014)
30. Holzinger, A., Jurisica, I.: Knowledge discovery and data mining in biomedical informatics: the future is in integrative, interactive machine learning solutions. In: Holzinger, A., Jurisica, I. (eds.) Interactive Knowledge Discovery and Data Mining in Biomedical Informatics. LNCS, vol. 8401, pp. 1–18. Springer, Heidelberg (2014)
31. Holzinger, A., Malle, B., Giuliani, N.: On graph extraction from image data. In: Ślęzak, D., Tan, A.-H., Peters, J.F., Schwabe, L. (eds.) BIH 2014. LNCS, vol. 8609, pp. 552–563. Springer, Heidelberg (2014)
32. Holzinger, A., Ofner, B., Dehmer, M.: Multi-touch graph-based interaction for knowledge discovery on mobile devices: state-of-the-art and future challenges. In: Holzinger, A., Jurisica, I. (eds.) Interactive Knowledge Discovery and Data Mining in Biomedical Informatics. LNCS, vol. 8401, pp. 241–254. Springer, Heidelberg (2014)
33. Holzinger, A., Ofner, B., Stocker, C., Calero Valdez, A., Schaar, A.K., Ziefle, M., Dehmer, M.: On graph entropy measures for knowledge discovery from publication network data. In: Cuzzocrea, A., Kittl, C., Simos, D.E., Weippl, E., Xu, L. (eds.) CD-ARES 2013. LNCS, vol. 8127, pp. 354–362. Springer, Heidelberg (2013)
34. Holzinger, A., Stocker, C., Dehmer, M.: Big complex biomedical data: towards a taxonomy of data. In: Obaidat, M.S., Filipe, J. (eds.) Communications in Computer and Information Science CCIS 455, pp. 3–18. Springer, Heidelberg (2014)
35. Huppertz, B., Holzinger, A.: Biobanks – a source of large biological data sets: open problems and future challenges. In: Holzinger, A., Jurisica, I. (eds.) Interactive Knowledge Discovery and Data Mining in Biomedical Informatics. LNCS, vol. 8401, pp. 317–330. Springer, Heidelberg (2014)
36. Jacob, L., Obozinski, G., Vert, J.P.: Group Lasso with overlap and graph Lasso. In: Proceedings of the 26th International Conference on Machine Learning (ICML), pp. 433–440 (2009)
37. Javanmard, A., Montanari, A.: Model selection for high-dimensional regression under the generalized irrepresentability condition. BMC Bioinformatics 26, 3012–3020 (2013)
38. Joachims, T., Finley, T., Yu, C.N.: Cutting-plane training of structural SVMs. BMC Bioinformatics 77(1), 27–59 (2009)
39. Kleinberg, J.: Navigation in a small world. Nature 406(6798), 845–845 (2000)
40. Klopocki, E., Kristiansen, G., Wild, P.J., Klaman, I., Castanos-Velez, E., Singer, G., Stöhr, R., Simon, R., Sauter, G., Leibiger, H., Essers, L., Weber, B., Hermann, K., Rosenthal, A., Hartmann, A., Dahl, E.: Loss of SFRP1 is associated with breast cancer progression and poor prognosis in early stage tumors. Nature 25(3), 641–649 (2004)
41. Knight, K., Fu, W.: Asymptotics for Lasso-type estimators. Ann. Stat. 28(5), 1356–1378 (2000)
42. Koontz, W., Narendra, P., Fukunaga, K.: A graph-theoretic approach to nonparametric cluster analysis. Nature 100(9), 936–944 (1976)

43. Kumpulainen, S., Jarvelin, K.: Barriers to task-based information access in molecular medicine. Nature **63**(1), 86–97 (2012)
44. Kurgan, L.A., Musilek, P.: A survey of knowledge discovery and data mining process models. Nature **21**(01), 1–24 (2006)
45. Lauritzen, S.L.: Graphical Models. Oxford University Press, Oxford (1996)
46. Law, V., Knox, C., Djoumbou, Y., Jewison, T., Guo, A.C., Liu, Y.F., Maciejewski, A., Arndt, D., Wilson, M., Neveu, V., Tang, A., Gabriel, G., Ly, C., Adamjee, S., Dame, Z.T., Han, B.S., Zhou, Y., Wishart, D.S.: Drugbank 4.0: shedding new light on drug metabolism. Nature **42**(D1), D1091–D1097 (2014)
47. Lee, S.: Sparse inverse covariance estimation for graph representation of feature structure. In: Holzinger, A., Jurisica, I. (eds.) Interactive Knowledge Discovery and Data Mining in Biomedical Informatics. LNCS, vol. 8401, pp. 227–240. Springer, Heidelberg (2014)
48. Lee, S.: Signature selection for grouped features with a case study on exon microarrays. In: Stańczyk, U., Jain, L.C. (eds.) Feature Selection for Data and Pattern Classification, pp. 329–349. Springer, Heidelberg (2015)
49. Lee, S., Wright, S.J.: Manifold identification in dual averaging methods for regularized stochastic online learning. Nature **13**, 1705–1744 (2012)
50. Lilla, C., Koehler, T., Kropp, S., Wang-Gohrke, S., Chang-Claude, J.: Alcohol dehydrogenase 1B (ADH1B) genotype, alcohol consumption and breast cancer risk by age 50 years in a german case-control study. Nature **92**(11), 2039–2041 (2005)
51. Lodhi, H., Saunders, C., Shawe-Taylor, J., Watkins, N.C.C.: Text classification using string kernels. Nature **2**, 419–444 (2002)
52. Ma, K.L., Muelder, C.W.: Large-scale graph visualization and analytics. Nature **46**(7), 39–46 (2013)
53. Mattmann, C.A.: Computing: a vision for data science. Nature **493**(7433), 473–475 (2013)
54. McCall, M., Murakami, P., Lukk, M., Huber, W., Irizarry, R.: Assessing affymetrix genechip microarray quality. BMC Bioinformatics **12**(1), 137 (2011)
55. McCall, M.N., Bolstad, B.M., Irizarry, R.A.: Frozen robust multiarray analysis (fRMA). BMC Bioinformatics **11**(2), 242–253 (2010)
56. Meinshausen, N., Bühlmann, P.: High-dimensional graphs and variable selection with the Lasso. BMC Bioinformatics **34**, 1436–1462 (2006)
57. Meinshausen, N., Bühlmann, P.: Stability selection. BMC Bioinformatics **72**(4), 417–473 (2010)
58. Müller, R.: Medikamente und Richtwerte in der Notfallmedizin, 11th edn. Ralf Müller Verlag, Graz (2012)
59. Nesterov, Y.E.: A method of solving a convex programming problem with convergence rate $o(1/k^2)$. Soviet Math. Dokl. **27**(2), 372–376 (1983)
60. Niakšu, O., Kurasova, O.: Data mining applications in healthcare: research vs practice. In: Databases and Information Systems Baltic DB & IS 2012, p. 58 (2012)
61. Otasek, D., Pastrello, C., Holzinger, A., Jurisica, I.: Visual data mining: effective exploration of the biological universe. In: Holzinger, A., Jurisica, I. (eds.) Interactive Knowledge Discovery and Data Mining in Biomedical Informatics. LNCS, vol. 8401, pp. 19–33. Springer, Heidelberg (2014)
62. Preuß, M., Dehmer, M., Pickl, S., Holzinger, A.: On terrain coverage optimization by using a network approach for universal graph-based data mining and knowledge discovery. In: Ślęzak, D., Tan, A.-H., Peters, J.F., Schwabe, L. (eds.) BIH 2014. LNCS, vol. 8609, pp. 564–573. Springer, Heidelberg (2014)

63. Schoenauer, M., Akrour, R., Sebag, M., Souplet, J.C.: Programming by feedback. In: Proceedings of the 31st International Conference on Machine Learning (ICML 2014), pp. 1503–1511 (2014)
64. Spinrad, N.: Google car takes the test. Nature **514**(7523), 528–528 (2014)
65. Strogatz, S.: Exploring complex networks. Nature **410**(6825), 268–276 (2001)
66. Tibshirani, R.: Regression shrinkage and selection via the Lasso. Nature **58**, 267–288 (1996)
67. Tseng, P.: Convergence of a block coordinate descent method for nondifferentiable minimization. Nature **109**(3), 475–494 (2001)
68. Vandenberghe, L., Boyd, S., Wu, S.P.: Determinant maximization with linear matrix inequality constraints. Nature **19**(2), 499–533 (1998)
69. Wagner, H., Dłotko, P., Mrozek, M.: Computational topology in text mining. In: Ferri, M., Frosini, P., Landi, C., Cerri, A., Di Fabio, B. (eds.) CTIC 2012. LNCS, vol. 7309, pp. 68–78. Springer, Heidelberg (2012)
70. Washio, T., Motoda, H.: State of the art of graph-based data mining. Nature **5**(1), 59 (2003)
71. Wishart, D.S., Knox, C., Guo, A.C., Shrivastava, S., Hassanali, M., Stothard, P., Chang, Z., Woolsey, J.: Drugbank: a comprehensive resource for in silico drug discovery and exploration. Nature **34**, D668–D672 (2006)
72. Wittkop, T., Emig, D., Truss, A., Albrecht, M., Boecker, S., Baumbach, J.: Comprehensive cluster analysis with transitivity clustering. Nature **6**(3), 285–295 (2011)
73. Yoshida, K., Motoda, H., Indurkhya, N.: Graph-based induction as a unified learning framework. Nature **4**(3), 297–316 (1994)
74. Yuan, M., Lin, Y.: Model selection and estimation in regression with grouped variables. Nature **68**, 49–67 (2006)
75. Yuan, M., Lin, Y.: Model selection and estimation in the Gaussian graphical model. Biometrika **94**(1), 19–35 (2007)
76. Zhao, P., Yu, B.: On model selection consistency of Lasso. Biometrika **7**, 2541–2563 (2006)
77. Zhengxiang, Z., Jifa, G., Wenxin, Y., Xingsen, L.: Toward domain-driven data mining. In: International Symposium on Intelligent Information Technology Application Workshops, pp. 44–48 (2008)
78. Zhu, X.: Persistent homology: an introduction and a new text representation for natural language processing. In: IJCAI, IJCAI/AAAI (2013)
79. Zou, H.: The adaptive Lasso and its Oracle properties. Biometrika **101**(476), 1418–1429 (2006)
80. Zou, H., Hastie, T.: Regularization and variable selection via the elastic net. Biometrika **67**, 301–320 (2005)
81. Zudilova-Seinstra, E., Adriaansen, T.: Visualisation and interaction for scientific exploration and knowledge discovery. Biometrika **13**(2), 115–117 (2007)

Local Pattern Detection in Attributed Graphs

Jean-François Boulicaut[1], Marc Plantevit[2], and Céline Robardet[1(✉)]

[1] Université de Lyon, CNRS, INSA de Lyon, LIRIS UMR5205,
69621 Villeurbanne, France
{jean-francois.boulicaut,celine.robardet}@insa-lyon.fr
[2] Université de Lyon, CNRS, Univ. Lyon1, LIRIS UMR5205,
69622 Villeurbanne, France
marc.plantevit@univ-lyon1.fr

Abstract. We propose to mine the topology of a large attributed graph by finding regularities among vertex descriptors. Such descriptors are of two types: (1) the vertex attributes that convey the information of the vertices themselves and (2) some topological properties used to describe the connectivity of the vertices. These descriptors are mostly of numerical or ordinal types and their similarity can be captured by quantifying their co-variation. Mining topological patterns relies on frequent pattern mining and graph topology analysis to reveal the links that exist between the relation encoded by the graph and the vertex attributes. In this paper, we study the network of authors who have cooperated at some time with Katharina Morik according to the data available in DBLP database. This is a nice occasion for formalizing different questions that can be considered when an attributed graph describes both a type of interaction and node descriptors.

Keywords: Attributed graph mining · Katharina Morik co-authorship

1 Introduction

A timely challenge concerns enriched graph mining to support knowledge discovery. We recently proposed the topological pattern domain [25], a kind of gradual pattern that extends the rank-correlated sets from [6] to support attributed graph analysis. In such graphs, the binary relation encoded by the graph is enriched by vertex numerical attributes. However, existing methods that support the discovery of local patterns in graphs mainly focus on the topological structure of the patterns, by extracting specific subgraphs while ignoring the vertex attributes (cliques [21], quasi-cliques [20,29]), or compute frequent relationships between vertex attribute values (frequent subgraphs in a collection of graphs [16] or in a single graph [5]), while ignoring the topological status of the vertices within the whole graph, e.g., the vertex connectivity or centrality. The same limitation holds for the methods proposed in [18,24,27,28], which identify sets of vertices that have similar attribute values and that are close neighbors. Such approaches only focus on a local neighborhood of the vertices and do not consider the connectivity of the vertex in the whole graph.

© Springer International Publishing Switzerland 2016
S. Michaelis et al. (Eds.): Morik Festschrift, LNAI 9580, pp. 168–183, 2016.
DOI: 10.1007/978-3-319-41706-6_8

To investigate the relations that may exist between the position of the vertices within the graph and their attribute values, we proposed to extract topological patterns that are sets made of vertex attributes and topological measures. Such measures quantify the topological status of each vertex within the graph. Some of these measures are based on the close neighborhood of the vertices (e.g., the vertex degree), while others describe the connectivity of a vertex by considering its relationship with all other vertices (e.g., the centrality measures). Combining such microscopic and macroscopic properties characterizes the connectivity of the vertices and it may be a sound basis to explain why some vertices have similar attribute values.

Topological patterns of interest are composed of vertex properties that behave similarly over the vertices of the graph. The similarity among vertex properties can be captured by quantifying their correlation, which may be positive or negative. To that end, we extend the Kendall rank correlation coefficient to any number of variables, as well as to negative correlation. Whereas this measure is rather theoretically sounded, its evaluation is computationally demanding as it requires to consider all vertex pairs to estimate the proportion of which that supports the pattern. The well known optimization techniques that are used for evaluating the correlation between two variables (and that leads to a theoretical complexity in $O(n \log n)$) do not extend directly when a higher number of variables is considered. We tackled this issue and proposed several optimization and pruning strategies that makes it possible to use this approach on large graphs. We also introduced several interestingness measures of topological patterns that differ by the pairs of vertices that are considered while evaluating the correlation between descriptors: (1) While all the vertex pairs are considered, patterns that are true all over the graph are extracted; (2) When including only the vertex pairs that are in a specific order regarding to a selected numerical or ordinal attribute reveals the topological patterns that emerge with respect to this attribute; (3) Examining the vertex pairs that are connected in the graph makes it possible to identify patterns that are *structurally correlated* to the relationship encoded by the graph. Besides, we designed an operator that identifies the top k representative vertices of a topological pattern.

In this paper, we study the network of authors who have cooperated at some time with Katharina Morik according to the data available in the DBLP database. Doing so, we emphasize powerful mechanisms for detecting new types of local patterns in interaction graphs. Indeed, we formalize different questions that can be considered when an attributed graph describes both interactions and vertex descriptors. This has not yet been studied systematically. It enables also to discuss the need for new post-processing techniques that exploit both the patterns and the graph data. Finally, detecting local patterns in various data types has motivated a lot of research in our group and writing a chapter at this Festschrift occasion is also an implied reference to the domain that gave us the first occasion to spend time and work with our smart colleague [22].

2   Related Work

Graph mining is an active topic in Data Mining. In the literature, there exist two main trends to analyze graphs. On the one hand, graphs are studied at a macroscopic level by considering statistical graph properties (e.g., diameter, degree distribution) [2,7]. On the other hand, sophisticated graph properties are discovered by using a local pattern mining approach. Recent approaches mine attributed graphs which convey more information. In such graphs, information is locally available on vertices by means of attribute values. As argued by Moser et al. [23], "*often features and edges contain complementary information, i.e., neither the relationships can be derived from the feature vectors nor vice versa*".

Attributed graphs are extensively studied, by means of clustering techniques (see e.g., [1,8,13,15,19,32]) whereas pattern mining techniques in such graphs have been less investigated. The pioneering work [23] proposes a method to find dense homogeneous subgraphs (i.e., subgraphs whose vertices share a large set of attributes). Similar to this work, Günnemann et al. [14] propose a method based on subspace clustering and dense subgraph mining to extract non redundant subgraphs that are homogenous with respect to vertex attributes. Silva et al. [28] extract pairs of dense subgraphs and Boolean attribute sets such that the Boolean attributes are strongly associated with the dense subgraphs. In [24], the authors propose the task of finding the collections of homogeneous k-clique percolated components (i.e., components made of overlapping cliques sharing a common set of true valued attributes) in Boolean attributed graphs. Another approach is presented in [18], where a larger neighborhood is considered. This pattern type relies on a relaxation of the accurate structure constraint on subgraphs. Roughly speaking, they propose a probabilistic approach to both construct the neighborhood of a vertex and propagate information into this neighborhood. Following the same motivation, Sese et al. [26] extract (not necessarily dense) subgraphs with common itemsets. Note that these approaches use a single type of topological information based on the neighborhood of the vertices. Furthermore, they do not handle numerical attributes as in our proposal. However, global statistical analysis [11] of a single graph considers several measures to describe the graph topology, but does not benefit from vertex attributes. Besides, current local pattern mining techniques on attributed graphs do not consider numerical attributes nor macroscopic topological properties. To the best of our knowledge, our paper represents a first attempt to combine both microscopic and macroscopic analysis on graphs by means of (emerging) topological pattern mining. Indeed, several approaches aim at building global models from local patterns [12], but none of them tries to combine information from different graph granularity levels.

Co-variation patterns are also known as gradual patterns [9] or rank-correlated itemsets [6]. Do et al. [9] use a support measure based on the length of the longest path between ordered objects. This measure has some drawbacks w.r.t. computational and semantics aspects. Calders et al. [6] introduce a support measure based on the Kendall's τ statistical measure. However, their approach is not defined to simultaneously discover up and down co-variation patterns as does our approach. Another novelty of our work is the definition of other inter-estingness measures to capture emerging co-variations. Finally, this work is also the first attempt to use co-variation pattern mining in attributed graphs.

3   Topological Vertex Properties

Let us consider a non-directed attributed graph $G = (V, E, L)$, where V is a set of n vertices, E a set of m edges, and $L = \{l_1, \cdots, l_p\}$ a set of p numerical or ordinal attributes associated with each vertex of V. Important properties of the vertices are encoded by the edges of the graph. From this relation, we can compute some topological properties that synthesize the role played by each vertex in the graph. The topological properties we are interested in range from a microscopic level – those that described a vertex based on its direct neighborhood – to a macroscopic level – those that characterize a vertex by considering its relationship to all other vertices in the graph. Statistical distributions of these properties are generally used to depict large graphs (see, e.g., [2, 17]). We propose here to use them as vertex descriptors.

3.1   Microscopic Properties

Let us consider here only three topological properties to describe the direct neighborhood of a vertex v:

- The degree of v is the number of edges incident to v ($deg(v) = |\{u \in V, \{u, v\} \in E\}|$). When normalized by the maximum number of edges a vertex can have, it is called the degree centrality coefficient: $\text{DEGREE}(v) = \frac{deg(v)}{n-1}$.
- The clustering coefficient evaluates the connectivity of the neighbors of v and thus its local density:

$$\text{CLUST}(v) = \frac{2|\{\{u, w\} \in E, \{u, v\} \in E \wedge \{v, w\} \in E\}|}{deg(v)(deg(v) - 1)}$$

3.2   Mesoscopic Property

We also consider the position of each vertex to the center of the graph, that is the distance – the number of edges of a shortest path – to a peculiar vertex. In the following, we call this property the $\text{MORIK\_NUMBER}(v)$ as we consider the relative position of the vertices to the vertex that corresponds to Katharina Morik.

3.3   Macroscopic Properties

We consider five macroscopic topological properties to characterize a vertex while taking into account its connectivity to all other vertices of the graph.

- The relative importance of vertices in a graph can be obtained through centrality measures [11]. Closeness centrality $\text{CLOSE}(v)$ is defined as the inverse of the average distance between v and all other vertices that are reachable from it. The distance between two vertices is defined as the number of edges of the shortest path between them: $\text{CLOSE}(v) = \frac{n}{\sum_{u \in V} |shortest\_path(u,v)|}$.
- The betweenness centrality $\text{BETW}(v)$ of v is equal to the number of times a vertex appears on a shortest path in the graph. It is evaluated by first computing all the shortest paths between every pair of vertices, and then counting the number of times a vertex appears on these paths: $\text{BETW}(v) = \sum_{u,w} \mathbb{1}_{shortest\_path(u,w)}(v)$.
- The eigenvector centrality measure (EGVECT) favours vertices that are connected to vertices with high eigenvector centrality. This recursive definition can be expressed by the following eigenvector equation $Ax = \lambda x$ which is solved by the eigenvector x associated to the largest eigenvalue λ of the adjacency matrix A of the graph.
- The PAGERANK index [4] is based on a random walk on the vertices of the graph, where the probability to go from one vertex to another is modelled as a Markov chain in which the states are vertices and the transition probabilities are computed based on the edges of the graph. This index reflects the probability that the random walk ends at the vertex itself:

$$\text{PAGERANK}(v) = \alpha \sum_u \mathbb{1}_E(\{u,v\}) \frac{\text{PAGERANK}(u)}{deg(u)} + \frac{1-\alpha}{n}$$

where the parameter α is the probability that a random jump to vertex v occurs.
- Network constraint [30] evaluates to what extent person's contacts are redundant

$$\text{NETWORK}(v) = \sum_{u|(u,v) \in E} [\frac{1}{deg(v)} + \sum_{w|(u,w) \ and \ (v,w) \in E} (\frac{1}{deg(v)} \frac{1}{deg(u)})]^2$$

When its value is low, the contacts are rather disconnected, whereas when it is high, the contacts are close or strongly tied.

These 9 topological properties characterizes the graph relationship encoded by E. These properties, along with the set of vertex attributes L, constitutes the set of vertex descriptors $\mathcal{D}$ used in this paper.

4   Topological Patterns

Let us now consider topological patterns as a set of vertex attributes and topological properties that behave similarly over a large part of the vertices of the graph. We assume that all topological properties and vertex attributes are of numerical or ordinal type, and we propose to capture their similarity by quantifying their co-variation over the vertices of the graph. Topological patterns are defined as $P = \{D_1{}^{s_1}, \cdots, D_\ell{}^{s_\ell}\}$, where D_j, $j = 1 \ldots \ell$, is a vertex descriptor from $\mathcal{D}$ and $s_j \in \{+, -\}$ is its co-variation sign. In the following, we propose three pattern interestingness measures that differ in the pairs of vertices considered for their evaluation.

4.1   Topological Patterns over the Whole Graph

Several signed vertex descriptors co-vary if the orders induced by each of them on the set of vertices are consistent. This consistency is evaluated by the number of vertex pairs ordered the same way by all descriptors. The number of such pairs constitutes the so-called support of the pattern. This measure can be seen as a generalization of the Kendall's τ measure. When we consider all possible vertex pairs, this interestingness measure is defined as follows:

Definition 1 (*Supp_{all}*). *The support of a topological pattern P over all possible pairs of vertices is:*

$$Supp_{all}(P) = \frac{|\{(u, v) \in V^2 \mid \forall D_j^{s_j} \in P : D_j(u) \rhd_{s_j} D_j(v)\}|}{\binom{n}{2}}$$

where $\rhd_{s_j}$ denotes $<$ when s_j is equal to $+$, and $\rhd_{s_j}$ denotes $>$ when s_j is equal to $-$.

This measure gives the number of vertex pairs (u, v) such that u is strictly lower than v on all descriptors with sign $+$, and u is strictly higher than v on descriptors with sign $-$.

As mentioned in [6], $Supp_{all}$ is an anti-monotonic measure for positively signed descriptors. This is still true when considering negatively signed ones: adding D_{l+1}^- to a pattern P leads to a support lower than or equal to that of P since the pairs (u, v) that support P must also satisfy $D_{l+1}(u) > D_{l+1}(v)$. Besides, when adding descriptors with negative sign, the support of some patterns can be deduced from others, the latter referred to as symmetrical patterns.

Property 1 (Support of symmetrical patterns). Let P be a topological pattern and $\overline{P}$ be its symmetrical, that is, $\forall D_j^{s_j} \in P$, $D_j^{\overline{s_j}} \in \overline{P}$, with $\overline{s_j} = \{+, -\} \setminus \{s_j\}$. If a pair (u, v) of V^2 contributes to the support of P, then the pair (v, u) contributes to the support of $\overline{P}$. Thus, we have $Supp_{all}(P) = Supp_{all}(\overline{P})$.

Topological patterns and their symmetrical patterns are semantically equivalent. To avoid the irrelevant computation of duplicate topological patterns, we exploit Property 1 and enforce the first descriptor of a pattern P to be signed by $+$.

Mining frequent topological patterns consists in computing all sets of signed descriptors P, but not their symmetrical ones, such that $Supp_{all}(P) \geq minsup$, where $minsup$ is a user-defined minimum support threshold.

4.2   Other Interestingness Measures

To identify most interesting topological patterns, we propose to give to the end-user the possibility of guiding its data mining process by querying the patterns with respect to their correlation with the relationship encoded by the graph or with a selected descriptor. Therefore, we revisit the notion of emerging patterns [10] by identifying the patterns whose support is significantly greater (i.e., according to a growth-rate threshold) in a specific subset of vertex pairs than in the remaining ones. This subset can be defined in different ways according to the end-user's motivations: either it is defined by the vertex pairs that are ordered with respect to a selected descriptor called the class descriptor, or it is equal to E, the set of edges. Whereas the former highlights the correlation of a pattern with the class descriptor, the latter enables to characterize the importance of the graph structure within the support of the topological pattern.

Emerging Patterns w.r.t. a Selected Descriptor. Let us consider a selected descriptor $C \in \mathcal{D}$ and a sign $r \in \{+, -\}$. The set of pairs of vertices that are ordered by C^r is

$$\mathcal{C}_{C^r} = \{(u, v) \in V^2 \mid C(u) \rhd_r C(v)\}$$

The support measure based on the vertex pairs of $\mathcal{C}_{C^r}$ is defined below.

Definition 2 (*$Supp_{C^r}$*). *The support of a topological pattern P over C^r is:*

$$Supp_{C^r}(P) = \frac{|\{(u, v) \in \mathcal{C}_{C^r} \mid \forall D_j^{s_j} \in P : D_j(u) \rhd_{s_j} D_j(v)\}|}{|\mathcal{C}_{C^r}|}$$

Analogously, the support of P over the pairs of vertices that do not belong to $\mathcal{C}_{C^r}$ is denoted $Supp_{C^{\overline{r}}}(P)$. To evaluate the impact of C^r on the support of P, we consider the growth rate of the support of P over the partition of vertex pairs $\{\mathcal{C}_{C^r}, \mathcal{C}_{C^{\overline{r}}}\}$: $Gr(P, C^r) = \frac{Supp_{C^r}(P)}{Supp_{C^{\overline{r}}}(P)}$

If $Gr(P, C^r)$ is greater than a minimum growth-rate threshold, then P is referred to as emerging with respect to C^r. If $Gr(P, C^r) \approx 1$, P is as frequent in $\mathcal{C}_{C^r}$ as in $\mathcal{C}_{C^{\overline{r}}}$. If $gr(P, C^r) \gg 1$, P is much more frequent in $\mathcal{C}_{C^r}$ than in $\mathcal{C}_{C^{\overline{r}}}$. For example, $Gr(\{h^+, i^-, \text{BETW}^+\}, t^+) = 2.31$. The intuition behind this definition is to identify the topological patterns that are mostly supported by pairs of vertices that are also ordered by the selected descriptor.

Emerging Patterns w.r.t. the Graph Structure. It is interesting to measure if the graph structure plays an important role in the support of a topological pattern P. To this end, we define a similar support measure based on pairs that belongs to E, the set of edges of the graph:

$$\mathcal{C}_E = \{(u, v) \in V^2 \mid \{u, v\} \in E\}$$

Based on this set of pairs, we define the support of P as:

Definition 3 ($Supp_E$). *The support of a topological pattern P over the pairs of vertices that are linked in G is:*

$$Supp_E(P) = \frac{2|\{(u,v) \in \mathcal{C}_E \mid \forall D_j^{s_j} \in P : D_j(u) \triangleright_{s_j} D_j(v)\}|}{|\mathcal{C}_E|}$$

The maximum value of the numerator is $\frac{|\mathcal{C}_E|}{2}$ since: (1) if $(u,v) \in \mathcal{C}_E$ then $(v,u) \in \mathcal{C}_E$, and (2) it is not possible that $\forall D_j^{s_j} \in P$, $D_j(u) \triangleright_{s_j} D_j(v)$ and $D_j(v) \triangleright_{s_j} D_j(u)$ at the same time. For instance, the pattern $\{h^+, i^-\}$ is supported by all the twenty possible pairs that are edges, its support is thus equal to 1. The support of P over the pairs of vertices that do not belong to $\mathcal{C}_E$ is denoted $Supp_{\overline{E}}(P)$.

As before, to evaluate the impact of E on the support of P, we consider the growth rate of the support of P over the partition of vertex pairs $\{\mathcal{C}_E, \mathcal{C}_{\overline{E}}\}$:
$Gr(P, E) = \frac{Supp_E(P)}{Supp_{\overline{E}}(P)}$.

$Gr(P, E)$ enables to assess the impact of the graph structure on the pattern. Therefore, if $Gr(P, E) \gg 1$, P is said to be *structurally* correlated. If $Gr(P, E) \ll 1$, the graph structure tends to inhibit the support of P.

5   Top k Representative Vertices

The user may be interested in identifying the vertices that are the most representative of a given topological pattern, thus enabling the projection of the patterns back into the graph. For example, the representative vertices of the pattern $\{t^+, \text{BETW}^-\}$ would be researchers with a relatively large number of IEEE TKDE papers and a low betweenness centrality measure.

We denote by $S(P)$ the set of vertex pairs (u,v) that constitutes the support of a topological pattern P:

$$S(P) = \{(u,v) \in V^2 \mid \forall D_j^{s_j} \in P : D_j(u) \triangleright_{s_j} D_j(v)\}$$

which forms, with V, a directed graph $G_P = (V, S(P))$. This graph satisfies the following property.

Property 2. The graph $G_P = (V, S(P))$ is transitive and acyclic.

Proof. Let us consider $(u,v) \in V^2$ and $(v,w) \in V^2$ such that, $\forall D_j^{s_j} \in P :$ $D_j(u) \triangleright_{s_j} D_j(v)$ and $D_j(v) \triangleright_{s_j} D_j(w)$. Thus, $D_j(u) \triangleright_{s_j} D_j(w)$ and $(u,w) \in S(P)$. Therefore, G_P is transitive.

As $\triangleright_s \in \{<, >\}$, it stands for a strict inequality. Thus, if $(u,v) \in S(P)$, $(v,u) \notin S(P)$. Furthermore, as G_P is transitive, if there exists a path between u and v, there is also an arc $(u,v) \in S(P)$. Therefore, $(v,u) \notin S(P)$ and we can conclude that G_P is acyclic.

As G_P is acyclic, it admits a topological ordering of its vertices, which is, in the general case, not unique. The top k representative vertices of a topological pattern P are identified on the basis of such a topological ordering of V and are the k last vertices with respect to this ordering. Considering that an arc $(u, v) \in S(P)$ is such that v dominates u on P, this vertex set contains the most dominant vertices on P. The top k representative vertices of P can be easily identified by ordering the vertices by their incoming degree.

Although the support of topological patterns is an anti-monotonic measure, its computation is quadratic in the number of vertices of the graph which prevents the extraction of such patterns on large graph using classical pattern mining algorithms. To overcome this problem, we proposed in [25] an upper bound on this measure that can be computed linearly in the number of vertices. This upper bound takes advantages of the presence of ties in the descriptor values. By pre-computing some indexes on the descriptors, almost all non frequent patterns are pruned without computing their support when the minimum support is high.

The computation of topological patterns is done in an ECLAT-based way [31]. More precisely, all the subsets of a pattern P are always evaluated before P itself. In this way, by storing all frequent patterns in the hash-tree $\mathcal{M}$, the anti-monotonic frequency constraint is fully-checked on the fly. We compute the upper bound on the support to prune non-promising topological patterns. When this upper bound is greater than the minimum threshold, the exact support is computed. Another optimization is based on the deduction of the support from already evaluated patterns: A pair of vertices that supports a pattern P can support pattern PA^+ or pattern PA^-, or none of them. Thus, another upper bound on $Supp_{all}(PA^-)$ is $Supp_{all}(P) - Supp_{all}(PA^+)$. Note that these patterns have already been considered before the evaluation of PA^-. So, to be stringent, we bound the support by taking the minimum between this value and the upper bound. When computing the support of the pattern, the top k representative vertices are also identified.

6 Studying Katharina Morik's Network

In the following, we propose to use TopGraphMiner to study the scientific co-authorship network of Katharina Morik. After presenting the attributed graph we generate from the DBLP digital library[1], we provide qualitative results that show the implication of Katharina in the machine learning community.

6.1 Katharina Morik's Co-authorship Network

The co-authorship graph is built from the DBLP digital library. Regarding Katharina's bibliography, we select all the conference venues and journals in which Katharina has at least one DBLP entry[2]. We gather all the publications in these conference venues and journals since their foundation, and derived a graph where the vertices stand for the authors and edges link two authors who

---

[1] http://dblp.uni-trier.de/.

[2] http://www.dblp.org/search/index.php?query=author:katharina_morik.

co-authored at least one paper in this corpus. To each vertex, we associate the number of publications in each of these 53 selected conferences or journals as vertex properties. We then removed isolated vertices, that is to say, authors who has no co-author in the selected publications. The resulting attributed graphs involves 81 222 vertices and 466 152 undirected edges. Notice that, even if this attributed graph is generated based on Katharina's publications, her co-authors only represent 0.1 % of the vertices of the whole graph, while the vertices whose distance to Katharina is at most 2 represent less than 2 % of the whole set of vertices. The average Morik number is 4.05 and 4033 authors have no path to Katharina (infinite Morik number). There are 1428 connected components.

Figure 1 presents this co-authorship graph restricted to the authors that are at most at a distance of 2 from Katharina and that have a degree value greater than 20. Applying the community detection Chinese Whisper algorithm [3], we obtain 68 communities whose most salient are represented on the figure. The purple community, that gathers 177 authors including Katharina, is very dense (1096 edges). It brings together well identified researchers in data mining, machine learning and data bases. The other main communities are labeled on the graph. Our goal is to analyse this graph with regard to several questions:

– Are there any interesting patterns among publications?

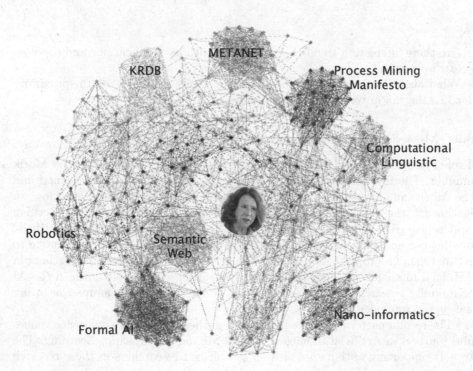

Fig. 1. Research domains associated to Katharina's co-authors. (Color figure online)

Table 1. Emerging patterns w.r.t. morik_number$^-$.

Pattern	Top 20
IJCAI$^+$, KI$^+$, GWAI$^+$, Informatik_Spektrum$^+$, morik_number$^-$	Katharina Morik, Wolfgang Wahlster, Bernhard Nebel, Thomas Christaller, Wolfgang Hoeppner, Jörg H. Siekmann, Günther Görz, Frank Puppe, Udo Hahn, Hans-Hellmut Nagel, Franz Baader, Christopher Habel, Bernd Neumann, Ulrich Furbach, Joachim Hertzberg
IJCAI$^+$, ICML$^+$, Machine_Learning$^+$, Knowl._Inf._Syst.$^+$, Data_Min._Knowl._Discov.$^+$, morik_number$^-$	Katharina Morik, Wray L. Buntine, Kristian Kersting, Floriana Esposito, Xindong Wu, Eamonn J. Keogh, Zhi-Hua Zhou, Siegfried Nijssen, Hiroshi Motoda, João Gama, Jie Tang, Salvatore J. Stolfo, Dacheng Tao, Michael J. Pazzani, Wei Liu, Chris H. Q. Ding, Tao Li, Bin Li

– Are there interesting trends between some authors' publications and topological properties?
– What about Katharina's role in this graph? Can we characterize the proximity to Katharina in terms of co-authorship?

6.2 Most Emerging Pattern with Respect to Morik Number

Table 1 presents two interesting patterns that strongly emerge with the Morik number. The first pattern gathers 4 conferences that are positively signed and the Morik number that is negatively signed: The more authors are close to Katharina, the more they publish in IJCAI as well as in three other German conferences (KI - Künstliche Intelligenz, GWAI - German workshop on artificial intelligence and Informatik_Spektrum) Notice that GWAI changes its name to KI in 1993. The top 20 supporting authors gathers the German researchers in Artificial Intelligence. They are close to Katharina who is wellknown in the AI community research, and she also actively contributes to the animation of her national community.

The second pattern presented in Table 1 gathers the major conference venues and journals in Artificial Intelligence, Data Mining and Machine Learning. The top 20 supporting authors are all well established researchers in these research areas.

The first pattern with the Morik number positively signed is presented in Table 2. It gathers the conference ICASSP in signal processing that is positively

Table 2. Emerging pattern w.r.t. morik_number[+].

Pattern	Top 20
ICASSP[+], IJCAI[−], KR[−], KI[−], morik_number[+]	Gyula Hermann, Victor Lazzarini, Joseph Timoney, Fred Kitson, Manuel Duarte Ortigueira, Abbas Mohammadi, Riwal Lefort, Jean-Marc Boucher, Artur Przelaskowski, Kenichi Miyamoto, Emiru Tsunoo, Olaf Schreiner, Murtaza Taj, Salim Chitroub, Saptarshi Das, Ales Procházka, Amrane Houacine, Yasuyuki Ichihashi, Pablo Javier Alsina, Valeri Mladenov

Table 3. Emerging patterns w.r.t. Morik number that mix vertex and topological attributes.

Pattern	Top 20
ICASSP[+], IJCAI[−], Degree[−], Closeness[−], Betweennes[−], NetworkConstraint[+], morik_number[+]	Jacob Ninan, Marc Beacken, Hinrich R. Martens, Jyun-Jie Wang, William H. Haas, J. G. Cook, Lawrence J. Ziomek, José R. Nombela, T. J. Edwards, Judith G. Claassen, Shigekatsu Irie, Alberto R. Calero, Takaaki Ueda, Hisham Hassanein, Peter Strobach, Liubomire G. Iordanov, N. A. M. Verhoeckx, Guy R. L. Sohie, Sultan Mahmood, Matt Townsend
KI[+], Degree[+], Closeness[+], NetworkConstraint[−], morik_number[−]	Bernhard Nebel, Katharina Morik, Deborah L. McGuinness, Mark A. Musen, Rudi Studer, Steffen Staab, Hans W. Guesgen, Bamshad Mobasher, Simon Parsons, Thorsten Joachims, Alex Waibel, Kristian Kersting, Matthias Jarke, Manuela M. Veloso, Wolfgang Nejdl, Alfred Kobsa, Virginia Dignum, Alessandro Saffiotti, Hans Uszkoreit, Antonio Krüger

signed and 3 conferences in Machine Learning that are negatively signed: The farther the authors from Katharina, the more they published at ICASSP and the less they contribute to AI conferences IJCAI, KI and KR (Principles of knowledge representation and reasoning). The support of this pattern is rather low.

The most emerging patterns w.r.t. Morik number that mix vertex and topological attributes are presented in Table 3. The first pattern is similar to the pattern of Table 2 and the additional topological attributes corroborate the eccentricity of the pattern relative to the graph. The second pattern brings together confirmed researchers in artificial intelligence and machine learning, who have

Table 4. Emerging patterns involving the French-speaking data mining conference EGC.

Pattern	Top 20
EGC$^+$, Data_Min._Knowl._Discov.$^+$, morik_number$^-$	Katharina Morik, Bart Goethals, Céline Robardet, Didier Dubois, Michèle Sebag, Luc De Raedt, Mohammed Javeed Zaki, Einoshin Suzuki, Heikki Mannila, Jian Pei, Élisa Fromont, Toon Calders, Adriana Prado, Gilles Venturini, Szymon Jaroszewicz, João Gama, Alice Marascu, Osmar R. Zaïane, Pascal Poncelet, Jean-François Boulicaut
EGC$^+$, Knowl._Inf._Syst.$^+$, Data_Min._Knowl._Discov.$^+$, morik_number$^-$	Katharina Morik, Bart Goethals, João Gama, Mohammed Javeed Zaki, Jian Pei, Heikki Mannila, Osmar R. Zaïane, Toon Calders, Szymon Jaroszewicz, Einoshin Suzuki, Pascal Poncelet, Christophe Rigotti, Jean-François Boulicaut, Marie-Christine Rousset, Maguelonne Teisseire, Florent Masseglia, Gregory Piatetsky-Shapiro
EGC$^+$, Knowl._Inf._Syst.$^+$, morik_number$^-$	Katharina Morik, Bart Goethals, Fosca Giannotti, Mohand-Said Hacid, Toon Calders, Mohammed Javeed Zaki, Osmar R. Zaïane, Heikki Mannila, João Gama, Dominique Laurent, Jian Pei, Szymon Jaroszewicz, Einoshin Suzuki, Patrick Gallinari, David Genest, Mohand Boughanem, François Scharffe, Marc Plantevit, Laure Berti-Equille, Zbigniew W. Ras

published at Künstliche Intelligenz. They are very central in the graph and their neighborhood is not so much connected.

6.3   Where Are We in Katharina's Network? An Interactive Exploration of the Patterns

After considering the patterns that maximize the growth rate w.r.t. Morik number, we now look for patterns supported by the authors of this paper. Many of these patterns involve the French-speaking conference EGC (see Table 4) and journals in data mining. This is due to the fact that Katharina gave a keynote at EGC in 2009. The top 20 supporting authors are either French or prestigious invited speakers at this conference.

The first pattern of Table 5 can be interpreted thanks to a Dagstuhl seminar organized by Katharina called *Local Pattern Detection*. The goal of this seminar was to bring together prominent European researchers in the field of local pattern discovery. The Data Mining and Knowledge Discovery journal is the most important one that publishes results in that area. The second one is around

Table 5. Patterns related to Dagstuhl seminars.

Pattern	Top 20
LocalPatternDetection[+], Data_Min._Knowl._Discov.[+], morik_number[-]	Katharina Morik, Stefan Rüping, Francesco Bonchi, Niall M. Adams, Marko Grobelnik, David J. Hand, Dunja Mladenic, Frank Höppner, Saso Dzeroski, Einoshin Suzuki, Nada Lavrac, Jean-François Boulicaut, Myra Spiliopoulou, Ruggero G. Pensa, Johannes Fürnkranz, Filip Zelezny
Parallel_Universes_and_Local_Patterns[+], morik_number[-]	Katharina Morik, Arno Siebes, Michael R. Berthold, Michael Wurst, David J. Hand, Bernd Wiswedel, Frank Höppner, Emmanuel Müller, Élisa Fromont, Claus Weihs, Niall M. Adams, Mirko Böttcher, Ralph Krieger, Bruno Crémilleux, Ira Assent, Marie-Odile Cordier, Thomas Seidl, Heike Trautmann, Rene Quiniou, Arnaud Soulet

the seminar *Parallel Universes and Local Patterns* that was also organized by Katharina and colleagues.

7   Conclusion

We have been using an algorithm that supports network analysis by finding regularities among vertex topological properties and attributes. It mines frequent topological patterns as up and down co-variations involving both attributes and topological properties of graph vertices. In addition, we defined two interestingness measures to capture the significance of a pattern with respect to either a given descriptor, or the relationship encoded by the graph edges. Furthermore, by identifying the top k representative vertices of a topological pattern, we support a better interaction with end-users. While [25] has given details about the whole methodology and has sketched several case studies, we decided in this chapter to analyze co-authorship network of our colleague Katharina Morik. We have shown that it supports the discovery of sensible patterns.

Acknowledgments. We thank Adriana Prado for her help. We also gratefully acknowledge support from the CNRS/IN2P3 Computing Center.

References

1. Akoglu, L., Tong, H., et al.: PICS: parameter-free identification of cohesive sub-groups in large graphs. In: SIAM DM, pp. 439–450 (2012)
2. Albert, R., Barabási, A.L.: Topology of complex networks: local events and universality. Phys. Rev. **85**, 5234–5237 (2000)
3. Biemann, C.: Chinese whispers: an efficient graph clustering algorithm and its application to natural language processing problems. In: Proceedings of the First Workshop on Graph Based Methods for Natural Language Processing, pp. 73–80. Association for Computational Linguistics (2006)
4. Brin, S., Page, L.: The anatomy of a large-scale hypertextual web search engine. Comput. Netw. **30**(1–7), 107–117 (1998)
5. Bringmann, B., Nijssen, S.: What is frequent in a single graph? In: Washio, T., Suzuki, E., Ting, K.M., Inokuchi, A. (eds.) PAKDD 2008. LNCS (LNAI), vol. 5012, pp. 858–863. Springer, Heidelberg (2008)
6. Calders, T., Goethals, B., Jaroszewicz, S.: Mining rank-correlated sets of numerical attributes. In: KDD, pp. 96–105 (2006)
7. Chakrabarti, D., Zhan, Y., Faloutsos, C.: R-MAT: a recursive model for graph mining. In: SIAM SDM (2004)
8. Cheng, H., Zhou, Y., Yu, J.X.: Clustering large attributed graphs. TKDD **5**(2), 12 (2011)
9. Do, T., Laurent, A., Termier, A.: Efficient parallel mining of closed frequent gradual itemsets. In: IEEE ICDM, pp. 138–147 (2010)
10. Dong, G., Li, J.: Efficient mining of emerging patterns: discovering trends and differences. In: KDD, pp. 43–52 (1999)
11. Freeman, L.C.: A set of measures of centrality based on betweenness. Sociometry **40**(1), 35–41 (1977)
12. Fürnkranz, J., Knobbe, A.J.: Guest editorial: global modeling using local patterns. DMKD **21**, 1–8 (2010)
13. Ge, R., Ester, M., Gao, B.J., et al.: Joint cluster analysis of attribute data and relationship data. TKDD **2**(2), 1–35 (2008)
14. Günnemann, S., et al.: Subspace clustering meets dense subgraph mining: a synthesis of two paradigms. In: IEEE ICDM, pp. 845–850 (2010)
15. Günnemann, S., et al.: A density-based approach for subspace clustering in graphs with feature vectors. In: PKDD, pp. 565–580 (2011)
16. Jiang, D., Pei, J.: Mining frequent cross-graph quasi-cliques. ACM TKDD **2**(4), 1–42 (2009)
17. Kang, U., Tsourakakis, C.E., Appel, A.P., Faloutsos, C., Leskovec, J.: Hadi: mining radii of large graphs. ACM TKDD **5**(2), 8 (2011)
18. Khan, A., Yan, X., Wu, K.L.: Towards proximity pattern mining in large graphs. In: SIGMOD, pp. 867–878 (2010)
19. Liao, Z.X., Peng, W.C.: Clustering spatial data with a geographic constraint. Knowl. Inf. Syst. **31**, 1–18 (2012)
20. Liu, G., Wong, L.: Effective pruning techniques for mining quasi-cliques. In: Daelemans, W., Goethals, B., Morik, K. (eds.) ECML PKDD 2008, Part II. LNCS (LNAI), vol. 5212, pp. 33–49. Springer, Heidelberg (2008)
21. Makino, K., Uno, T.: New algorithms for enumerating all maximal cliques. In: Hagerup, T., Katajainen, J. (eds.) SWAT 2004. LNCS, vol. 3111, pp. 260–272. Springer, Heidelberg (2004)

22. Morik, K., Boulicaut, J.-F., Siebes, A. (eds.): Local Pattern Detection. LNCS (LNAI), vol. 3539. Springer, Heidelberg (2005)
23. Moser, F., Colak, R., Rafiey, A., Ester, M.: Mining cohesive patterns from graphs with feature vectors. In: SIAM SDM, pp. 593–604 (2009)
24. Mougel, P.N., Rigotti, C., Gandrillon, O.: Finding collections of k-clique percolated components in attributed graphs. In: PAKDD (2012)
25. Prado, A., Plantevit, M., Robardet, C., Boulicaut, J.-F.: Mining graph topological patterns: finding covariations among vertex descriptors. IEEE Trans. Knowl. Data Eng. 25(9), 2090–2104 (2013)
26. Sese, J., Seki, M., Fukuzaki, M.: Mining networks with shared items. In: CIKM, pp. 1681–1684 (2010)
27. Silva, A., Meira, W., Zaki, M.: Structural correlation pattern mining for large graphs. In: Workshop on Mining and Learning with Graphs (2010)
28. Silva, A., Meira, W., Zaki, M.J.: Mining attribute-structure correlated patterns in large attributed graphs. PVLDB 5(5), 466–477 (2012)
29. Uno, T.: An efficient algorithm for solving pseudo clique enumeration problem. Algorithmica 56(1), 3–16 (2010)
30. Wang, D.J., Shi, X., McFarland, D.A., Leskovec, J.: Measurement error in network data: a re-classification. Soc. Netw. 34(4), 396–409 (2012)
31. Zaki, M.J.: Scalable algorithms for association mining. IEEE Trans. Knowl. Data Eng. 12(3), 372–390 (2000)
32. Zhou, Y., Cheng, H., Yu, J.: Graph clustering based on structural/attribute similarities. PVLDB 2(1), 718–729 (2009)

Advances in Exploratory Pattern Analytics on Ubiquitous Data and Social Media

Martin Atzmüller[✉]

Research Center for Information System Design, University of Kassel,
Wilhelmshöher Allee 73, 34121 Kassel, Germany
atzmueller@cs.uni-kassel.de

Abstract. Exploratory analysis of ubiquitous data and social media includes resources created by humans as well as those generated by sensor devices. This paper reviews recent advances concerning according approaches and methods, and provides additional review and discussion. Specifically, we focus on exploratory pattern analytics implemented using subgroup discovery and exceptional model mining methods, and put these into context. We summarize recent work on description-oriented community detection, spatio-semantic analysis using local exceptionality detection, and class association rule mining for activity recognition. Furthermore, we discuss results and implications.

1 Introduction

In ubiquitous and social environments, a variety of heterogenous data is generated, e.g., by sensors and social media, cf. [3]. For obtaining first insights into the data, description-oriented *exploratory* data mining approaches can then be applied.

Subgroup discovery [2,8,50,87,88] is such an exploratory approach for discovering interesting subgroups – as an instance of *local pattern detection* [52,67,68]. The interestingness is usually defined by a certain property of interesting formalized by a quality function. In the simplest case, a binary target variable is considered, where the share in a subgroup can be compared to the share in the dataset in order to detect (exceptional) deviations. More complex target concepts consider sets of target variables. In particular, *exceptional model mining* [8,55] focuses on more complex quality functions, considering complex *target models*, e.g., given by regression models or Bayesian networks with a deviating behavior for certain subgroups, cf. [37,38]. In the context of ubiquitous data and social media [3–5], interesting target concepts are given, e.g., by densely connected graph structures (communities) [17], exceptional spatio-semantic distributions [24], or class association rules [18].

This paper summarizes recent work on community detection, behavior characterization and spatio-temporal analysis using subgroup discovery and exceptional model mining. We start with the introduction of necessary foundational concepts in Sect. 2. After that, Sect. 3 provides a compact overview of recent

© Springer International Publishing Switzerland 2016
S. Michaelis et al. (Eds.): Morik Festschrift, LNAI 9580, pp. 184–207, 2016.
DOI: 10.1007/978-3-319-41706-6_9

scientific advances summarizing our recent work [4,6,7,17,21,24]. Furthermore, we describe exemplary results, and conclude with a discussion of implications and future directions in Sect. 4.

2   Background

Below, we first introduce some basic notation. After that, we provide a brief summary of basic concepts with respect to subgroup discovery

2.1   Basic Notation

Formally, a *database* $D = (I, A)$ is given by a set of individuals I and a set of attributes A. A *selector* or *basic pattern* $sel_{a_i=v_j}$ is a Boolean function $I \to \{0, 1\}$ that is true if the value of attribute $a_i \in A$ is equal to v_j for the respective individual. The set of all basic patterns is denoted by S.

For a numeric attribute a_{num} selectors $sel_{a_{num} \in [min_j; max_j]}$ can be defined analogously for each interval $[min_j; max_j]$ in the domain of a_{num}. The Boolean function is then set to true if the value of attribute a_{num} is within the respective range.

2.2   Patterns and Subgroups

Basic elements used in subgroup discovery are patterns and subgroups. Intuitively, a *pattern* describes a *subgroup*, i.e., the subgroup consists of instances that are covered by the respective pattern. It is easy to see, that a pattern describes a fixed set of instances (subgroup), while a subgroup can also be described by a set of patterns, if there are different options for covering the subgroup' instances. In the following, we define these concepts more formally.

Definition 1. *A* subgroup description *or (complex) pattern sd is given by a set of basic patterns* $sd = \{sel_1, \ldots, sel_l\}$, *where* $sel_i \in S$, *which is interpreted as a conjunction, i.e.,* $sd(I) = sel_1 \land \ldots \land sel_l$, *with* $length(sd) = l$.

Without loss of generality, we focus on a conjunctive pattern language using nominal attribute – value pairs as defined above in this paper; internal disjunctions can also be generated by appropriate attribute – value construction methods, if necessary. We call a pattern sd' a *superpattern* (or *refinement*) of a *subpattern* sd, iff $sd \subset sd'$.

Definition 2. *A* subgroup (extension)

$$sg_{sd} := ext(sd) := \{i \in I \,|\, sd(i) = true\}$$

is the set of all individuals which are covered by the pattern sd.

As search space for subgroup discovery the set of all possible patterns 2^S is used, that is, all combinations of the basic patterns contained in S. Then, appropriate efficient algorithms, e.g., [19,27,58] can be applied.

2.3   Interestingness of a Pattern

A large number of quality functions has been proposed in literature, cf. [41] for estimating the interestingness of a pattern – selected according to the analysis task.

Definition 3. *A quality function* $q \colon 2^S \to \mathbb{R}$ *maps every pattern in the search space to a real number that reflects the interestingness of a pattern (or the extension of the pattern, respectively).*

Many quality functions for a single target concept (e.g., binary [8,50] or numerical [8,56]), trade-off the size $n = |ext(sd)|$ of a subgroup and the deviation $t_{sd} - t_0$, where t_{sd} is the average value of a given target concept in the subgroup identified by the pattern sd and t_0 the average value of the target concept in the general population. In the binary case, the averages relate to the *share* of the target concept. Thus, typical quality functions are of the form

$$q_a(sd) = n^a \cdot (t_{sd} - t_0), \, a \in [0; 1]. \tag{1}$$

For binary target concepts, this includes, for example, the *weighted relative accuracy* for the size parameter $a = 1$ or a simplified binomial function, for $a = 0.5$. *Multi-target concepts*, e.g., [24,51,88] that define a target concept captured by a set of variables can be defined similarly, e.g., by extending an univariate statistical test to the multivariate case, e.g., [24]: Then, the multivariate distributions of a subgroup and the general population are compared in order to identify interesting (and exceptional) patterns.

While a quality function provides a *ranking* of the discovered subgroup patterns, often also a statistical assessment of the patterns is useful in data exploration. Quality functions that directly apply a statistical test, for example, the Chi-Square quality function, e.g., [8] provide a p-Value for simple interpretation. However, the Chi-Square quality function estimates deviations in two directions. An alternative, which can also be directly mapped to a p-Value is given by the *adjusted residual* quality function q_r, since the values of q_r follow a large standard normal distribution, cf. [1]:

$$q_r = n(p - p_0) \cdot \frac{1}{\sqrt{np_0(1 - p_0)(1 - \frac{n}{N})}} \tag{2}$$

The result of top-k subgroup discovery is the set of the k patterns $sd_1, \ldots, sd_k$, where $sd_i \in 2^S$ with the highest interestingness according to the applied quality function. A subgroup discovery task can now be specified by the 5-tuple: (D, c, S, q, k), where c indicates the target concept; the search space 2^S is defined by set of basic patterns S. In addition, we can consider constraints with respect to the *complexity* of the patterns. We can restrict the length l of the descriptions to a certain maximal value, e.g., with length $l = 1$ we only consider subgroup descriptions containing one selector, etc.

For several quality functions *optimistic estimates* [8,19,43,56] can be applied for determining upper quality bounds: Consider the search for the k best subgroups: If it can be proven, that no subset of the currently investigated hypothesis is interesting enough to be included in the result set of k subgroups, then we can skip the evaluation of any subsets of this hypothesis, but can still guarantee the optimality of the result. More formally, an optimistic estimate $oe(q)$ of a quality function q is a function such that $p \subseteq p' \to oe(q(p)) \geq q(p')$, i.e., such that no refinement p' of the pattern p can exceed the quality obtained by $oe(q(p))$.

3   Methods

With the rise of ubiquitous and mobile devices, social software and social media, a wealth of user-generated data is being created covering the according interactions in the respective systems an environments. In the following, we focus on social media and ubiquitous data: We adopt an intuitive definition of social media, regarding it as online systems and services in the ubiquitous web, which create and provide social data generated by human interaction and communication, cf. [4,7].

In this context, exploratory analytics provides the means to get insights into a number of exemplary analysis options, e.g., focusing on social behavior in mobile social networks. In the context of ubiquitous and social environments, exploratory data analysis is therefore a rather important approach, e.g., for getting first insights into the data: Here, subgroup discovery and exceptional model mining are prominent methods that can be configured and adapted to various analytical tasks. As outlined above, subgroup discovery [2,8,50,87,88] has been established as a general and broadly applicable technique for descriptive and exploratory data mining: It aims at identifying descriptions of subsets of a dataset that show an interesting behavior with respect to certain interestingness criteria, formalized by a quality function. Standard subgroup discovery approaches commonly focus on a *single* target concept as the property of interest [87], that can already be applied for common analytical questions like deviations of some parameters. Furthermore, since the quality function framework also enables *multi-target concepts*, e.g., [8,24,51,88] these enable even more powerful approaches for data analytics.

Figure 1 shows an overview on methods adapted and extended to the specific analytical tasks in the context of social media and ubiquitous data. Below, we discuss these in more detail, summarizing our recent work [17,18,24].

3.1   Description-Oriented Community Detection Using Subgroup Discovery

Important inherent structures in ubiquitous and social environments are given by communities, cf. [72,86]. Typically, these are seen as certain subsets of nodes of a graph with a dense structure. Classic community detection, e.g., [39] for a survey, just identifies subgroups of nodes with a dense structure, lacking an interpretable description. That is, no concise nor easily interpretable community description is provided.

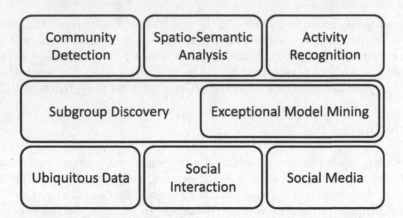

Fig. 1. Overview on the applied subgroup discovery and exceptional model mining approaches: We focus on the exploratory mining and analysis of social interaction in ubiquitous data and social media, tackling communities, human activities and behavior, and spatial-temporal characteristics, e.g., relating to events.

In [17], we focus on· *description-oriented community detection* using subgroup discovery. For providing both structurally valid and interpretable communities we utilize the graph structure as well as additional descriptive features of the graph's nodes. Using additional descriptive features of the nodes contained in the network, we approach the task of identifying communities as sets of nodes together with a *description*, i.e., a logical formula on the values of the nodes' descriptive features. Such a *community pattern* then provides an intuitive description of the community, e.g., by an easily interpretable conjunction of attribute-value pairs. Basically, we aim at identifying communities according to standard community quality measures, while providing characteristic descriptions at the same time.

As a simple example, we can consider a friendship graph common in online social systems. In the social bookmarking system BibSonomy[1] [35], for example, users can declare their friendship toward other users. This results in a directed graph, where nodes are denoted by users, and edges denote the friendship relations. Furthermore, in BibSonomy each user can tag resources like publications and web pages, i.e., assign a set of descriptive tags to certain resources. Then, the set of tags of a user can be considered as a description of that user's interests. Thus, the description-oriented community detection task in this context is to find user groups, where users are well connected given the friendship link structure, and also share a set of tags, as common features. Description-oriented community detection thus both needs to mine the graph-space and the description-space in an efficient way. In the following, we summarize the approach presented in [17], outlining the COMODO algorithm for fast description-oriented community detection, and present exemplary results.

---

[1] http://www.bibsonomy.org.

Overview. The COMODO algorithm for description-oriented community detection aims at discovering the top-k communities (described by community patterns) with respect to a number of standard community evaluation functions. The method is based on a generalized subgroup discovery approach [23,57] adapted to attributed graph data, and also tackles typical problems that are not addressed by standard approaches for community detection such as pathological cases like small community sizes.

In [17] the approach is demonstrated on data sets from three social systems namely, i.e., from the social bookmarking systems BibSonomy and delicious[2], and from the social media platform last.fm[3]. However, the presented approach is not limited to such systems and can be applied to any kind of graph-structured data for which additional descriptive features (node labels) are available, e.g., certain activity in telephone networks, interactions in face-to-face contacts [16], and according edge-attributed graphs.

Algorithm. COMODO is a fast branch-and-bound algorithm utilizing optimistic estimates [43,87] which are efficient to compute. This allows COMODO to prune the search space significantly, as we will see below.

As outlined above, COMODO utilizes both the graph structure, as well as descriptive information of the attributed graph, i.e., the label information of the nodes. This information is contained in two data structures: The graph structure is encoded in graph G while the attribute information is contained in database D describing the respective attribute values of each node. In a preprocessing step, we merge these data sources. Since the communities considered in our approach do not contain isolated nodes, we can describe them as sets of edges. We transform the data (of the given graph G and the database D containing the nodes' descriptive information) into a new data set focusing on the edges of the graph G: Each data record in the new data set represents an edge between two nodes. The attribute values of each such data record are the common attributes of the edge's two nodes. For a more detailed description, we refer to [17].

The FP-growth algorithm (cf. [44]) for mining association rules, the SD-Map* algorithm for fast exhaustive subgroup discovery [19], as well as quality functions operating on the graph structure form the basis of COMODO. COMODO utilizes an extended FP-tree structure, called the *community pattern tree* (CP-tree) to efficiently traverse the solution space. The tree is built in two scans of the graph data set and is then mined in a recursive divide-and-conquer manner, cf. [19,57]. The CP-tree contains the frequent nodes in a header table, and links to all occurrences of the frequent basic patterns in the tree structure.

The main algorithmic procedure of COMODO is shown in Algorithm 1. First, patterns containing only one basic pattern are mined. Then recursively, patterns conditioned on the occurrence of a (prefixed) complex pattern (as a set of basic patterns, chosen in the previous recursion step) are considered. For each following recursive step, a conditional CP-tree is constructed, given the

---

[2] http://www.delicious.com.

[3] http://last.fm.

Algorithm 1. COMODO

---

procedure COMODO-Mine (cf. [17] for an extended description)

Input: Current community pattern tree CPT, pattern $\hat{p}$, priority queue $top\text{-}k$, int k (max. number of patterns), int $maxLength$ (max. length of a pattern), int τ_n (min. community size)

1: $COM =$ new dictionary: $basicpattern \rightarrow pattern$
2: $minQ = minQuality(top\text{-}k)$
3: **for all** b in CPT.**getBasicPatterns do**
4:    $p = createRefinement(\hat{p}, b)$
5:    $COM[b] = p$
6:    **if** $size(p, CPT) \geq \tau_n$ **then**
7:      **if** $quality(p, F) \geq minQ$ **then**
8:        $addToQueue(top\text{-}k, p)$
9:        $minQ = minQuality(top\text{-}k)$
10: **if** $length(\hat{p}) + 1 < maxLength$ **then**
11:    $refinements = sortBasicPatternsByOptimisticEstimateDescending(COM)$
12:    **for all** b in $refinements$ **do**
13:      **if** $optimisticEstimate(COM[b]) \geq minQ$ **then**
14:        $CCPT = getConditionalCPT(b, CPT, minQ)$
15:        Call COMODO-Mine($CCPT$, $COM[b]$, $top\text{-}k$)

---

conditional pattern base of a frequent basic pattern (CP-node). The conditional pattern base consists of all the prefix paths of such a CP-node, i.e., all the paths from the root node to the CP-node.

Given the conditional pattern base, a (smaller) CP-tree is generated: the *conditional CP-tree* of the respective CP-Node. If the conditional CP-tree just consists of one path, then the community descriptions can be generated by considering all the combinations of the nodes contained in the path. Otherwise, the new tree is subjected to the next recursion step. We refer to [44] for more details on CP-trees and FP-growth.

As shown in the algorithm, we consider three options for pruning and sorting according to the current optimistic estimates:

1. **Sorting:** During the iteration on the currently active basic pattern queue when processing a (conditional) CP-tree, we can dynamically reorder the basic patterns that have not been evaluated so far by their optimistic estimate value. In this way, we evaluate the *more promising* basic patterns first. This heuristic can help to obtain and to propagate higher values for the pruning threshold early in the process, thus, helping to prune larger portions of the search space (line 11).
2. **Pruning:** If the optimistic estimate for the conditioning basic pattern is below the threshold given by the k best community pattern qualities (line 13), then we can omit a branch.
3. **Pruning:** When building a (conditional) community pattern tree, we can omit all the CP-nodes with an optimistic estimate below the mentioned quality threshold (line 14).

To efficiently compute the community evaluation functions together with their optimistic estimates COMODO stores additional information in the *community pattern nodes* (CP-nodes) of the CP-tree, depending on the used quality function. Each CP-node of the CP-tree captures information about the aggregated edge information concerning the database D and the respective graph. For each node, we store the following information:

- The basic pattern (selector) corresponding to the attribute value of the CP-node. This selector describes the community (given by a set of edges) covering the CP-node.
- The edge count of the (partial) community represented by the CP-node, i.e., the aggregated count of all edges that are accounted for by the CP-node and its basic pattern, respectively.
- The set of nodes that are connected by the set of edges of the CP-node, i.e., the nodes making up the respective subgroup.

Each edge data record also stores the contributing nodes and their degrees (in- and out-degree in the directed case). Then, as outlined in [17] we can compute standard quality functions efficiently, e.g., for the *Modularity* [71–73] or the *Segregation Index* [40].

Exemplary Evaluation Results. In our evaluation, we focused on two aspects: The efficiency of the proposed optimistic estimates, and the validity of the obtained community patterns. In order to evaluate the efficiency, we count the number of search steps, i.e., community allocations that are considered by the COMODO algorithm. We compared the total number of search steps (no optimistic estimate pruning) to optimistic estimate pruning using different community quality measures. Additionally, we measured the impact of using different minimal community size thresholds. Exemplary results are shown in Figs. 2 and 3 for the BibSonomy click graph and the delicious friend graph, for $k = 10, 20, 50$ and minimal size thresholds $\tau_n = 10, 20$. We consider a number of standard community quality functions: The *segregation index* [40], the *inverse average ODF (out degree fraction)* [59], and the *modularity* [71].

The large, exponential search space can be exemplified, e.g., for the click graph with a total of about $2 \cdot 10^{10}$ search steps for a minimal community size threshold $\tau_n = 10$. The results demonstrate the effectiveness of the proposed descriptive mining approach applying the presented optimistic estimates. The implemented pruning scheme makes the approach scalable for larger data sets, especially when the local modularity quality function is chosen to assess the communities' quality. Concerning the validity of the patterns, we focused on structural properties of the patterns and the subgraphs induced by the respective community patterns. We applied the significance test described in [53] for testing the statistical significance of the density of a discovered subgraph. Furthermore, we compared COMODO to three baseline community detection algorithms [42, 63,75], where COMODO shows a significantly better performance concerning validity and description length (for more details, we refer to [17]).

Overall, the results of the structural evaluations indicate statistically valid and significant results. Also, these show that COMODO does not exhibit the typical problems and pathological cases such as small community sizes that are

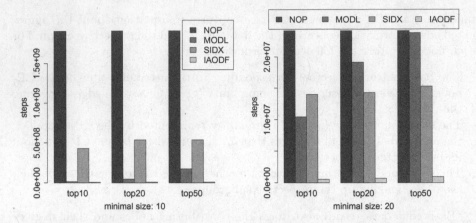

Fig. 2. Runtime performance of COMODO on BibSonomy click graph [17]: Search steps with no optimistic estimate pruning (*NOP*) vs. community quality functions with optimistic estimate pruning: MODL (Local Modularity), SIDX (Segregation Index) and IAODF (Inverse Average-ODF), for minimal size thresholds $\tau_n = 10, 20$.

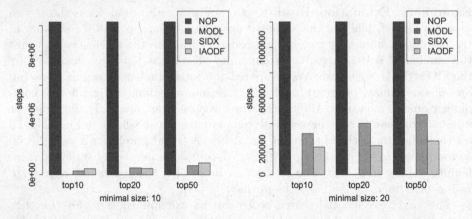

Fig. 3. Runtime performance of COMODO on the Delicious friend graph [17]: Search steps with no optimistic estimate pruning (*NOP*) vs. community quality functions with optimistic estimate pruning: MODL (Local Modularity), SIDX (Segregation Index) and IAODF (Inverse Average-ODF), for minimal size thresholds $\tau_n = 10, 20$.

often encountered when using typical community mining methods. Furthermore, COMODO is able to detect communities that are typically captured by shorter descriptions leading to a lower description complexity, compared to the baselines, cf. [17].

3.2   Exceptional Model Mining for Spatio-Semantic Analysis

Ubiquitous data mining has many facets including descriptive approaches: These can help for obtaining a first overview on a dataset, for summarization, for uncovering a set of interesting patterns, analyzing their inter-relations [6,24,65, 66], and refinement [9]. Exploratory analysis on ubiquitous data needs to handle different heterogenous and complex data types, e.g., considering a combination of a dataset containing attributive and context information about certain data points with spatial and/or temporal information, cf. [46,62,79,80]. Then, also semantic aspects concerning attributes, locations, and time can be considered.

In [6,24], we present an adaptation of subgroup discovery using exceptional model mining formalizations on ubiquitous data – focusing the on spatio-semantic analysis in [24]: We consider subgroup discovery and assessment approaches for obtaining interesting descriptive patterns, cf. [28,29,32]. The proposed exploratory approach enables to obtain first insights into the spatio-semantic space. In the context of an environmental application, the presented approach provides for the detailed inspection and analysis of objective and subjective data and according measurements. Below, we sketch the approach presented in [24] and summarize illustrating results.

Overview. The approach for exploratory subgroup analytics utilizes concepts of exceptional model mining in order to analyze complex target concepts on ubiquitous data. In particular, we focus on the interrelation between sensor measurements, subjective perceptions, and descriptive tags. Here, we propose a novel multi-target quality function for ranking the discovered subgroups, based on the Hotelling's T-squared test [45], see [24] for a detailed discussion.

Our application context is given by the *WideNoise Plus* smartphone application for measuring environmental noise. The individual data points include the measured noise in decibel (dB), associated subjective perceptions (feeling, disturbance, isolation, and artificiality) and a set of tags (free text) for providing an extended semantic context for the individual measurements. For the practical implementation, we utilize the VIKAMINE[4] tool [20] for subgroup discovery and analytics; it is complemented by methods of the R environment for statistical computing [77] in order to implement a semi-automatic pattern discovery process[5] based on automatic discovery and visual analysis methods.

Dataset – *WideNoise Plus*. For the analysis, we utilize real-world data from the EveryAware project[6], specifically, on collectively organized noise measurements collected using the *WideNoise Plus* application between December 14, 2011 and June 6, 2014. *WideNoise Plus* allows the collection of noise measurements using smartphones, including noise level (dB) measured using the microphone, location (latitude/longitude), as well as a timestamp when the measurement was taken. In addition, when taking a measurement the user can add subjective information about the context perceptions, encoded in the interval $[-5; 5]$ for *feeling*: [hate;love], *disturbance*: [hectic;calm], *isolation*: [alone;social], *artificiality*: [man-made;nature]. Furthermore, the user can assign tags to the

---

[4] http://vikamine.org.

[5] http://rsubgroup.org.

[6] http://www.everyaware.eu.

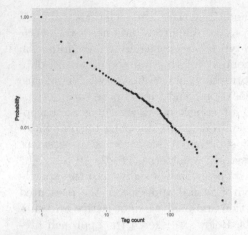

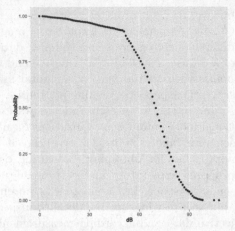

Fig. 4. Cumulated tag count distribution in the dataset. The y-axis provides the probability of observing a tag count larger than a certain threshold on the x-axis, cf. [24].

Fig. 5. Cumulated distribution of noise measurement (dB). The y-axis provides the probability for observing a measurement with a dB value larger than a certain threshold on the x-axis, cf. [24].

measurement for additional descriptive information, e.g., "noisy", "indoor", or "calm", providing the semantic context of the specific measurement. The data are stored and processed by the backend based on the UBICON platform [13,14].[7]

The applied dataset contains 6,600 data records and 2,009 distinct tags: The available tagging information was cleaned such that only tags with a length of at least three characters were considered. Only data records with valid tag assignments were included. Furthermore, we applied stemming and split multi-word tags into distinct single word tags.

Exemplary Analysis Results. In our experiments, we initially performed some basic statistical analysis of the observed distributions as well as experiments on correlating the subjective and objective data. Doing that, we observed typical phenomena in the domain of tagging data, while the correlations are expressed on a medium level. This directly motivated the development and application of the proposed advanced techniques using our subgroup analytics approach. This allows us to focus on the relation between objective and subjective data given patterns of tagging data in more detail.

Figures 4, 5, 6 and 7 provide basic statistics about the tag count and measured noise distributions, as well as the value distributions of the perceptions and the number of tags assigned to a measurement. Figure 5 shows the distribution of the collected dB values, with a mean of 67.42 dB.

---

[7] http://www.ubicon.eu.

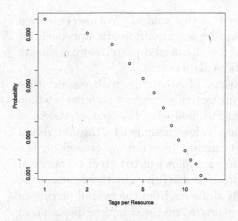

Fig. 6. Cumulated tag per record distribution in the dataset. The y-axis provides the probability of observing a tag per record count larger than a certain threshold on the x-axis, cf. [24].

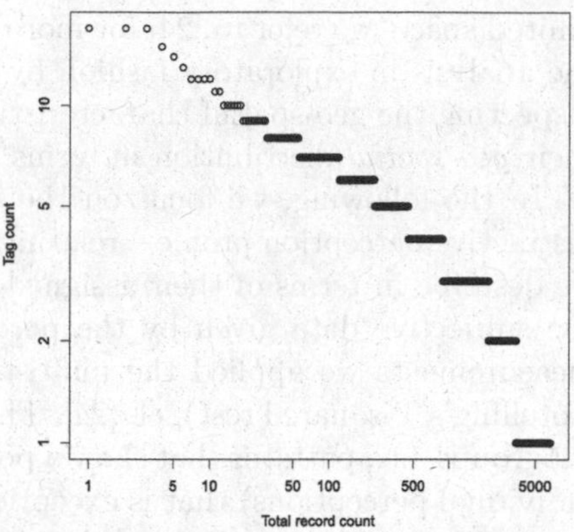

Fig. 7. Distribution of assigned tags per resource/data record, cf. [24].

In Fig. 6, we observe a typical heavy-tailed distributions of the tag assignments. Also, as can be observed in Figs. 4 and 7, the tag assignment data is rather sparse, especially concerning larger sets of assigned tags. However, it already allows to draw some conclusions on the tagging semantics and perceptions. In this context, the relation between (subjective) perceptions and (objective) noise measurements is of special interest. Table 1 shows the results of analyzing the correlation between the subjective and objective data. As shown in the table, we observe the expected trend that higher noise values correlate with the subjective "hate", "hectic" or "man-made" situations. While the individual correlation values demonstrate only medium correlations, they are nevertheless statistically significant.

Table 1. Correlation analysis between subjective (perceptions) and objective (dB) measurements; all values are statistically significant ($p < 0.01$).

	Feeling	Disturbance	Isolation	Artificiality
dB	−0.27	−0.32	−0.32	0.19

For a detailed analysis, we first focused on subgroup patterns for hot-spots of low or high noise levels, i.e., on patterns that are characteristic for areas with low or high noise. We were able to identify several characteristic tags for noisy environments, for example, *north AND runway*, *heathrow*, and *aeroplan*, which relate to Heathrow noise monitoring case study, cf. [13] for more details. For more quiet environments, we also observed typical patterns, e.g., focusing on the tags *park*, *forest*, *outdoor*, and *room*, and combinations of these. Due to the limited space, we refer to [24] for more details on this analysis. We also extended the analysis in exploratory fashion by providing a semi-automatic approach for inspecting the geo-spatial characteristics of the discovered patterns by assessing their *geo-spatial* distribution in terms of its *peakiness* [90].

In the following, we focus on the discovery of subgroups with respect to a distinctive perception profile – relating to subjective *perception patterns* – which we describe in terms of their assigned tags. For analyzing the characteristics of the subjective data given by the perception values assigned to the individual measurements we applied the multi-target quality function q_H (based on the Hotelling's T-squared test), cf. [24]. This function allows us to detect exceptional subgroups, i.e., patterns that show a perception profile (given by the means of the individual perceptions) that is exceptionally different from the overall picture of the perceptions (respectively, their means estimated on the complete dataset). In addition, we also analyzed, which patterns show a rather "conforming" behavior to the overall mean values. For that, we applied the quality function $q'_H = \frac{1}{q_H}$. Using the reciprocal of q_H we could then identify patterns for which their deviation was quite small, i.e., close to the general trend in the complete dataset. Table 2 presents the obtained results, where the rows 1–10 in the table denote deviating patterns (q_H), while rows 11–20 show conforming patterns.

For comparison, the overall means of the perceptions are given by: feeling $= -0.83$, disturbance $= -0.64$, isolation $= -0.19$, artificiality $= -2.33$. As we can observe in the table, the deviating patterns tend to correspond to more *noisy* patterns; the majority of the patterns shows a dB value above the mean in the complete dataset (67.42 dB). Furthermore, most of the patterns relate to the Heathrow case study, e.g., *north AND runway*, *plane AND south*; an interesting pattern is given by *plane AND runway AND garden* – people living close to Heathrow obviously tend to measure noise often in their garden. For the *conforming* patterns we mostly observe patterns with a mean dB close to the general mean. However, interestingly there are some patterns that show an increased mean and also "unexpected" patterns, e.g., *street AND traffic* or *airport*.

Overall, these results confirm the trends that we observed in the statistical analysis above indicating a medium correlation of the perceptions with the noise patterns. However, combinations of descriptive tags, and the contributions of individual perceptions is only provided using advanced techniques, like the proposed subgroup discovery approach using a complex multi-target concept for the detection of local exceptional patterns. While the initial statistical analysis of the perceptions provides some initial insights on subjective and objective data, again these results motivate our proposed approach as a flexible and powerful tool for the analysis of subgroups and their relations in this spatio-semantic context. Further steps then include appropriate visualization and introspection techniques, e.g., [2, 8, 25, 28].

Table 2. Exemplary perception patterns [24]: rows 1–10 show deviating patterns, while rows 11–20 show conforming patterns. Overall means (perceptions): feeling $= -0.83$, disturbance $= -0.64$, isolation $= -0.19$, artificiality $= -2.33$. The table shows the size of the subgroups, their quality according to the applied quality function, the mean of the measured dB values, and the means of the individual perceptions.

id	description	size	quality	mean dB	feeling	disturbance	isolation	artificiality
1	north AND runway	31	6223.79	80.32	-4.87	-4.97	-4.32	-4.97
2	heathrow	635	3609.66	69.71	-4.84	-4.79	-4.21	-4.90
3	aeroplan	550	3345.64	67.29	-4.79	-4.71	-4.70	-4.79
4	north	32	1813.34	79.59	-4.69	-4.69	-4.31	-4.97
5	esterno	548	1660.91	69.86	0.99	1.34	1.55	-1.89
6	plane AND runway AND garden	33	1237.88	79.45	-2.21	-2.27	1.09	-2.24
7	nois	648	1214.25	66.34	-4.39	-4.14	-4.20	-4.29
8	plane AND south	65	1186.62	79.54	-3.29	-3.12	-0.35	-3.29
9	voci	270	1138.21	71.80	0.93	1.32	2.10	-2.32
10	plane AND runway	91	999.63	79.96	-3.74	-3.66	-1.45	-3.77
11	park	26	0.72	66.69	-0.19	0.12	-0.81	-0.85
12	san	27	0.50	70.74	-0.15	-0.22	0.04	-1.37
13	lorenzo AND outdoor	22	0.29	70.77	0.00	-0.14	0.32	-1.27
14	street AND traffic	33	0.25	70.12	-1.55	-0.88	0.61	-3.45
15	univers	25	0.24	57.20	-0.32	0.32	0.88	-2.16
16	lorenzo	25	0.23	71.00	0.04	0.00	0.32	-1.16
17	land AND nois	20	0.20	75.80	-2.70	-1.15	0.10	-1.65
18	work	92	0.20	56.27	-0.40	0.23	-0.32	-1.67
19	room	25	0.19	50.52	1.08	1.36	-1.16	-1.96
20	airport	23	0.17	72.57	-0.04	-1.35	1.96	-3.26

3.3 Class Association Rule Mining Using Subgroup Discovery

With more and more ubiquitous devices, sensor data capturing human activities is becoming a universal data source for the analysis of human behavioral patterns. In particular, *activity recognition* has become a prominent research field with many successful methods for the classification of human activities. However, often the learned models are either "black-box" models such as neural networks, or are rather complex, e.g., in the case of random forests or large decision trees. In this context, we propose exploratory pattern analytics for constructing rule-based models in order to aid interpretation by humans, supported using appropriate quality and complexity measures [11,12].

Below, we summarize a novel approach for *class association rule mining* [60,61,84,89] presented in [18]. We propose an *adaptive framework* for mining such rules using *subgroup discovery*, and demonstrate the effectiveness of our approach using real-world activity data collected using mobile phone sensors. We summarize the proposed approach and algorithmic framework, before we provide exemplary results of an evaluation using real world activity data obtained by mobile phone sensors. The effectiveness of the approach is demonstrated by a comparison with typical *descriptive* models, i.e., using a rule-based (*Ripper* [36]) and a decision tree classifier (*C4.5* [76]) as a baseline.

Overview. Associative classification approaches integrate association rule mining and classification strategies. Basically, class association rules are special association rules with a fixed class attribute in the rule consequent. In order to mine such rules, we apply subgroup discovery. In the case of class association rules, the respective class can be defined as the target concept (i.e., the rule head) of the subgroups. Then, subgroup discovery can be adapted as a rule generator for class association rule mining.

In summary, in [18] we adapt subgroup discovery to class association rule mining, and embed it into an adaptive approach for obtaining a rule set that aims to target a simple rule base with an adequate level of predictive power, i.e., combining simplicity and accuracy. We utilize standard methods of rule selection and evaluation, that can be integrated into our framework: Liu et al. [61], for example, propose the *CBA* algorithm, which includes association rule mining and subsequent rule selection. It applies a covering strategy, selecting rules one by one, minimizing the total error. In addition to the rule mining and selection techniques, there are several strategies for the final decision of how to combine rules for classification ("voting" of the matching rules), e.g., [81].

Algorithmic Framework. For our adaptive framework, we distinguish the *learning phase* that constructs the model, and the *classification phase* that applies the model.

Model Construction. For the construction of the model, we apply the steps described in Algorithm 2. Basically, CARMA starts with discovering class association rules for each class c contained in the dataset. Using subgroup discovery, we collect a set of class association rules for the specific class, considering a maximal length of the concerned patterns. After that, we apply a boolean *ruleset assessment* function a in order to check, if the quality of the ruleset is good enough. If the outcome of this test is positive, we continue with the next class. Otherwise, we increase the maximal *length* of a rule (up to a certain user-definable threshold T_l). After the final set of all class association rules for all classes has been determined, we apply the *rule selection function* r in order to obtain a set of class association rules that optimizes predictive power on the trainingset. That is, the rule selection function aims to estimate classification error and should select the rules according to coverage and accuracy of the rules on the trainingset.

Classification. In the classification phase, we apply the rules contained in a model. For aggregating the predictions of the matching rules, we apply a specific *rule combination* strategy, cf. [81]. Examples include *unweighted voting* (majority vote according to the matching rules for the respective class), *weighted voting* (including weights for the matching rules), or *best rule* (classification according to the matching rule with the highest confidence).

Summary. In contrast to existing approaches, the CARMA framework is based on subgroup discovery for class association rule mining. This allows for selection

Algorithm 2. CARMA: Framework for Adaptive Class Association Rule Mining [18]

Input: Database D, set of classes C, parameter k specifying the cardinality of top-k pattern set, parameter T_l denoting the maximal possible length of a subgroup pattern, quality function q, ruleset assessment function a, rule selection function r.

1: Patterns $P = \emptyset$
2: **for all** $c \in C$ **do**
3:    Current length threshold $length = 1$
4:    **while** true **do**
5:       Obtain candidate patterns P^* by $SubgroupDiscovery(D, c, S, q, k)$
6:       **if** Current candidate patterns are good enough, i.e., $a(P^*) = true$ **then**
7:          $P = P \cup P^*$
8:          break
9:       **else if** $length > T_l$ **then**
10:          break
11:       **else**
12:          $length = length + 1$
13: Add a default pattern (rule) for the most frequent class to P
14: Apply rule selection function: $P = r(P)$
15: **return** P

---

of a suitable quality function for generating the rules, in contrast to (simple) confidence/support-based approaches. Then, e.g., significance criteria can be easily integrated. Furthermore, CARMA applies an adaptive strategy for balancing rule complexity (size) with predictive accuracy by applying a ruleset assessment function, in addition to the rule selection function. The framework itself does not enforce a specific strategy, but leaves this decision to a specific configuration. In our implementation in [18], for example, we follow the rule selection strategy of CBA; the ruleset assessment is done by a median-based ranking of the according confidences of the rules, i.e., estimated by the respective shares of the class contained in the subgroup covered by the respective rule. Here, we test if the median of the rules' confidences is above a certain threshold $\tau_c = 0.5$.

Exemplary Evaluation Results. In [18] we compared an instantiation of the CARMA framework against two baselines: The Ripper algorithm [36] as a rule-based learner, and the C4.5 algorithm [76] for learning decision trees. For the subgroup discovery step in the CARMA framework, we apply the BSD algorithm [58], utilizing the *adjusted residual* quality function, cf. Sect. 2, which directly maps to significance criteria. Furthermore, we apply an adaptation of the CBA algorithm [61] for the rule selection function, We opted for interpretable patterns with a maximal length of 7 conditions, and set the respective threshold $T_l = 7$ accordingly. In the evaluation, we used three different *TopK* values: 100, 200 and 500. For the rule combination strategy, we experimented with four strategies: taking the best rule according to confidence and Laplace value, the unweighted voting strategy, and the weighted voting (Laplace) method, cf. [18,81] for a detailed discussion. All experiments were performed using 10-fold cross-validation on an activity dataset with 27 activities (classes) and 116 features, cf. [18] for details.

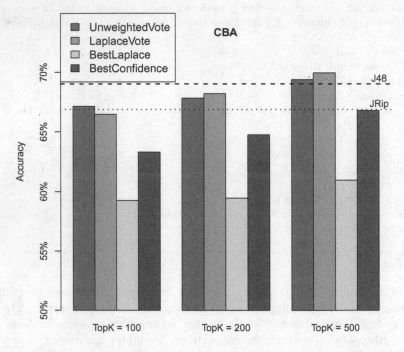

Fig. 8. Comparison of the accuracy of CARMA using the standard CBA method for rule selection, with different rule combination strategies to the baselines, cf. [18]. (Color figure online)

Figure 8 shows the accuracy of CARMA using these parametrizations. Overall, it is easy to see that the proposed approach is able to outperform the baselines in accuracy. Furthermore, it outperformed both as well in complexity, since it always had a significantly lower average complexity regarding the average number of conditions in a rule. For the baselines, C4.5 showed a better performance than Ripper, however, with a more complex model (1394 rules) that were also more complex themselves; Ripper had a slightly lower accuracy but a significantly lower number of rules and average rule length. The proposed CARMA approach outperforms both concerning the combination of accuracy and simplicity. Considering the voting functions, we observe that the functions (unweighted voting, and weighted Laplace) always outperforms the rest. In our experiments, using larger values of k indicates a higher accuracy – here also the complexity (in the number of rules) can be tuned. We observe a slight trade-off between accuracy and complexity. Basically, the parameter k seems to have an influence on the complexity, while the remaining instantiations do not seem to have a strong influence.

In summary, the proposed framework always provides a more compact model than the baseline algorithms. In our experiments, it is at least in the same range or even better than the baselines. Considering the best parameter instantiation, the proposed approach is able to outperform both baselines concerning the accuracy and always provides a more compact model concerning rule complexity, cf. [18] for more details.

4 Conclusions and Outlook

Subgroup discovery and exceptional model mining provide powerful and comprehensive methods for knowledge discovery and exploratory analysis. In this paper, we summarized recent advances concerning according approaches and methods in the context of ubiquitous data and social media. Specifically, we focused on exploratory pattern analytics implemented using subgroup discovery and exceptional model mining methods, summarizing recent work on description-oriented community detection, spatio-semantic analysis using local exceptionality detection, and class association rule mining using subgroup discovery. The methods were embedded into evaluations and case studies demonstrating their theoretical as well as practical impact and implications.

Interesting future directions include the adaptation and extension of knowledge-intensive approaches, e.g., [22, 26, 30, 31, 54, 69, 85]. This also concerns the incorporation of multiple relations, e.g., in the form of *partitioning knowledge* [10], or making use of multiplex and multi-modal networks [47, 64, 70, 78, 82], for modeling complex relations on ubiquitous data and social media and the analysis of emerging semantics [15, 65, 66]. Furthermore, the extended analysis of sequential data can be applied in both spatio-temporal dimensions [74], also concerning dynamics in the spatio-temporal space, e.g., for an extended temporal modeling of ubiquitous relations [48, 49]: Here, possible methods for extension and adaptation include temporal pattern mining for event detection [34], or temporal subgroup analytics [83], especially considering sophisticated exceptional model classes in that area. In addition, for including dynamics of spatial and temporal properties, for example, Markov chain approaches can be extended towards exceptional model mining, e.g., for modeling and analyzing sequential hypotheses and trails [33] in order to detect exceptional sequential transition patterns.

References

1. Agresti, A.: An Introduction to Categorical Data Analysis. Wiley-Blackwell, Hoboken (2007)
2. Atzmueller, M.: Knowledge-Intensive Subgroup Mining - Techniques for Automatic and Interactive Discovery, Dissertations in Artificial Intelligence-Infix (Diski), vol. 307. IOS Press, March 2007
3. Atzmueller, M.: Mining social media. Informatik Spektrum **35**(2), 132–135 (2012)
4. Atzmueller, M.: Mining social media: key players, sentiments, and communities. WIREs: Data Min. Knowl. Disc. **2**, 411–419 (2012)

5. Atzmueller, M.: Onto collective intelligence in social media: exemplary applications and perspectives. In: Proceedings of 3rd International Workshop on Modeling Social Media Hypertext (MSM 2012). ACM, New York (2012)

6. Atzmueller, M.: Data mining on social interaction networks. J. Data Min. Digital Humanit. **1**, June 2014

7. Atzmueller, M.: Social behavior in mobile social networks: characterizing links, roles and communities. In: Chin, A., Zhang, D. (eds.) Mobile Social Networking: An Innovative Approach. Computational Social Sciences, pp. 65–78. Springer, Heidelberg (2014)

8. Atzmueller, M.: Subgroup discovery - advanced review. WIREs: Data Min. Knowl. Disc. **5**(1), 35–49 (2015)

9. Atzmüller, M., Baumeister, J., Hemsing, A., Richter, E.-J., Puppe, F.: Subgroup mining for interactive knowledge refinement. In: Miksch, S., Hunter, J., Keravnou, E.T. (eds.) AIME 2005. LNCS (LNAI), vol. 3581, pp. 453–462. Springer, Heidelberg (2005)

10. Atzmueller, M., Baumeister, J., Puppe, F.: Evaluation of two strategies for case-based diagnosis handling multiple faults. In: Proceedings of 2nd Conference Professional Knowledge Management (WM 2003), Luzern, Switzerland (2003)

11. Atzmüller, M., Baumeister, J., Puppe, F.: Quality measures and semi-automatic mining of diagnostic rule bases. In: Seipel, D., Hanus, M., Geske, U., Bartenstein, O. (eds.) INAP/WLP 2004. LNCS (LNAI), vol. 3392, pp. 65–78. Springer, Heidelberg (2005)

12. Atzmueller, M., Baumeister, J., Puppe, F.: Semi-automatic learning of simple diagnostic scores utilizing complexity measures. Artif. Intell. Med. **37**(1), 19–30 (2006). Special Issue on Intelligent Data Analysis in Medicine

13. Atzmueller, M., Becker, M., Doerfel, S., Kibanov, M., Hotho, A., Macek, B.E., Mitzlaff, F., Mueller, J., Scholz, C., Stumme, G.: Ubicon: observing social and physical activities. In: Proceedings of IEEE International Conference on Cyber, Physical and Social Computing (CPSCom), pp. 317–324. IEEE Computer Society, Washington, D.C. (2012)

14. Atzmueller, M., Becker, M., Kibanov, M., Scholz, C., Doerfel, S., Hotho, A., Macek, B.E., Mitzlaff, F., Mueller, J., Stumme, G.: Ubicon and its applications for ubiquitous social computing. New Rev. Hypermedia Multimedia **20**(1), 53–77 (2014)

15. Atzmueller, M., Benz, D., Hotho, A., Stumme, G.: Towards mining semantic maturity in social bookmarking systems. In: Proceedings of Workshop on Social Data on the Web, 10th International Semantic Web Conference, ISWC, Bonn, Germany (2011)

16. Atzmueller, M., Doerfel, S., Hotho, A., Mitzlaff, F., Stumme, G.: Face-to-face contacts at a conference: dynamics of communities and roles. In: Atzmueller, M., Chin, A., Helic, D., Hotho, A. (eds.) MUSE 2011 and MSM 2011. LNCS, vol. 7472, pp. 21–39. Springer, Heidelberg (2012)

17. Atzmueller, M., Doerfel, S., Mitzlaff, F.: Description-oriented community detection using exhaustive subgroup discovery. Inf. Sci. **329**, 965–984 (2016)

18. Atzmueller, M., Kibanov, M., Hayat, N., Trojahn, M., Kroll, D.: Adaptive class association rule mining for human activity recognition. In: Proceedings of 6th International Workshop on Mining Ubiquitous and Social Environments (MUSE), ECML/PKDD, Porto, Portugal (2015)

19. Atzmueller, M., Lemmerich, F.: Fast subgroup discovery for continuous target concepts. In: Rauch, J., Raś, Z.W., Berka, P., Elomaa, T. (eds.) ISMIS 2009. LNCS, vol. 5722, pp. 35–44. Springer, Heidelberg (2009)

20. Atzmueller, M., Lemmerich, F.: VIKAMINE – open-source subgroup discovery, pattern mining, and analytics. In: Flach, P.A., Bie, T., Cristianini, N. (eds.) ECML PKDD 2012, Part II. LNCS, vol. 7524, pp. 842–845. Springer, Heidelberg (2012)

21. Atzmueller, M., Lemmerich, F.: Exploratory pattern mining on social media using geo-references and social tagging information. Int. J. Web Sci. **2**(1–2), 80–112 (2013)

22. Atzmueller, M., Lemmerich, F., Reutelshoefer, J., Puppe, F.: Wiki-enabled semantic data mining - task design, evaluation and refinement. In: Proceedings of 2nd International Workshop on Design, Evaluation and Refinement of Intelligent Systems (DERIS 2009), Krakow, Poland, vol. 545. CEUR-WS (2009)

23. Atzmueller, M., Mitzlaff, F.: Efficient descriptive community mining. In: Proceedings of 24th International FLAIRS Conference, pp. 459–464. AAAI Press, Palo Alto (2011)

24. Atzmueller, M., Mueller, J., Becker, M.: Exploratory subgroup analytics on ubiquitous data. In: Atzmueller, M., Chin, A., Scholz, C., Trattner, C. (eds.) MUSE/MSM 2013, LNAI 8940. LNCS (LNAI), vol. 8940, pp. 1–20. Springer, Heidelberg (2015)

25. Atzmueller, M., Puppe, F.: Semi-automatic visual subgroup mining using VIKAMINE. J. Univ. Comput. Sci. **11**(11), 1752–1765 (2005)

26. Atzmüller, M., Puppe, F.: A methodological view on knowledge-intensive subgroup discovery. In: Staab, S., Svátek, V. (eds.) EKAW 2006. LNCS (LNAI), vol. 4248, pp. 318–325. Springer, Heidelberg (2006)

27. Atzmüller, M., Puppe, F.: SD-Map – a fast algorithm for exhaustive subgroup discovery. In: Fürnkranz, J., Scheffer, T., Spiliopoulou, M. (eds.) PKDD 2006. LNCS (LNAI), vol. 4213, pp. 6–17. Springer, Heidelberg (2006)

28. Atzmueller, M., Puppe, F.: A case-based approach for characterization and analysis of subgroup patterns. J. Appl. Intell. **28**(3), 210–221 (2008)

29. Atzmueller, M., Puppe, F.: Semi-automatic refinement and assessment of subgroup patterns. In: Proceedings of 21st International Florida Artificial Intelligence Research Society Conference, pp. 518–523. AAAI Press, Palo Alto (2008)

30. Atzmueller, M., Puppe, F., Buscher, H.P.: Towards knowledge-intensive subgroup discovery. In: Proceedings of LWA 2004, Workshop KDML, Germany, pp. 117–123 (2004)

31. Atzmueller, M., Puppe, F., Buscher, H.P.: Exploiting background knowledge for knowledge-intensive subgroup discovery. In: Proceedings of 19th International Joint Conference on Artificial Intelligence (IJCAI), Edinburgh, Scotland, pp. 647–652 (2005)

32. Atzmueller, M., Puppe, F., Buscher, H.P.: Profiling examiners using intelligent subgroup mining. In: Proceedings of 10th International Workshop on Intelligent Data Analysis in Medicine and Pharmacology (IDAMAP-2005), Aberdeen, Scotland, pp. 46–51 (2005)

33. Atzmueller, M., Schmidt, A., Kibanov, M.: DASHTrails: an approach for modeling and analysis of distribution-adapted sequential hypotheses and trails. In: Proceedings of WWW 2016 (Companion). IW3C2/ACM (2016)

34. Batal, I., Fradkin, D., Harrison, J., Moerchen, F., Hauskrecht, M.: Mining recent temporal patterns for event detection in multivariate time series data. In: Proceedings of ACM SIGKDD International Conference on Knowledge Discovery and Data Mining, KDD 2012, pp. 280–288. ACM, New York (2012)

35. Benz, D., Hotho, A., Jäschke, R., Krause, B., Mitzlaff, F., Schmitz, C., Stumme, G.: The social bookmark and publication management system bibsonomy. VLDB **19**, 849–875 (2010)

36. Cohen, W.W.: Fast effective rule induction. In: Twelfth International Conference on Machine Learning, pp. 115–123. Morgan Kaufmann (1995)
37. Duivesteijn, W., Knobbe, A., Feelders, A., van Leeuwen, M.: Subgroup discovery meets bayesian networks - an exceptional model mining approach. In: Proceedings of International Conference on Data Mining (ICDM), pp. 158–167. IEEE, Washington, D.C. (2010)
38. Duivesteijn, W., Feelders, A., Knobbe, A.J.: Different slopes for different folks: mining for exceptional regression models with Cook's distance. In: Proceedings of ACM SIGKDD International Conference on Knowledge Discovery and Data Mining, pp. 868–876. ACM, New York (2012)
39. Fortunato, S.: Community detection in graphs. Phys. Rep. **486**(3–5), 75–174 (2010)
40. Freeman, L.: Segregation in social networks. Sociol. Methodol. Res. **6**(4), 411 (1978)
41. Geng, L., Hamilton, H.J.: Interestingness measures for data mining: a survey. ACM Comput. Surv. **38**(3), 1–32 (2006)
42. Gregory, S.: Finding overlapping communities in networks by label propagation. New J. Phys. **12**, 103018 (2010)
43. Grosskreutz, H., Rüping, S., Wrobel, S.: Tight optimistic estimates for fast subgroup discovery. In: Daelemans, W., Goethals, B., Morik, K. (eds.) ECML PKDD 2008, Part I. LNCS (LNAI), vol. 5211, pp. 440–456. Springer, Heidelberg (2008)
44. Han, J., Pei, J., Yin, Y.: Mining frequent patterns without candidate generation. In: Chen, W., Naughton, J., Bernstein, P.A. (eds.) Proceedings of SIGMOD, pp. 1–12. ACM Press, May 2000
45. Hotelling, H.: The generalization of student's ratio. Ann. Math. Statist. **2**(3), 360–378 (1931)
46. Hotho, A., Pedersen, R.U., Wurst, M.: Ubiquitous data. In: May, M., Saitta, L. (eds.) Ubiquitous Knowledge Discovery. LNCS, vol. 6202, pp. 61–74. Springer, Heidelberg (2010)
47. Kibanov, M., Atzmueller, M., Illig, J., Scholz, C., Barrat, A., Cattuto, C., Stumme, G.: Is web content a good proxy for real-life interaction? A case study considering online and offline interactions of computer scientists. In: Proceedings of IEEE/ACM International Conference on Advances in Social Networks Analysis and Mining (ASONAM). IEEE Press, Boston (2015)
48. Kibanov, M., Atzmueller, M., Scholz, C., Stumme, G.: On the evolution of contacts and communities in networks of face-to-face proximity. In: Proceedings of IEEE International Conference on Cyber, Physical and Social Computing (CPSCom). IEEE Computer Society, Boston (2013)
49. Kibanov, M., Atzmueller, M., Scholz, C., Stumme, G.: Temporal evolution of contacts and communities in networks of face-to-face human interactions. China Inf. Sci. **57**, 1–17 (2014)
50. Klösgen, W.: Explora: a multipattern and multistrategy discovery assistant. In: Fayyad, U.M., Piatetsky-Shapiro, G., Smyth, P., Uthurusamy, R. (eds.) Advances in Knowledge Discovery and Data Mining, pp. 249–271. AAAI Press (1996)
51. Klösgen, W.: Subgroup discovery. In: Handbook of Data Mining and Knowledge Discovery, chap. 16.3. Oxford University Press, New York (2002)
52. Knobbe, A.J., Cremilleux, B., Fürnkranz, J., Scholz, M.: From local patterns to global models: the LeGo approach to data mining. In: Proceedings of the ECML/PKDD-08 Workshop (LeGo-2008), pp. 1–16 (2008)
53. Koyuturk, M., Szpankowski, W., Grama, A.: Assessing significance of connectivity and conservation in protein interaction networks. J. Comput. Biol. **14**(6), 747–764 (2007)

54. Lavrač, N., Vavpetič, A., Soldatova, L., Trajkovski, I., Kralj Novak, P.: Using ontologies in semantic data mining with SEGS and g-SEGS. In: Proceedings of the 14th International Conference on Discovery Science (DS) (2011)

55. Leman, D., Feelders, A., Knobbe, A.J.: Exceptional model mining. In: Daelemans, W., Goethals, B., Morik, K. (eds.) ECML PKDD 2008, Part II. LNCS (LNAI), vol. 5212, pp. 1–16. Springer, Heidelberg (2008)

56. Lemmerich, F., Atzmueller, M., Puppe, F.: Fast exhaustive subgroup discovery with numerical target concepts. Data Min. Knowl. Disc. **30**(3), 711–762 (2015). http://dx.doi.org/10.1007/s10618-015-0436-8

57. Lemmerich, F., Becker, M., Atzmueller, M.: Generic pattern trees for exhaustive exceptional model mining. In: Flach, P.A., Bie, T., Cristianini, N. (eds.) ECML PKDD 2012, Part II. LNCS, vol. 7524, pp. 277–292. Springer, Heidelberg (2012)

58. Lemmerich, F., Rohlfs, M., Atzmueller, M.: Fast discovery of relevant subgroup patterns. In: Proceedings of International FLAIRS Conference, pp. 428–433. AAAI Press, Palo Alto (2010)

59. Leskovec, J., Lang, K.J., Dasgupta, A., Mahoney, M.W.: Community structure in large networks: natural cluster sizes and the absence of large well-defined clusters. CoRR abs/0810.1355 (2008)

60. Li, W., Han, J., Pei, J.: CMAR: accurate and efficient classification based on multiple class-association rules. In: Cercone, N., Lin, T.Y., Wu, X. (eds.) Proceedings of International Conference on Data Mining (ICDM), pp. 369–376. IEEE Computer Society (2001)

61. Liu, B., Hsu, W., Ma, Y.: Integrating classification and association rule mining. In: Proceedings of ACM SIGKDD International Conference on Knowledge Discovery and Data Mining, pp. 80–86. AAAI Press, August 1998

62. May, M., Berendt, B., Cornuéjols, A., Gama, J., Giannotti, F., Hotho, A., Malerba, D., Menesalvas, E., Morik, K., Pedersen, R., et al.: Research challenges in ubiquitous knowledge discovery. In: Next Generation of Data Mining, pp. 131–150 (2008)

63. McDaid, A., Hurley, N.: Detecting highly overlapping communities with model-based overlapping seed expansion. In: Proceedings of the 2010 International Conference on Advances in Social Networks Analysis and Mining, ASONAM 2010, pp. 112–119. IEEE Computer Society, Washington, D.C. (2010)

64. Mitzlaff, F., Atzmueller, M., Benz, D., Hotho, A., Stumme, G.: Community assessment using evidence networks. In: Atzmueller, M., Hotho, A., Strohmaier, M., Chin, A. (eds.) MUSE/MSM 2010. LNCS, vol. 6904, pp. 79–98. Springer, Heidelberg (2011)

65. Mitzlaff, F., Atzmueller, M., Hotho, A., Stumme, G.: The social distributional hypothesis. J. Soc. Netw. Anal. Min. **4**(216), 1–14 (2014)

66. Mitzlaff, F., Atzmueller, M., Stumme, G., Hotho, A.: Semantics of user interaction in social media. In: Ghoshal, G., Poncela-Casasnovas, J., Tolksdorf, R. (eds.) Complex Networks IV. SCI, vol. 476, pp. 13–25. Springer, Heidelberg (2013)

67. Morik, K.: Detecting interesting instances. In: Hand, D.J., Adams, N.M., Bolton, R.J. (eds.) Pattern Detection and Discovery. LNCS (LNAI), vol. 2447, pp. 13–23. Springer, Heidelberg (2002)

68. Morik, K., Boulicaut, J.-F., Siebes, A. (eds.): Local Pattern Detection. LNCS (LNAI), vol. 3539, pp. 115–134. Springer, Heidelberg (2005)

69. Morik, K., Potamias, G., Moustakis, V., Charissis, G.: Knowledgeable learning using MOBAL: a medical case study. Appl. Artif. Intell. **8**(4), 579–592 (1994)

70. Mucha, P.J., Richardson, T., Macon, K., Porter, M.A., Onnela, J.P.: Community structure in time-dependent, multiscale, and multiplex networks. Science **328**(5980), 876–878 (2010)

71. Newman, M.E., Girvan, M.: Finding and evaluating community structure in networks. Phys. Rev. E Stat. Nonlin. Soft Matter Phys. **69**(2), 1–15 (2004)
72. Newman, M.E.J.: Detecting community structure in networks. Eur. Phys. J. **38**, 321–330 (2004)
73. Newman, M.E.J.: Modularity and community structure in networks. Proc. Nat. Acad. Sci. U.S.A. **103**(23), 8577–8582 (2006)
74. Piatkowski, N., Lee, S., Morik, K.: Spatio-temporal random fields: compressible representation and distributed estimation. Mach. Learn. **93**(1), 115–139 (2013)
75. Pool, S., Bonchi, F., van Leeuwen, M.: Description-driven community detection. Trans. Intell. Syst. Technol. **5**(2), 1–21 (2014)
76. Quinlan, R.: C4.5: Programs for Machine Learning. Morgan Kaufmann Publishers, San Mateo (1993)
77. R Development Core Team: R: A Language and Environment for Statistical Computing. R Foundation for Statistical Computing, Vienna (2009). ISBN: 3-900051-07-0. http://www.R-project.org
78. Scholz, C., Atzmueller, M., Barrat, A., Cattuto, C., Stumme, G.: New insights and methods for predicting face-to-face contacts. In: Kiciman, E., Ellison, N.B., Hogan, B., Resnick, P., Soboroff, I. (eds.) Proceedings of International AAAI Conference on Weblogs and Social Media. AAAI Press, Palo Alto (2013)
79. Scholz, C., Atzmueller, M., Stumme, G.: Unsupervised and hybrid approaches for on-line RFID localization with mixed context knowledge. In: Andreasen, T., Christiansen, H., Cubero, J.-C., Raś, Z.W. (eds.) ISMIS 2014. LNCS, vol. 8502, pp. 244–253. Springer, Heidelberg (2014)
80. Scholz, C., Doerfel, S., Atzmueller, M., Hotho, A., Stumme, G.: Resource-aware on-line RFID localization using proximity data. In: Gunopulos, D., Hofmann, T., Malerba, D., Vazirgiannis, M. (eds.) ECML PKDD 2011, Part III. LNCS, vol. 6913, pp. 129–144. Springer, Heidelberg (2011)
81. Sulzmann, J.N., Fnkranz, J.: A comparison of techniques for selecting and combining class association rules. In: Baumeister, J., Atzmueller, M. (eds.) Proceedings of LWA, Technical report, Department of Computer Science, University of Wzburg, Germany, vol. 448, pp. 87–93 (2008)
82. Symeonidis, P., Perentis, C.: Link prediction in multi-modal social networks. In: Calders, T., Esposito, F., Hüllermeier, E., Meo, R. (eds.) ECML PKDD 2014, Part III. LNCS, vol. 8726, pp. 147–162. Springer, Heidelberg (2014)
83. Sez, C., Rodrigues, P., Gama, J., Robles, M., Garcmez, J.: Probabilistic change detection and visualization methods for the assessment of temporal stability in biomedical data quality. Data Min. Knowl. Disc. **29**, 950–975 (2014)
84. Thabtah, F.: A review of associative classification mining. Knowl. Eng. Rev. **22**(1), 37–65 (2007)
85. Vavpetič, A., Lavrač, N.: Semantic subgroup discovery systems and workflows in the SDM-toolkit. Comput. J. **56**(3), 304–320 (2013)
86. Wasserman, S., Faust, K.: Social Network Analysis: Methods and Applications. Structural Analysis in the Social Sciences, 1st edn. Cambridge University Press, Cambridge (1994)
87. Wrobel, S.: An algorithm for multi-relational discovery of subgroups. In: Komorowski, J., Żytkow, J.M. (eds.) PKDD 1997. LNCS, vol. 1263, pp. 78–87. Springer, Heidelberg (1997)
88. Wrobel, S., Morik, K., Joachims, T.: Maschinelles Lernen und Data Mining. Handbuch der Künstlichen Intelligenz **3**, 517–597 (2000)

89. Yin, X., Han, J.: CPAR: classification based on predictive association rules. In: Barbar, D., Kamath, C. (eds.) Proceedings of SIAM International Conference on Data Mining (SDM), pp. 331–335. SIAM (2003)
90. Zhang, H., Korayem, M., You, E., Crandall, D.J.: Beyond co-occurrence: discovering and visualizing tag relationships from geo-spatial and temporal similarities. In: Proceedings of International Conference on Web Search and Data Mining (WSDM), pp. 33–42. ACM, New York (2012)

Understanding Human Mobility with Big Data

Fosca Giannotti[1], Lorenzo Gabrielli[1], Dino Pedreschi[2],
and Salvatore Rinzivillo[1]([✉])

[1] KDDLab, ISTI - CNR, Pisa, Italy
{fosca.giannotti,lorenzo.gabrielli,salvatore.rinzivillo}@isti.cnr.it
[2] KDDLab, Dipartimento di Informatica, Università di Pisa, Pisa, Italy
pedre@di.unipi.it
http://kdd.isti.cnr.it

Abstract. The paper illustrates basic methods of mobility data mining, designed to extract from the big mobility data the patterns of collective movement behavior, i.e., discover the subgroups of travelers characterized by a common purpose, profiles of individual movement activity, i.e., characterize the routine mobility of each traveler. We illustrate a number of concrete case studies where mobility data mining is put at work to create powerful analytical services for policy makers, businesses, public administrations, and individual citizens.

Keywords: Mobility data mining · Big data analytics

1 Introduction

The large availability of location aware services allows the collection of huge repositories of movement data. These new sources of data give an unprecedented opportunity to have a social microscope of individual collective and global behaviours. Here we focus on mobility data, such as the call detail records from mobile phones and the GPS tracks from car navigation devices, which represent society-wide proxies of human mobile activities. These big mobility data help us understand human mobility, and discover the hidden patterns and profiles that characterize the trajectories we follow during our daily activity. The paper illustrates the basic methods of mobility data mining, designed to extract from the big mobility data the patterns of collective movement behavior, i.e., discover the subgroups of travelers characterized by a common purpose, profiles of individual movement activity, typical path followed by many travellers. These methods are the basic breaks to support analytical questions such as:

- What are the most popular itineraries followed from the origin to the destination of people's travels? What routes, what timing, what volume for each such itinerary?
- How do people leave the city toward suburban areas (or vice-versa)? What is the spatio-temporal distribution of such trips?

© Springer International Publishing Switzerland 2016
S. Michaelis et al. (Eds.): Morik Festschrift, LNAI 9580, pp. 208–220, 2016.
DOI: 10.1007/978-3-319-41706-6_10

- How to understand the accessibility to key mobility attractors, such as large facilities, railway stations or airports? How do people behave when approaching an attractor?
- Are there geographic borders that emerge from the way people use the territory for their daily activities? If so, how do we define such borders? Are these borders matching the administrative ones?

This paper shortly illustrates the system Urban Mobility Atlas, that visually synthesizes the complex analytical processes in a toolset of measures for various mobility dimensions of a geographical area. We focus on the challenge of constructing novel mobility indicators from Big Data, capable of capturing the mobility vocation of a territory: what is the relationship between systematic and non systematic behavior? Is a territory amenable for adopting a new mobility behavior such as car-pooling or for massive diffusion of electric vehicles? In the following we will consider a big dataset of GPS traces of private vehicles circulating in central Italy, in the region of Tuscany. The owners of these cars are subscribers of a *pay-as-you-drive* car insurance contract, under which the tracked trajectories of each vehicle are periodically sent (through the GSM network) to a central server for anti-fraud and anti-theft purposes. This dataset has been donated for research purposes by *Octo Telematics Italia S.r.l* (oct), the leader for this sector in Europe. The whole dataset describes about 150,000 cars tracked during a month (May, 2012) in Tuscany.

2 Mobility Data Analysis

The collection of different sources of data is the very first step of an analytical process that involves several transformations of data to gather useful and novel knowledge from it. In particular, for mobility data it is necessary, at first, to process the raw observations of each position at a given time into a trajectory, i.e. an higher level object representing the movement of an individual from an origin to a destination.

This process is very complex and depends mainly on the spatio-temporal granularity of the raw data. In [3] it is settled as the ground for a new research field, namely Mobility Data Mining (MDM). The majority of methods of MDM are centred around the concept of trajectory. In [2] the authors present the advantages of combining different MDM algorithms and methods to derive more complex analytical processes that can be easily deployed as stand-alone services.

The Origin Destination matrix model provides a very compact representation of traffic demands, by abstracting the single trajectories to flows between any two regions. The large availability of sensed tracks enables us to automatically extract OD matrices, by dividing the territory into a partition of cells and by counting the movements from each origin to each destination. A visual interface allows the analyst to browse and select interesting flows. The combination of the OD matrix model with clustering methods, for example, may enable an analyst to understand which are the most popular itineraries followed by vehicles entering (or exiting) to a given territory. The extraction of OD matrix models strictly depends on the spatio-temporal granularity of the movement data. When using a data source with a variable sampling rate, like for example GSM data, it is necessary to plan reconstruction strategies for missing data. In [4] the authors use data mining methods and frequency analysis to derive the OD matrix from a large collection of GSM data. In this approach, the systematic movements are first identified within the raw data and the OD matrix is derived from this subset. The outcoming OD model is then fed to a traffic simulator for traffic assignment to the road network.

Using the OD Matrix Exploration we may select relevant flows that may be further investigated. For example, we can focus on the set of trajectories leaving the center of the city and moving towards North-East (Fig. 1 (top)). Despite all these trips originating in the city center and ending in the NE suburbs, a broad diversity is still evident, as we can notice from the temporal distributions of two main clusters in Fig. 3. In order to understand which are the most popular itineraries followed by the selected travels, we apply an algorithm that automatically detects significant groups of similar trips.

In particular we use the density-based clustering algorithm with the Spatial Route distance function. The clustering algorithm produces a set of clusters, each of which can be visualized by means of a thematic rendering where the trajectories in the same cluster are drawn with the same color. Figure 2 (bottom) shows how the most popular clusters highlight the main routes used by drivers to leave the center towards NE. The frequent behaviors identified by the clustering processing may be also analyzed by the temporal dimension. For example, in Fig. 3, the temporal distribution of the trips within the two major clusters, along a typical day is presented. We can use such statistics to interpret the semantic of each group of trips.

From the analysis of collective movements we are also able to identify the actual borders that emerge from daily behaviours of people [6]. The aim of discovering borders at a meso-scale is motivated by providing decision-support tools for policy makers, capable of suggesting optimal administrative borders for the government of the territory. This analytical process is based on techniques developed for complex network analysis. Figure 4 shows the resulting borders by analyzing the movements of around 40k vehicles for one month in the center-west part of Tuscany. From the figure it is possible to recognize the relevant socio-cultural district of the region. For comparison, we also draw the current administrative borders of each municipality.

Shifting from the collective behaviors to the individuals, we can investigate the daily routines of each person. The daily mobility of each user can be essentially summarized by a set of single trips that the user performs during the day. When trying to extract a mobility profile of users [7] our interest is in the trips that are part of their habits, therefore neglecting occasional variations that divert from their typical behavior. Therefore in order to identify the individual mobility profiles of users from their GPS traces, the following steps will be performed - see Fig. 5: *(1)* divide the whole history of the user into trips (Fig. 5(a)); *(2)* group

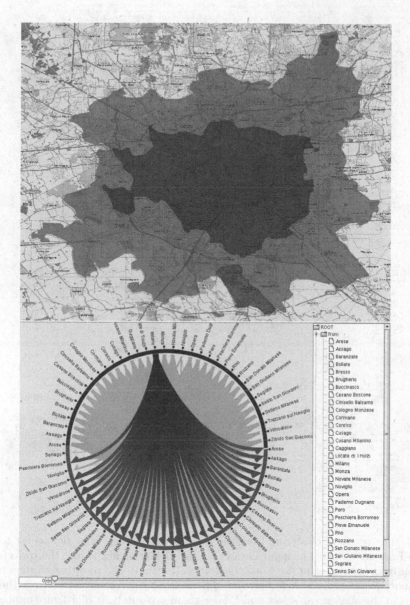

Fig. 1. An example of Matrix model exploration with GPS data. (top) Spatial partition of the territory: the highlighted cell is linked with the selection on the visual explorer. (bottom) The visual explorer to browse the OD Matrix: each region is represented with a node, nodes are displayed in a circular layout. The arc connecting two nodes represents the flow, i.e., the number of trips from the origin to the destination node; the arc width is proportional to the flow.

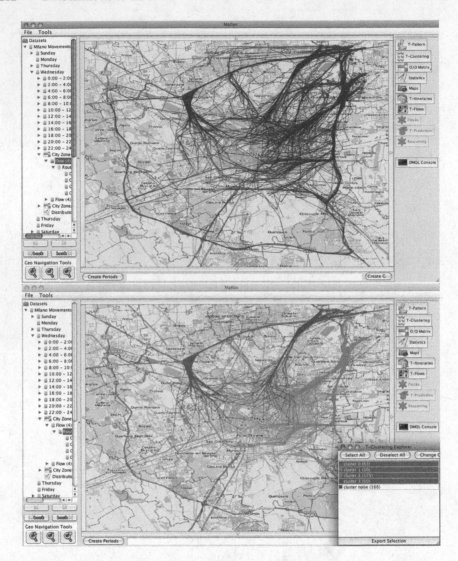

Fig. 2. The result of clustering from the trajectories moving from the center to the North-East area. (top) The input dataset for the clustering algorithm: the trajectories moving from the center to the North-East area. (bottom) The resulting clusters using the Route Similarity distance function. The clusters are visualized using a themed color, and the analyst can select and browse them separately. (Color figure online)

trips that are similar, discarding the outliers (Fig. 5(b)); *(3)* from each group, extract a set of representative trips, to be used as mobility profiles (Fig. 5(c)).

The concept of individual profiles is also exploited in [1] where the authors reconstruct the calling habits of customers of a mobile operator to determine a class for each individual among a set of predefined ones: resident, visitors,

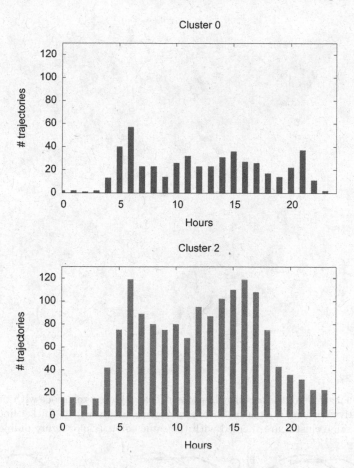

Fig. 3. Temporal distribution of the trajectories in the clusters of Fig. 2 (bottom). Cluster 0 (top) do not exhibit significant peaks, while cluster 2 (bottom) has a peak in the morning and one in the afternoon. The temporal profile of Cluster 2 captures two commuting behaviors: a group leaving the city in the morning (commuters going to work outside), and a larger group leaving the city in the late afternoon (commuters coming back home in the suburbs after work.)

commuters. In this case, the relations between the individual and the collectivity are exploited.

3   Human Mobility Indicators for the Urban Mobility Atlas

Starting from the analytical processes presented in the previous section, it is interesting to derive quantitative estimations of the main characteristics that distinguish the mobility of a territory. We call such estimators *mobility indicators* and, for each territory, we compute and aggregate several measurement to

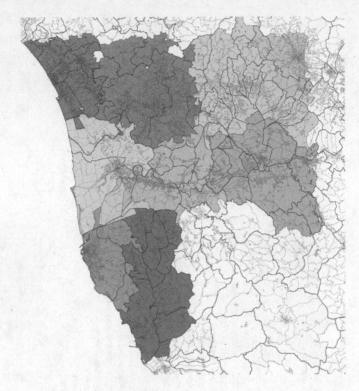

Fig. 4. Visualization of the mobility borders in Tuscany. As a reference with the existing administrative borders, the perimeter of each town is drawn with a thicker line. Regions within the same cluster are themed with the same color. (Color figure online)

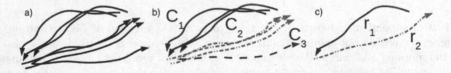

Fig. 5. Mobility profile extraction process: (a) trip identification; (b) group detection/outlier removal; (c) selection of representative mobility profiles

have an overview of the whole mobility. As a first step, it is crucial to identify the mobility analysis that may be relevant for our objectives. In particular, we describe here two complex mobility indicators that are based on individual characteristics of the observed population: the *systematicity index* measures the tendency of a population of following frequent routines to move among places; the *radius of gyration* measures the tendency of a population of exploring places far away from their home.

Both these measures are associated to an individual: we need to extend them from a single user to a group of users. We can imagine the mobility of a user as composed of several locations she has visited during her movements. Some of these locations are more relevant to the individual, for example since she spend the majority of her time in a few places. By analysing the individual history, we can determine the *Most Frequent Location* (MFL) and we may assume this as the *home location*. Given the MFL of a user u, we may univocally associate u to the area where her MFL is located. This association can be exploited to define a derived estimator for an area A as an aggregation of the measures of all users whose MFL is located in A.

To determine the systematicity of a user, we exploit the mobility profile extraction method [7]. This method allows us to separate systematic movements from the infrequent ones. The individual systematicity index of a user u is given by the ratio of the systematic trips over the non-systematic ones. This index is extended to a territory A by averaging the individual indexes of each individual linked to A.

Moreover, by means of cluster analysis, it is possible to aggregate individual profiles of users to determine common routes to access a territory. We call such routes *access patterns*. The access patterns can be exploited to reason about the shared strategies of users to access a destination. In combination with the systematicity index, it is possible to give a qualitative measure of such routes. We have designed a visual widget to represent the combination of access patterns for a specific municipality. Figure 6 shows a compact representation of access patterns arriving to Florence: linestrings represent the individual mobility profiles, similar profiles are rendered with the same color, and for each group we have a pie-chart representing the ratio of systematic mobility over the whole movements.

From the comparison of the two plots in Fig. 6 we can grasp easily the differences between two different cities: on the top we have a large metropolitan area capable of attracting mobility for services and work; on the bottom we have a small town whose mobility is mainly influenced by the neighbouring municipalities. Besides the differences in volume, we can notice which access patterns are used for systematic movements. In Florence the ratio of systematic trips is very high for every access path, in Montepulciano, on the contrary, the differences among the incoming routes. This result may highlight the relationship and exchange of commuters with neighbouring cities.

At a glance, it is possible to notice the main routes of systematic movements arriving from North. On the contrary, the trajectories coming from South are equally populated by systematic and non-systematic trajectories.

The concept of Radius of Gyration gives a measure of the spreadness of visited locations of a user over the geography. It is based on the concept of *center of mass* r_{cm} of a user, defined as a two-dimensional vector representing the weighted mean point of the locations visited by an individual. We can measure the mass associated to a location with its visitation frequency or the time spent in the location, obtaining the following definition:

$$r_{cm} = \frac{1}{N} \sum_{i \in L} n_i r_i \tag{1}$$

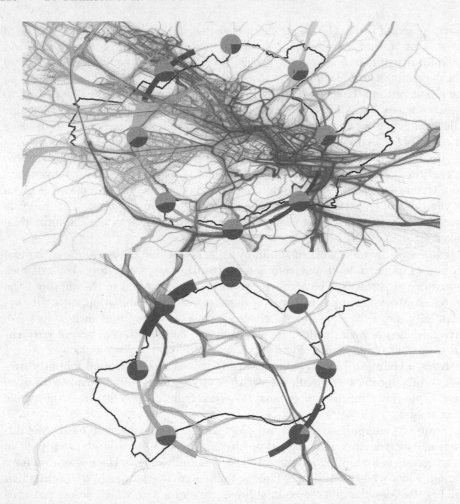

Fig. 6. Incoming traffic to Florence (top) and Montepulciano (bottom). Movements are aggregated by similarity of access point to the city. Each route is annotated with the *Systematicity Indicator*, representing the ratio between systematic and regular movements. (Color figure online)

where L is the set of locations visited by the user, r_i is a two-dimensional vector describing the geographic coordinates of location i; n_i is the visitation frequency or the total time spent; and N is the sum over all the locations n_i (i.e. total number of visits or time spent). The radius of gyration rg of a user characterizes how spread out the visited locations are from the center of mass. In mathematical terms, it is defined as the root mean square distance of the locations from the center of mass:

$$r_g = \sqrt{\frac{1}{N} \sum_{i \in L} n_i (r_i - r_{cm})^2} \tag{2}$$

where r_i and r_{cm} are the vector of coordinates of location i and center of mass respectively.

The radius of gyration provides us with a measure of mobility volume, indicating the typical distance traveled by an individual and providing an estimation of her tendency to move.

By analysing the distribution of the values of r_g in a city, we can understand which are the characteristic movements of a city. In Fig. 7 (left) we may notice a skew distribution of r_g in the observed population in Florence.[1] The majority of people has a very limited value of radius, mainly within the city boundaries. However, there are still a consistent part of people travelling long distances.

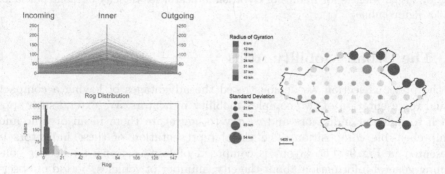

Fig. 7. Visual widgets for Radius of Gyration indicator for the city of Florence. (left) distribution of the r_g on the resident population. (right) Spatial distribution of mean and standard deviation of r_g per cell. (Color figure online)

Since each individual is linked to her MFL, we design also a spatial distribution of the r_g. In particular, we consider a spatial partition of the territory (we have used a regular grid in Fig. 7 (right)) and for each cell we aggregate the corresponding values of r_g: the mean value is coded with a color scale (linked also to the distribution histogram); the standard deviation is represented as the circle radius. From the widget in Fig. 7 (right) we can notice how the internal areas of the city have both a low mean value of r_g and also a low standard deviation. On the boundary, instead, we can notice higher values of r_g and also a very high variance.

If we consider another city, like for example Montepulciano in Fig. 8, we can highlight the different behaviours in mobility of the respective residents. In this case, people tend to travel longer distances, as it emerges from the r_g distribution. This may be justified with the need of finding services that are not directly accessible in a small town.

---

[1] An high resolution version of the graphics is available online at http://kdd.isti.cnr. it/uma.

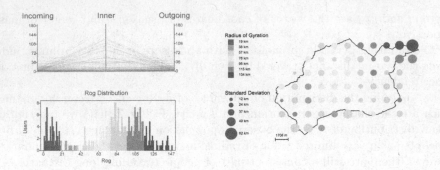

Fig. 8. Visual widgets for Radius of Gyration indicator for the city of Montepulciano. (Color figure online)

4   The Urban Mobility Atlas

In the previous section we demonstrated the advantages of having a compact visual representation of two complex mobility indicators. We extend such approach to different indicators and we try to aggregate them in an organic and comprehensible composition. The visual representation of these indicators is presented in Fig. 9. The layout is composed of four sections: an header containing general information about the city (number of vehicles, period of observation, etc.); a series of mobility statistics (distribution of length of trajectories,

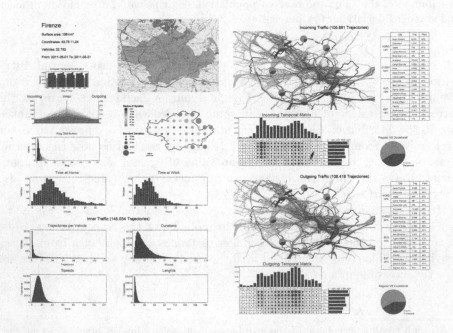

Fig. 9. Urban Mobility Atlas for the city of Florence. (Color figure online)

distribution of duration, etc.); distribution of individual mobility characteristics (radius of gyration, time spent per location,etc.); incoming traffic statistics; outgoing traffic statistics. The Urban Mobility Atlas can be created on a territory given the availability of a large dataset of vehicular trajectory. We created an instance of the Urban Mobility Atlas in Tuscany, exploiting the GPS dataset provided by *OCTO Telematics* covering around the 5 % of the circulating vehicles. An interesting question is to assess if the dimension òf such a dataset, as well as the spatial coverage that it shows, are enough to support the accuracy and representativeness of the overall vehicular mobility; several studies on the literature [5] have tackled this problem by building an extremely accurate predictor that, given GPS data observation, estimates the real traffic values as measured by road sensors (ground truth). The user can navigate through a visual interface, where she can browse each city on a map.

5   Conclusions

This paper introduces the *Urban Mobility Atlas*, a visual summary of mobility indicators, having the objective of synthesizing the mobility of a city. The proposed visualization is based on a set of mobility data mining processes, linked to specific territories, i.e. city in our case. The indicators are computed at the individual level, by estimating relevant locations visited by each vehicle and the complete set of movements observed. The set of individual indicators is then linked to the territory and aggregated by the chosen spatial partition. The system has been implemented and deployed on the set of municipalities in Tuscany. The current version of the system is accessible at the URL http://kdd.isti.cnr.it/uma with a static visual interface. We plan to extend the approach by transforming the analytical processes into a set of API (Application Programming Interface) in order to enable external developer to ensemble new configuration or new application.

References

1. Furletti, B., Gabrielli, L., Renso, C., Rinzivillo, S.: Identifying users profiles from mobile calls habits. In: Proceedings of the ACM SIGKDD International Workshop on Urban Computing, UrbComp 2012, pp. 17–24. ACM, New York (2012)
2. Giannotti, F., Nanni, M., Pedreschi, D., Pinelli, F., Renso, C., Rinzivillo, S., Trasarti, R.: Unveiling the complexity of human mobility by querying and mining massive trajectory data. VLDB J. **20**(5), 695–719 (2011)
3. Giannotti, F., Pedreschi, D. (eds.): Mobility, Data Mining and Privacy - Geographic Knowledge Discovery. Springer, Heidelberg (2008)
4. Nanni, M., Trasarti, R., Furletti, B., Gabrielli, L., Mede, P.V.D., Bruijn, J.D., de Romph, E., Bruil, G.: MP4-A project: mobility planning for Africa. In: In D4D Challenge @ 3rd Conference on the Analysis of Mobile Phone datasets (NetMob 2013), Cambridge, USA (2013)
5. Pappalardo, L., Rinzivillo, S., Qu, Z., Pedreschi, D., Giannotti, F.: Understanding the patterns of car travel. Eur. Phys. J. Spec. Top. **215**(1), 61–73 (2013)

6. Rinzivillo, S., Mainardi, S., Pezzoni, F., Coscia, M., Pedreschi, D., Giannotti, F.: Discovering the geographical borders of human mobility. KI - Künstliche Intelligenz **26**(3), 253–260 (2012)
7. Trasarti, R., Pinelli, F., Nanni, M., Giannotti, F.: Mining mobility user profiles for car pooling. In: Apté, C., Ghosh, J., Smyth, P. (eds.) Proceedings of the 17th ACM SIGKDD International Conference on Knowledge Discovery and Data Mining, San Diego, CA, USA, 21–24 August 2011, pp. 1190–1198. ACM (2011)

On Event Detection from Spatial Time Series for Urban Traffic Applications

Gustavo Souto[1] and Thomas Liebig[2(✉)]

[1] Fraunhofer Institute for Software and Systems Engineering, Fraunhofer ISST,
Dortmund, Germany
gustavo.souto@isst.fraunhofer.de
[2] Artificial Intelligence Group, TU Dortmund University, Dortmund, Germany
thomas.liebig@tu-dortmund.de

Abstract. Since the last decades the availability and granularity of location-based data has been rapidly growing. Besides the proliferation of smartphones and location-based social networks, also crowdsourcing and voluntary geographic data led to highly granular mobility data, maps and street networks. In result, location-aware, smart environments are created. The trend for personal self-optimization and monitoring named by the term 'quantified self' will speed-up this ongoing process. The citizens in conjunction with their surrounding smart infrastructure turn into 'living sensors' that monitor all aspects of urban living (traffic load, noise, energy consumption, safety and many others). The "Big Data"-based intelligent environments and smart cities require algorithms that process these massive amounts of spatio-temporal data. This article provides a survey on event processing in spatio-temporal data streams with a special focus on urban traffic.

1 Introduction

Early detection of anomalies in spatio-temporal data streams provides many applications for smart cities and is a major research topic since the availability and granularity of location-based data has been rapidly growing in the last decades.

Besides, the proliferation of smartphones and location-based social networks, also crowdsourcing and voluntary geographic data led to highly granular mobility data, maps and street networks. In result, location-aware, smart environments are created. The trend for personal self-optimization and monitoring named by the term 'quantified self' will speed-up this ongoing process. The citizens in conjunction with their surrounding smart infrastructure turn into 'living sensors' that monitor all aspects of urban living (traffic load, noise, energy consumption, safety and many others).

The "Big Data"-based intelligent environments and smart cities require algorithms that process these massive amounts of spatio-temporal data in real-time. But key challenges for streaming analysis are (1) one-pass processing (2) limited amount of memory and (3) limited time to process [6].

© Springer International Publishing Switzerland 2016
S. Michaelis et al. (Eds.): Morik Festschrift, LNAI 9580, pp. 221–233, 2016.
DOI: 10.1007/978-3-319-41706-6_11

Spatio-temporal data comes in a variety of forms and representations, depending on the domain, the observed phenomenon, and the observation method. In principle, there are three types of spatio-temporal data streams [19]: *spatial time series*, *events*, and *trajectories*.

- A *spatial time series* consists of tuples $(attribute, object, time, location)$.
- An *event* of a particular type $event_i$ is triggered from a spatial time series under certain conditions and contains the tuples verifying these conditions $(event_i, object_n, time_n, location_n)$.
- A *trajectory* is a spatial time series for a particular $object_i$. It contains the location per time and is a series of tuples $(object_i, time_n, location_n)$.

The increasing availability of massive heterogeneous streaming data for public organizations, governments and companies pushes their inclusion in incident recognition systems. Leveraging insights from these data streams offers a more detailed and real-time picture of traffic, communication, or social networks, to name a few, which still is a key challenge for early response and disaster management. Detecting events in spatio-temporal data is a widely investigated research area (see e.g. [1] for an overview). Depending on the application, the event detection can analyze single trajectories (e.g. of persons or vehicles), group movements, spatio-temporal measurements, or heterogeneous data streams. Following examples highlight capabilities of these approaches:

- *Individual Mobility:* Within airports (or other security region) it is valuable to monitor whether individuals enter some restricted area. The analysis of stops or of sudden decelerations allows detection of unusual behaviour. Sequences of such events can be matched against predefined mobility patterns [12], e.g. to identify commuters.
- *Group Movement:* During public events the early detection of hazardous pedestrian densities gains much attention. The patterns one could distinguish and detect in group movement are *encounter*, *flock* or *leadership* pattern [10].
- *Spatio Temporal Measurements:* A spatio-temporal value spans a whole region. This could be traffic flow, air pollution, noise, etc. The sudden rise or decline of these values indicates an anomaly.
- *Heterogeneous Data Streams:* The combination of previously described types of anomalies provides event filters in an urban environment based on heterogeneous data (e.g. GPS data of pedestrians, traffic loop data, mobile phone network data).

In the paper at-hand we provide a introductory survey on (1) functions on heterogeneous spatio-temporal data streams, Sect. 2, (2) pattern matching, Sect. 3, (3) anomaly detection in spatio-temporal time series, Sect. 4, and (4) streaming frameworks, Sect. 5. All four aspects are relevant for implementing real-world event detection systems that process heterogeneous data streams.

2 Function Classes on Heterogeneous Spatio-Temporal Time Series

In general functions for event detection from heterogeneous data streams can be classified using a former concept of raster-geography, namely *map-algebra* [5]. Both, raster geography and heterogeneous spatio-temporal data analysis consider data which is provided in multiple layers (i.e. one layer per data stream). Functions can be applied to one or multiple layers. Thus, spatial functions split into four groups: *local, focal, zonal* and *global* ones [5], illustrated in Fig. 1.

- *Local functions* operate on every single cell in a layer. And the cell is processed without reference to surrounding cells. An example is a map transformation, the multiplication with a constant, or the comparison with a threshold.
- *Focal functions* process cell data depending on the values of neighboring cells. The neighborhood can be defined by arbitrary shapes. Example functions are moving averages and nearest neighbor methods.
- *Zonal functions* process cells on the base of zones, these are cells that hold a common characteristic. Zonal functions allow the combination of heterogeneous data streams in various layers by application of functions to one layer if another layer already fulfills another condition.
- *Global functions* process the entire data. Examples are distance based operations.

For heterogeneous data streams analysis, expressiveness of these four function types is important to *derive* low-level events (incidents), to *combine* low-level events (e.g. aggregation, clustering, prediction etc.) and to *trigger* high-level events.

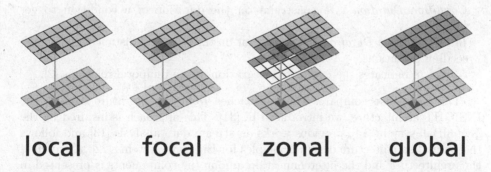

local   focal   zonal   global

Fig. 1. Function classes on Spatio-Temporal data, Dark blue highlights the currently processed location. Light blue cells indicate the regions whose values are used for computation. Best viewed in color. (Color figure online)

3   Event Pattern Matching

The exploitation of spatio-temporal event patterns is a major research field in mobility mining. Event pattern matching focuses on the task to match sequences of events against event patterns and to trigger another event (which is raised for further analysis) in case the sequence matches. Recently, pattern-graphs were introduced in [27], their pattern description is capable to express the temporal relations among various occurring events following the interval-calculus [2]. As an example the co-occurrence of two low-level events may trigger any high-level event. With spatio-temporal data streams also spatial relations are important to consider. The region connection calculus [28] lists relations of spatial events that are essential for a spatio-temporal pattern matcher.

Possible frameworks for event pattern matchers are the event calculus [32], finite automaton [12] and other pattern matcher [8,27] or even complex frameworks which allow application of local, focal, zonal and global functions e.g. [14,30]. The requirements for spatio-temporal pattern matcher in a smart city scenario are:

- to operate in real time,
- to incorporate spatial [28] and temporal [2] relations
- to provide local, focal, zonal, and global [5] predicates on the attributes, and
- to pose arbitrary queries formed of these elements (regular language [23], Kleene closure [18]).

In Table 1 we compare the features of state-of-the-art event detection frameworks. The temporal expressiveness is split into the following four categories:

- *Pattern Duration* is a constraint on the temporal distance of first and last condition in a pattern.
- *Condition Duration* is a constraint on the duration of a condition to get matched.
- *Inter-Condition Duration* is a constraint on the temporal distance among succeeding conditions.
- *Complete* indicates the complete integration of the temporal relations [2].

The table also compares the approaches from the literature against the INSIGHT architecture, we introduced in [31]. This approach is inspired by the TechniBall system [14], previous works on stream data analysis [13] and follows the Lambda architecture design principles for Big Data systems [22]. A sketch of the architecture and the interconnection among the components is presented in Fig. 2. Every data stream is analysed individually for anomalies. In this detection functions (e.g. clustering, prediction, thresholds, etc.) on the data streams can be applied. The resulting anomalies are joined at a round table. A final Complex Event Processing component allows the formulation of complex regular expressions on the function values derived from heterogeneous data streams.

Table 1. Comparison of Spatio-Temporal Event Detection Frameworks

Approach	Time Algebra [2]				Spatial Algebra [28]	Regular Expressions	Spatial Functions [5]				Stream Processing
	complete	condition duration	inter-condition duration	pattern duration			local	focal	zonal	global	
Mobility Pattern [23]	-	✓	-	-	-	✓	-	-	-	-	-
SASE [17]	-	✓	-	✓	-	✓	✓	✓	✓	✓	✓
SASE+ [9]	-	✓	-	-	-	✓	✓	✓	✓	✓	✓
Cayuga [8]	-	-	-	✓	-	✓	✓	✓	✓	✓	✓
Spatio-Temporal Pattern Queries [30]	✓	✓	✓	✓	✓	-	✓	✓	✓	✓	-
Mobility Pattern Stream Matching [12]	-	✓	-	✓	-	✓	-	-	-	-	✓
Event calculus [3,32]	✓	✓	-	✓	-	✓	✓	✓	✓	✓	✓
Temporal Pattern Graphs [27]	✓	✓	✓	✓	-	✓	✓	-	-	-	✓
INSIGHT architecture [31]	✓	✓	✓	✓	✓	✓	✓	✓	✓	✓	✓

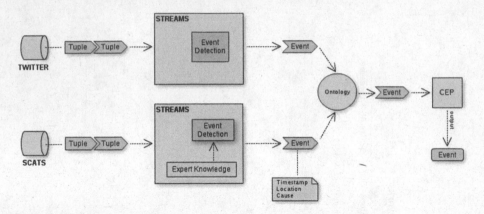

Fig. 2. INSIGHT architecture for event detection from heterogeneous data streams exemplified with two input streams Twitter and traffic loop data derived by SCATS, compare [4,31].

4   Anomaly Detection on Spatial Time Series

This section discusses state-of-the-art of anomaly detection in traffic condition data streams as this paper focuses on smart cities and traffic is a major aspect of a smart city. However, some techniques generalize also to other spatio-temporal phenomena as noise, pollution, etc. For a comprehensive survey on outlier detection from spatio-temporal data streams we point the reader to [16].

4.1   Statistical Approach

Pang et al. proposed an approach [25] which extends the Likelihood Ratio Test (LRT) framework to detect abnormal traffic patterns in taxi trajectory data (GPS trajectories). The approach partitions the road network of Beijing into a spatial grid, regions (R), to deal better with the problem of finding abnormal patterns. The extended LRT uses statistical models which are Persistent Spatiotemporal Model (PSTO) and Emerging Spatiotemporal Outlier Model (ESTO) to compute the likelihood of "anomalousness" of a region and detect the emerging spatio-temporal outliers, respectively. In addition, the proposed statical model works with the Maximum Likelihood Estimation (MLE) and Upper-bounding strategy to estimate the parameters of models and prune the non-outliers, respectively. However, this approach does not use other source of data (e.g. weather, list of events in the city, social network) to reduce the uncertainty of detected events, as well as it does not present a good ratio of adaptability to face natural changes in the data stream over time.

In [34], Yang et al. present a non-parametric Bayesian method, or Bayesian Robust Principal Component Analysis (RPCA) - BRPCA, to detect traffic events on a road. This method takes the traffic observations as one dimensional data and converts it into a matrix format which in turn decomposes it into a superposition of low-rank, sparse, and noise matrices. In addition, this method proposed an extended BRPCA to deal with multiple variables/time series/data streams. The idea of that extended BRPCA is to improve the traffic detection by sharing a sparsity structure among multiple data streams affected by the same events. Such an approach uses multiple homogeneous data streams and a static weather data source in the detection process.

In [26], although the major goal of this work is not detect outlier itself, the authors propose a novel adaptive Artificial Neural Network (ANN) based filter to detect and remove them to build a training data. The ANN filter uses the training set (i.e., usually the 3 months of historical data - information from street loops) as incoming and thus analyzes whether the readings are twice the maximum value, if it holds true, then the method marks it as anomaly, otherwise removed.

In [36], the authors propose an approach to estimate the traffic which uses mobile probes to detect outliers in Handover Data of a suburban freeway. The approach detects anomalies in 2 steps. The first step applies Least Squares Support Vector Machine (LS-SVM) ensemble classifier to identify whether each individual handover link is an outlier or not, and the second step employs a statistical-based algorithm which evaluates whether the detected outlier holds any locally handover link which is anomalous as well.

Trilles et al. [33] propose a variation of the CUmulative SUM (CUSUM) algorithm to detect anomalies in data streams near to real-time. This approach is only applied when the observations are in-control, that is, the data is normally distributed. In the anomaly detection process the CUSUM is obtained by computing $S_i = S_{i-1} \cdot z_i$, where z_i is a standard normal variable which is computed as follows $z_i = \frac{x_i - \bar{x}}{s}$, where the s is the standard deviation of the time series, and x_i is the i-th data point of the time series. The events are detected by the Eq. 1, if S_{H_i} exceeds a predefined threshold (CUSUM control charts) $\hat{A} \pm h\sigma_x$ ($h = 5$ and σ_x is the standard deviation), then it is an *Up-Event* due to its increase and if S_{L_i} is greater than threshold (CUSUM control charts) $\hat{A} \pm h\sigma_x$ ($h = 5$ and σ_x is the standard deviation), then it is an *Down-Event* due to its decrease. The variable k is a slack-variable and denotes the reference value which is usually set to be one half of the mean. The advantages of this work are the application of a simple approach for Real-Time anomaly detection and the dashboard application to visualize the detected events. However, the work does not present experiments with a data source which has high refresh rate such as SCATS data stream.

$$S_{H_i} = MAX[0, (z_i - k) + S_{H_i} - 1]$$
$$S_{L_i} = MIN[0, (z_i - k) + S_{L_i} - 1]$$

(1)

4.2 Human/Driver's Behavior

Pan [24] proposes a new method to detect disruptions in typical traffic patterns (traffic anomalies) using crowd-sourcing and social media. This approach detects anomalies according to drivers' routing behaviour instead of traffic volume-based and speed on roads. In addition, it provides a view of congested road segments and their relationships among these segments. It also provides to the end-user a detour router to avoid or escape the congestions. This method also makes use of a historical tweets associated with the spatial region to represent the normal occurrences of each region. In order to retrieve only the relevant contents, this approach applies a simple filtering technique which compares the frequency of current tweets with historical tweets and apply a weight to each term according to its frequency, as well as the location and time information.

4.3 Unsupervised

Yang [35] investigates the problem of outlier detection on large-scale collective behaviors. His work extracts features from high-dimensional data streams using K-Nearest Neighbors (KNN) method to detect the anomalies. This method performs the anomaly detection in 3 phases as follows: (1) observations from multiple sensors, this phase organizes more than 400 sensors as high-dimensional time series; (2) manifold learning, it applies Locally Linear Embedding (LLE) computes and Principal Component Analysis (PCA) to obtain a feature at a higher abstraction level; and (3) outlier detection, this phase performs the outlier detection through the K-Nearest Neighbors. The approach works good since special days, or holidays, which might generate an abnormal flow are known in advance. For instance, New Year and Independence Day. However, from this characteristics, it indicates that the method cannot handle historical data as well as adapt itself according to the changes.

Guo et al. [15] propose a traffic flow outlier detection approach which focuses on the pattern changing detection problem to detect anomalies in traffic conditional data streams. The traffic data comes from inductive loop sensors of four regions in United State and United Kingdom as well as this works makes use of a short-term traffic condition forecasting system to evaluate the proposed approach. This approach performs the analysis of the incoming data point after the data point be processed by Integrated Moving Average filter (IMA) which captures the seasonal effect on the level of traffic conditional series, and then Kalman filter picks up the local effect flow levels after IMA, and GARCH filter models and predict time-varying conditional variance of the traffic flow process. These filters constitute together the integrated forecast system aforementioned.

4.4 Tree Approach

Liu et al. [21] present an approach based on features analysis to detect outliers points as well as trees which detects the relationship among anomalies in traffic data stream. This work uses taxi trajectory data (GPS trajectories) on the road network of Beijing. The approach presents a model with 3 main steps which processes the traffic data to build a region graph. The 3 main steps are (1) Building a region graph, (2) Detect outliers from graph edges, and (3) Discover relations among outliers (building a tree). Then, this method partitions the map of traffic into regions by employing Connected Components Labeling.

Each region holds a link to other region and a link is anomalous whether its features have the largest difference from both their temporal and spatial neighbors, and the STOTree algorithm captures the causal relationship among outliers. Although this work presents an interesting work about correlated anomalies in traffic data streams, the work does not provide experiments under an online setting for the traffic anomaly detection. Instead, it describes a set of algorithms which could be applied in such a setting.

4.5   Discussion

Although these works present some substantial advances in the field of anomaly detection in data streams, the field is still in its early stage, and therewith it is possible to see that such works hold some drawbacks as well as open tasks. Examples of open tasks are incorporate heterogeneous data streams, keep tracking of historical data (local and global), apply adaptive data stream models, use expert knowledge, develop straightforward and lightweight approaches for data stream analysis. These open tasks aim to improve the anomaly detection in data streams (in general), that is, decreasing the uncertainty whether the detected event is a true anomaly.

The use of heterogeneous data streams improves detection of anomalies by reducing the uncertainty about the events veracity, this issue has been little exploited in the traffic conditions domain. Outlier detectors should take into account external factors (e.g., weather and social events), such an issue has been exploited more than heterogeneous data streams, but their applications only refer to sources which provide static information, or general information from online forecast sources (e.g., wind speed, amount of rainfall, humidity), instead of precise information about what is happening around the city by using local sensors (e.g., flood in a specific region of a city). The works [4,31] use heterogeneous data streams to detect anomalies in a smart city, but it is still some open questions which need answers such as *"How to merge the flow of heterogeneous data streams to obtain a good result?"* and *"How to join the result of the analysis of each flow to detect the true anomaly detection?"*.

Adaptive classification models react to the natural changes of data stream. The change of the target variable value in which the model is trying to predict is well-known as *Concept Drift*. A model which adapts itself over time holds more chances to find a true anomaly than another model without such characteristic. Therefore, this feature is also important to find true anomalies, for more details, see [11,29]. Except [4] which applies adaptive function in its complex event processing (CEP), none of the other works we discussed in this work holds this characteristic in their approaches.

The expert knowledge data issue addresses interesting challenges for the anomaly detection in traffic condition. The expert knowledge along with a base of knowledge acquired during the detection process in traffic conditions data stream is an interesting challenge which should receive more attention from now on, because this topic has not been well explored in traffic conditions data streams domain and its use can raise the rate of true anomalies by reducing the uncertainty in the data. None of works we present in this work approach such a concept, the exception are [20,31] which use traffic network data from OpenStreetMap[1] (OSM).

Straightforward and lightweight anomaly detection approaches lead to the data stream analysis in critical environments (e.g. old devices, or even mobile ones in smart cities). This open task is important in traffic conditions field since data emitters might apply some privacy constraints, and therewith the device next to the sensor (e.g. SCATS - region computer) around the city, or user mobile device (i.e., small agent running in smartphone), (pre-)processes part of the data stream before send it to a central server. Therefore, anomaly detection approaches must also satisfy such resource constraints on consumption of energy/battery, CPU and memory.

5   Streaming Frameworks for Anomaly Detection

The implementation of previously presented real-time event detection algorithms (Sect. 4) and event pattern matchers (Sect. 3) is usually done in a streaming framework. A streaming framework models the data flow in the analysis process and therefore the connections of the streams to the individual process steps. The data from one step to the next is transferred as messages. In general, a streaming framework is characterized by the following features [7]:

- *Message Processing Semantics* describes how often a message is processed in the framework, and which ordering of the messages is assumed by the framework.
- *State Handling and Fault Tolerance* describing how the streaming framework provides fault tolerance. Usually, a streaming framework provides fault tolerance by resending data that has not been acknowledged by the recipient.
- *Scalability* describes how the streaming frame work scales out in case of increasing resources.
- *Portability* describes whether the execution is bound to a specific platform, or whether it could also be executed in other, e.g. embedded, environments.

In [7] the state-of-the-art streaming frameworks are compared according to this feature list.

Acknowledgements. This research was supported by the National Council for Scientific and Technological Development (CNPq), the European Union's Seventh Framework Programme under grant agreement number FP7-318225, INSIGHT and from the European Union's Horizon 2020 Programme under grant agreement number H2020-ICT-688380, VaVeL. Additionally, this work has been supported by Deutsche Forschungsgemeinschaft (DFG) within the Collaborative Research Center SFB 876, project A1.

---

[1] openstreetmap.org.

References

1. Aggarwal, C.C.: Outlier Detection. Springer, New York (2013)
2. Allen, J.F.: Maintaining knowledge about temporal intervals. Commun. ACM **26**(11), 832–843 (1983). http://doi.acm.org/10.1145/182.358434
3. Artikis, A., Weidlich, M., Gal, A., Kalogeraki, V., Gunopulos, D.: Self-adaptive event recognition for intelligent transport management. In: 2013 IEEE International Conference on Big Data, pp. 319–325, October 2013
4. Artikis, A., Weidlich, M., Schnitzler, F., Boutsis, I., Liebig, T., Piatkowski, N., Bockermann, C., Morik, K., Kalogeraki, V., Marecek, J., Gal, A., Mannor, S., Gunopulos, D., Kinane, D.: Heterogeneous stream processing and crowdsourcing for urban traffic management. In: Proceedings of 17th International Conference on Extending Database Technology (EDBT), Athens, Greece, 24–28 March 2014, pp. 712–723 (2014). OpenProceedings.org
5. Berry, J.K.: Gis modeling and analysis. In: Madden, M. (ed.) Manual of Geographic Information Systems, pp. 527–585. American Society for Photogrammetry and Remote Sensing (2009). http://books.google.de/books?id=ek-IQAAACAAJ
6. Bifet, A., Kirkby, R.: Data stream mining a practical approach (2009)
7. Bockermann, C.: A survey of the stream processing landscape. Technical report 6, TU Dortmund University, May 2014. http://www-ai.cs.uni-dortmund.de/PublicPublicationFiles/bockermann_2014b.pdf
8. Demers, A., Gehrke, J., Panda, B., Riedewald, M., Sharma, V., White, W.: Cayuga: a general purpose event monitoring system, pp. 412–422 (2007)
9. Diao, Y., Immerman, N., Gyllstrom, D.: Sase+: an agile language for kleene closure over event streams. Analysis (UM-CS-07-03), 1–13 (2007)
10. Dodge, S., Weibel, R., Lautenschütz, A.K.: Towards a taxonomy of movement patterns. Inf. Vis. **7**(3–4), 240–252 (2008)
11. Dongre, P.B., Makik, L.G.: A review on real time data stream classification and adapting to various concept drift scenarios. In: IEEE International Advance Computing Conference, vol. 1, pp. 533–537, February 2014
12. Florescu, S., Körner, C., Mock, M., May, M.: Efficient mobility pattern stream matching on mobile devices. In: Proceedings of the Ubiquitous Data Mining Workshop (UDM 2012), pp. 23–27 (2012)
13. Fuchs, G., Andrienko, N., Andrienko, G., Bothe, S., Stange, H.: Tracing the German centennial flood in the stream of tweets: first lessons learned (2013)
14. Gal, A., Keren, S., Sondak, M., Weidlich, M., Blom, H., Bockermann, C.: Grand challenge: the techniball system. In: Proceedings of the 7th ACM International Conference on Distributed Event-Based Systems, DEBS 2013, pp. 319–324. ACM, New York (2013)
15. Guo, J., Huang, W., Williams, B.M.: Real time traffic flow outlier detection using short-term traffic conditional variance prediction. Transp. Res. Part C Emerg. Technol. **50**, 160–172 (2014)
16. Gupta, M., Gao, J., Aggarwal, C., Han, J.: Outlier detection for temporal data. Synth. Lect. Data Min. Knowl. Disc. **5**(1), 1–129 (2014)
17. Gyllstrom, D., Diao, Y., Wu, E., Stahlberg, P., Anderson, G.: SASE: complex event processing over streams. Science **1**, 407–411 (2007)
18. Gyllstrom, D., Agrawal, J., Diao, Y., Immerman, N.: On supporting kleene closure over event streams. In: Alonso, G., Blakeley, J.A., Chen, A.L.P. (eds.) ICDE, pp. 1391–1393. IEEE (2008)

19. Liebig, T., Morik, K.: Report on end-user requirements, test data, and on prototype definitions. Technical report FP7-318225 D5.1, TU Dortmund and Insight Consortium Members, August 2013

20. Liebig, T., Piatkowski, N., Bockermann, C., Morik, K.: Route planning with real-time traffic predictions. In: Proceedings of the 16th LWA Workshops: KDML, IR and FGWM, pp. 83–94 (2014)

21. Liu, W., Zheng, Y., Chawla, S., Yuan, J., Xing, X.: Discovering spatio-temporal causal interactions in traffic data streams. In: Proceedings of the 17th ACM SIGKDD International Conference on Knowledge Discovery and Data Mining, KDD 2011, NY, USA, pp. 1010–1018 (2011). http://doi.acm.org/10.1145/2020408.2020571

22. Marz, N.: Big Data: Principles and Best Practices of Scalable Realtime Data Systems. O'Reilly Media, Sebastopol (2013). http://www.amazon.de/Big-Data-Principles-Practices-Scalable/dp/1617290343

23. du Mouza, C., Rigaux, P., Scholl, M.: Efficient evaluation of parameterized pattern queries. In: Herzog, O., Schek, H., Fuhr, N., Chowdhury, A., Teiken, W. (eds.) CIKM, pp. 728–735. ACM (2005)

24. Pan, B., Zheng, Y., Wilkie, D., Shahabi, C.: Crowd sensing of traffic anomalies based on human mobility and social media. In: Proceedings of the 21st ACM SIGSPATIAL International Conference on Advances in Geographic Information Systems, SIGSPATIAL 2013, NY, USA, pp. 344–353 (2013). http://doi.acm.org/10.1145/2525314.2525343

25. Pang, L.X., Chawla, S., Liu, W., Zheng, Y.: On detection of emerging anomalous traffic patterns using GPS data. Data Knowl. Eng. **87**, 357–373 (2013). http://dx.doi.org/10.1016/j.datak.2013.05.002

26. Passow, B.N., Elizondo, D., Chiclana, F., Witheridge, S., Goodyer, E.: Adapting traffic simulation for traffic management: a neural network approach. In: IEEE Annual Conference on Intelligent Transportation Systems (ITSC 2013), pp. 1402–1407, October 2013

27. Peter, S., Höppner, F., Berthold, M.R.: Learning pattern graphs for multivariate temporal pattern retrieval. In: Hollmén, J., Klawonn, F., Tucker, A. (eds.) IDA 2012. LNCS, vol. 7619, pp. 264–275. Springer, Heidelberg (2012)

28. Randell, D.A., Cui, Z., Cohn, A.G.: A spatial logic based on regions and connection. In: Nebel, B., Rich, C., Swartout, W.R. (eds.) KR, pp. 165–176. Morgan Kaufmann (1992)

29. Bifet, A., Kirkby, R., Pfahringer, B.: Data Stream Mining: A Practical Approach. The University of Waikato, Hamilton (2011)

30. Sakr, M.A., Güting, R.H.: Spatiotemporal pattern queries. GeoInformatica **15**(3), 497–540 (2011)

31. Schnitzler, F., Liebig, T., Mannor, S., Souto, G., Bothe, S., Stange, H.: Heterogeneous stream processing for disaster detection and alarming. In: IEEE International Conference on Big Data, pp. 914–923. IEEE Press (2014)

32. Skarlatidis, A., Paliouras, G., Vouros, G.A., Artikis, A.: Probabilistic event calculus based on Markov logic networks. In: Palmirani, M. (ed.) RuleML - America 2011. LNCS, vol. 7018, pp. 155–170. Springer, Heidelberg (2011)

33. Trilles, S., Schade, S., Belmonte, Ó., Huerta, J.: Real-time anomaly detection from environmental data streams. In: Bacao, F., Santos, M.Y., Painho, M. (eds.) AGILE 2015. Lecture Notes in Geoinformation and Cartography, pp. 125–144. Springer International Publishing, Cham (2015). http://dx.doi.org/10.1007/978-3-319-16787-9_8

34. Yang, S., Kalpakis, K., Biem, A.: Detecting road traffic events by coupling multiple timeseries with a nonparametric Bayesian method. IEEE Trans. Intell. Transp. Syst. **15**(5), 1936–1946 (2014)
35. Yang, S., Liu, W.: Anomaly detection on collective moving patterns. In: IEEE International Conference on Privacy, Security, Risk, and Trust and IEEE International Conference on Social Computing, vol. 7, pp. 291–296 (2011)
36. Yuan, Y., Guan, W.: Outlier detection of handover data for innersuburban freeway traffic information estimation using mobile probes. In: IEEE Vehicular Technology Conference (VTC Spring), pp. 1–5, May 2011

Compressible Reparametrization
of Time-Variant Linear Dynamical Systems

Nico Piatkowski[1(✉)] and François Schnitzler[2]

[1] Artificial Intelligence Group, TU Dortmund, 44227 Dortmund, Germany
nico.piatkowski@tu-dortmund.de
[2] Technicolor, 35576 Cesson-sévigné, France
francois.schnitzler.ml@gmail.com

Abstract. Linear dynamical systems (LDS) are applied to model data
from various domains—including physics, smart cities, medicine, biology, chemistry and social science—as stochastic dynamic process. Whenever the model dynamics are allowed to change over time, the number
of parameters can easily exceed millions. Hence, an estimation of such
time-variant dynamics on a relatively small—compared to the number of
variables—training sample typically results in dense, overfitted models.
Existing regularization techniques are not able to exploit the temporal
structure in the model parameters. We investigate a combined reparametrization and regularization approach which is designed to detect redundancies in the dynamics in order to leverage a new level of sparsity. On
the basis of ordinary linear dynamical systems, the new model, called
ST-LDS, is derived and a proximal parameter optimization procedure
is presented. Differences to l_1-regularization-based approaches are discussed and an evaluation on synthetic data is conducted. The results
show, that the larger the considered system, the more sparsity can be
achieved, compared to plain l_1-regularization.

1 Introduction

Linear dynamical systems (LDS) describe relationships among multiple quantities. The system defines how the quantities evolve over time in response to past
or external values. They are important for analyzing multivariate time-series
in various domains such as economics, smart-cities, computational biology and
computational medicine. This work aims at estimating the transition matrices
of finite, time-variant high-dimensional vector time-series.

Large probabilistic models [8,17] are parameterized by millions of variables. Moreover, models of spatio-temporal data like dynamic Bayesian networks
(DBN) [3] become large when transition probabilities between time-slices are not
time-invariant. This induces problems in terms of tractability and overfitting.
A generic solution to these problems is a restriction to *sparse* models. Approaches
to find sparse models by penalizing parameter vectors with many non-zero weights
are available (e.g., the LASSO [5,15]). However, setting model parameters to zero
implies changes to the underlying conditional independence structure [8]. This is
not desired if specific relations between variables are to be studied.

© Springer International Publishing Switzerland 2016
S. Michaelis et al. (Eds.): Morik Festschrift, LNAI 9580, pp. 234–250, 2016.
DOI: 10.1007/978-3-319-41706-6_12

To overcome this issue for spatio-temporal data, a combination of reparametrization and regularization has been proposed, called spatio-temporal random fields (STRF) [12], which enables sparse models while keeping the conditional independence structure intact. Although the model in the aforementioned work is presented entirely for discrete data, the underlying concept can be extended to continuous data as well. Here, this idea is investigated and evaluated for multivariate Gaussian data where the conditional independence structure is encoded by the entries of the inverse covariance matrix [8] and a set of transition matrices. It is assumed, that the spatial structure is known and the goal is to find a sparse *representation* of the model's dynamics.

Related Work. In the literature, known approaches that aim at the reduction of model parameters are based on the identification of sparse conditional independence structures which in turn imply sparse parameter vectors. The basic ideas of these approaches can be applied to both, (inverse) covariance matrices and transition matrices. Some important directions are discussed in the following.

General regularization-based methods for sparse estimation may be considered [5,15], but several approaches for dynamic systems arose in the last decades. In time-varying dynamic Bayesian networks [14], Song et al. describe how to find the conditional independence structure of continuous, spatio-temporal data by performing a kernel reweighting scheme for aggregating observations across time and applying ℓ_1-regularization for sparse structure estimation. In subsequent work, it is shown how to transfer their ideas to spatio-temporal data with discrete domains [7]. The objective function that is used in the latter approach contains a regularization term for the difference of the parameter vectors of consecutive time-slices. Therefore, it is technically the most similar to STRF. However, the estimation is performed locally for each vertex and the resulting local models are heuristically combined to arrive at a global model. It can be shown that this is indeed enough to consistently estimate the neighborhood of each vertex [13].

Statistical properties of conditional independence structure estimation in undirected models are presented in [20]. In particular, the authors investigate (i) the risk consistency and rate of convergence of the covariance matrix and its inverse, (ii) large deviation results for covariance matrices for non-identically distributed observations, and (iii) conditions that guarantee smoothness of the covariances.

Han and Liu [6] present the first analysis of the estimation of transition matrices under a high-dimensional doubly asymptotic framework in which the length and the dimensionality of the time-series are allowed to increase. They provide explicit rates of convergence between the estimator and the population transition matrix under different matrix norms.

ℓ_1-regularization is indeed not the only way for inducing sparsity into the model. Wong et al. [18] show how to incorporate the non-informative Jeffreys' hyperprior into the estimation procedure. The main benefits of their approach are the absence of any regularization parameter and approximate unbiasedness of the estimate. However, the resulting posterior function is non-convex and their simulation results indicate that the proposed method tends to underestimate the number of non-zero parameters.

Instead of regularization, score-based methods deliver a combinatorial alternative for structure learning. Therein, multiple independence tests are performed to detect local structures which are finally merged to a global conditional independence structure. Since a large number of tests has to be performed, the approach might not be applicable whenever the number of variables is high. Local search heuristics [10,16] can leverage such complexity issues by restricting the test-space to neighboring structures.

Approaches mentioned so far assume that a specific segmentation of the data in suitable time-slices is already available. Fearnhead [4] developed efficient dynamic programming algorithms for the computation of the posterior over the number and location of changepoints in time-series. Based on this line of research, Xuan and Murphy [19] show how to generalize Fearnheads algorithms to multidimensional time-series. Specifically, they model the conditional independence structure using sparse, ℓ_1-regularized, Gaussian graphical models. The techniques presented therein can be used to identify the maximum a posteriori segmentation of time-series, which is required to apply any of the algorithms mentioned above.

Contribution and Organization. It is shown how to adapt the STRF model [12] to time-variant linear dynamical systems. Two alternatives are discussed, namely a reparametrization of the exponential family form of the system and a reparametrization of the transition matrices. Furthermore, a proximal-algorithm-based optimization procedure [1,11] for the joint estimation of the compressed transition matrices is presented. Finally, we evaluate the proposed procedure on synthetic data in terms of quality and complexity. The results are compared to ℓ_1-regularization and ordinary LDS.

2   Linear Dynamical Systems

Before our spatio-temporal reparametrization can be explained, we introduce time-variant linear dynamical systems and their estimation from data. Let $x_{1:T} := (x_1, x_2, \ldots, x_T)$ be a n-dimensional real valued time-series. We assume that its autonomous dynamics are fully specified by a finite, discrete-time, affine matrix equation

$$x_t = A_{t-1}x_{t-1} + \varepsilon_t \qquad \text{for } 1 < t \leq T \tag{1}$$

with *state* $x_t \in \mathbb{R}^n$, *transition matrix* $A_t \in \mathbb{R}^{n \times n}$ and *noise* $\varepsilon_t \in \mathbb{R}^n$. We call x_1 the *initial state* of the system. In total, there are $T - 1$ transition matrices $A := (A_1, A_2, \ldots, A_{T-1})^1$. Each ε_t is drawn from the same multivariate Gaussian distribution $\varepsilon_t \sim \mathcal{N}(0, \Sigma)$. Due to this stochasticity, each x_t with $t > 1$ is a multivariate Gaussian random variable given x_{t-1}, with

---

[1] Notice that A is a short notation for all transition matrices of the system.

$x_t | x_{t-1} \sim \mathcal{N}(A_{t-1} x_{t-1}, \Sigma)$. If the initial state is considered as a random variable too, e.g., $x_1 \sim \mathcal{N}(0, \Sigma)$, the full joint probability density of $x_{1:T}$ may be denoted as:

$$\mathbb{P}_{A,\Sigma}(x_{1:T}) = \mathbb{P}_{\Sigma}(x_1) \prod_{t=1}^{T-1} \mathbb{P}_{A_t, \Sigma}(x_{t+1} | x_t) \tag{2}$$

If x_1 is deterministic instead, one may simply drop the leading factor in (2).

2.1 Parameter Estimation

Estimating the parameters of an LDS is typically done by maximizing the likelihood $\mathcal{L}(A, \Sigma^{-1}, \mathcal{D})$ of a given dataset $\mathcal{D} = \{x_{1:T}^i\}_{i=1}^N$ that contains N realizations of the time-series $x_{1:T}$.

$$\mathcal{L}(A, \Sigma^{-1}, \mathcal{D}) = \prod_{i=1}^{N} \mathbb{P}_{A,\Sigma}(x_{1:T}^i) \tag{3}$$

Notice that we parameterize the likelihood directly in terms of the inverse covariance matrix. Since non-degenerate covariance matrices are positive definite, such an inverse is guaranteed to exist. Due to numerical convenience, it is common to minimize the average negative log-likelihood $\ell(A, \Sigma^{-1}, \mathcal{D}) = -\frac{1}{NT} \log \mathcal{L}(A, \Sigma^{-1}, \mathcal{D})$ instead. By plugging (2) into (3) and substituting the Gaussian density for $\mathbb{P}$, the resulting objective function is:

$$\ell(A, \Sigma^{-1}, \mathcal{D}) = -\frac{1}{NT} \log \prod_{i=1}^{N} \mathbb{P}_{A,\Sigma}(x_{1:T}^i)$$

$$= -\frac{1}{NT} \sum_{i=1}^{N} \left(\log \mathbb{P}_{\Sigma}(x_1^i) + \sum_{t=1}^{T-1} \log \mathbb{P}_{A,\Sigma}(x_{t+1}^i | x_t) \right)$$

$$= C - \frac{1}{2} \log \det \Sigma^{-1} + \frac{1}{2NT} \sum_{i=1}^{N} \sum_{t=1}^{T} r_t^{i\top} \Sigma^{-1} r_t^i \tag{4}$$

with *residual vector* $r_t^i = x_t^i - A_{t-1} x_{t-1}^i$, constant $C = \frac{1}{2} n \log 2\pi$ and $\top$ indicates the transpose of a vector or matrix. Here, $\mathbb{P}_{\Sigma}(x_1)$ is absorbed into the summation by setting $x_0 := 0$ and $A_0 := 0$. In the last equation, we made use of the fact that $(\det \Sigma^{-1}) = (\det \Sigma)^{-1}$ since any covariance matrix is positive definite. ℓ is a convex function of the transition matrices and the inverse noise covariance matrix, due to the convexity of $-\log \det \Sigma^{-1}$ and $(A_{t-1} x_{t-1})^2$. First or second order optimization procedures may be applied to find the global minimizer of (4) w.r.t. A or Σ. Hence, it is useful to know the derivatives.

We adopt the notation from [9] whenever an expression involves matrix differential calculus. Let the operator vec $: \mathbb{R}^{m \times n} \to \mathbb{R}^{mn}$ transform a matrix into a vector by stacking the columns of the matrix one underneath the other—vec(M)

represents the matrix M in column-major order. The partial derivative of ℓ w.r.t. A_t for $1 \leq t < T$ is then

$$\frac{\partial \ell}{\partial \operatorname{vec}(A_t)^\top} = \frac{1}{2NT} \sum_{i=1}^{N} \frac{\partial \left(r_{t+1}^{i\top} \Sigma^{-1} r_{t+1}^{i}\right)}{\partial \operatorname{vec}(A_t)^\top}$$

$$= -\frac{1}{NT} \operatorname{vec} \left(\Sigma^{-1} \sum_{i=1}^{N} (x_{t+1}^i - A_t x_t^i) x_t^{i\top} \right)^\top \tag{5}$$

and its partial derivative w.r.t. Σ^{-1} is

$$\frac{\partial \ell}{\partial \operatorname{vec}\left(\Sigma^{-1}\right)^\top} = -\frac{1}{2} \frac{\partial \log \det \Sigma^{-1}}{\partial \operatorname{vec}\left(\Sigma^{-1}\right)^\top} + \frac{1}{2NT} \sum_{i=1}^{N} \sum_{t=1}^{T} \frac{\partial \left(r_t^{i\top} \Sigma^{-1} r_t^i\right)}{\partial \operatorname{vec}\left(\Sigma^{-1}\right)^\top}$$

$$= -\frac{1}{2} \operatorname{vec} \left(\Sigma + \frac{1}{2NT} \sum_{i=1}^{N} \sum_{t=1}^{T} r_t^i r_t^{i\top} \right)^\top$$

Notice that the first order condition $\partial \ell / \partial \operatorname{vec}\left(\Sigma^{-1}\right)^\top = 0$ implies that the minimizer Σ^* must be equal to the empirical second moment of the transformed residual vector. A similar closed form can be derived for A_t^* whenever $\sum_{i=1}^{N} x_t^i x_t^{i\top}$ is invertible.

2.2 Sparse Estimation

Using closed-form expressions for A_t^* or Σ^{-1*} typically results in dense matrices, i.e., solutions with almost no zero entries. This might not be desired, either because sparse solutions allow for faster computation, or because the resulting matrices should reveal insights about the dependency between variables. A way to achieve this is to bias the solution towards sparse matrices by regularizing the objective function:

$$\ell^{\text{reg}}(A, \Sigma^{-1}, \mathcal{D}) = \ell(A, \Sigma^{-1}, \mathcal{D}) + g(A, \Sigma^{-1})$$

where g is an arbitrary non-negative function, the *regularizer*, that somehow measures the *complexity* that is induced by A and Σ^{-1}. Hence, minimizing ℓ^{reg} will produce solutions that trade off quality (in our case: likelihood) against complexity. It is common to choose a norm as regularizer. In particular, the l_1-norm is known to induce sparse solution [5,15]. For the LDS objective (4), this results in

$$\ell^{l_1\text{-LDS}}(A, \Sigma^{-1}, \mathcal{D}) = \ell(A, \Sigma^{-1}, \mathcal{D}) + \lambda \sum_{t=1}^{T-1} \|A_t\|_1 + \delta \|\Sigma^{-1}\|_1 \tag{6}$$

where $\| \cdot \|_1$ is the entry-wise matrix l_1-norm, i.e., $\|M\|_1 = \sum_{i=1}^{n} \sum_{j=1}^{m} |[M]_{i,j}|$ for any $n \times m$ matrix M. Here, λ and δ are positive weights which control the strength of the regularization. The larger λ (δ), the smaller will the norm of the resulting A_t (Σ^{-1}) be. That is, the larger λ or δ, the higher the number of zero entries in A_t or Σ^{-1}, respectively.

Remark 1. Zeros at the (i,j)-th entry of the inverse covariance matrix correspond to conditional independence between the variables $[\boldsymbol{x}_t]_i$ and $[\boldsymbol{x}_t]_j{}^2$, given all the other variables $\{[\boldsymbol{x}_t]_k : k \neq i \neq j\}$ [8]. Since $\boldsymbol{\Sigma}^{-1}$ is an inverse covariance matrix, it is symmetric. Hence, it may be interpreted as the (weighted) adjacency matrix of an undirected graphical structure $G(\boldsymbol{\Sigma}^{-1}) = (V, U)$ with $n = |V|$ vertices and an edge set E. If the estimation is carried out via numerical optimization, special care has to be taken to ensure that the estimated $\boldsymbol{\Sigma}^{-1}$ is symmetric and positive definite. Results on the estimation of sparse inverse covariance matrices may be found in [2,5,21]. In what follows, we assume that $\boldsymbol{\Sigma}$ is known. This is in line with the original STRF, where a spatial graphical structure is assumed to be given [12].

Due to the l_1 term, (6) can not be optimized by conventional numerical methods because $|x|$ is not differentiable at $x = 0$. However, if the gradient of (6) is Lipschitz continuous with modulus L, the proximal gradient method is guaranteed to converge with rate $\mathcal{O}(1/k)$ when a fixed stepsize $\eta \in (0, 1/L]$ is used [11].

Recall that we are interested in minimizing (6) w.r.t. all transition matrices $\boldsymbol{A}_t$. Hence, we consider block-wise minimization of the $\boldsymbol{A}_t$. The proximal alternating linearized minimization [1] is a variant of the general proximal gradient algorithm which is designed for a block-wise setting. A closer investigation of (5) shows, that each partial derivative of (6) w.r.t. $\boldsymbol{A}_t$ is indeed Lipschitz continuous. It's block Lipschitz constant is $L_t = \frac{1}{T}\|\boldsymbol{\Sigma}^{-1}\|_F \frac{1}{N}\sum_{i=1}^N \boldsymbol{x}_t^i \boldsymbol{x}_t^{i\top}\|_F = \|(\partial/\partial \operatorname{vec}(\boldsymbol{A}_t)^\top)(\partial\ell/\partial\operatorname{vec}(\boldsymbol{A}_t)^\top)\|_F$, which is the Frobenius norm of the gradient's Jacobian w.r.t. to $\boldsymbol{A}_t$. This is based on the fact that any differentiable vector-valued function whose gradient has bounded norm is Lipschitz continuous.

Using these moduli of continuity, the optimization consists of iteratively updating all transition matrices. Let $\gamma > 1$. In each iteration, the transition matrices are updated according to

$$\operatorname{vec}\left(\boldsymbol{A}_t^{\mathrm{new}}\right)^\top = \operatorname{prox}_{\gamma L_t}\left(\operatorname{vec}\left(\boldsymbol{A}_t\right)^\top - \frac{1}{\gamma L_t}\frac{\partial\ell}{\partial\operatorname{vec}\left(\boldsymbol{A}_t\right)^\top}\right)$$

with

$$\operatorname{prox}_\lambda^{\|\cdot\|_1}(\boldsymbol{x}) = \arg\min_{\boldsymbol{y}}\left(\|\boldsymbol{y}\|_1 + \frac{\lambda}{2}\|\boldsymbol{x} - \boldsymbol{y}\|_2^2\right).$$

Moreover, since $\|\cdot\|_1$ is fully separable, it can be shown (see, e.g., [11]) that

$$[\operatorname{prox}_\lambda^{\|\cdot\|_1}(\boldsymbol{x})]_j = \begin{cases} x_j - \lambda, & x_j > \lambda \\ 0, & |x_j| \leq \lambda \\ x_j + \lambda, & x_j < -\lambda \end{cases}.$$

---

^{2} $[\boldsymbol{x}]_i$ represents the i-th component of vector $\boldsymbol{x}$. Moreover, $[\boldsymbol{M}]_{i,j}$ represents the entry in row i and column j of matrix $\boldsymbol{M}$.

2.3 LDS and the Exponential Family

The STRF reparametrization for discrete state Markov random fields is formulated for exponential families [17]. We will now shortly recap the exponential family form of the multivariate Gaussian, which is also known as *information form*. An exponential family with natural parameter $\boldsymbol{\theta} \in \mathbb{R}^d$ may be denoted as

$$\mathbb{P}_{\boldsymbol{\theta}}(\boldsymbol{x}) = \exp\left(\langle \boldsymbol{\theta}, \phi(\boldsymbol{x}) \rangle - B(\boldsymbol{\theta})\right) \tag{7}$$

with log partition function $B(\boldsymbol{\theta}) = \log \int \exp\left(\langle \boldsymbol{\theta}, \phi(\boldsymbol{x}) \rangle\right) d\boldsymbol{x}^3$. In case of the multivariate Gaussian, parameter and sufficient statistic $\phi : \mathbb{R}^n \to \mathbb{R}^d$ are given by

$$\boldsymbol{\theta} = \begin{pmatrix} -\frac{1}{2}\text{vec}(\boldsymbol{\Sigma}^{-1}) \\ \boldsymbol{\Sigma}^{-1}\boldsymbol{\mu} \end{pmatrix} \text{ and } \phi(\boldsymbol{x}) = \begin{pmatrix} \text{vec}(\boldsymbol{x}\boldsymbol{x}^\top) \\ \boldsymbol{x} \end{pmatrix},$$

respectively. Moreover, the closed form of $B(\boldsymbol{\theta})$ can be computed by the n-dimensional Gaussian integral:

$$B(\boldsymbol{\theta}) = \log \int \exp\left(\langle \boldsymbol{\theta}, \phi(\boldsymbol{x}) \rangle\right) d\boldsymbol{x}$$

$$= \log \int \exp\left(-\frac{1}{2}\boldsymbol{x}^\top \boldsymbol{\Sigma}^{-1}\boldsymbol{x} + \boldsymbol{x}^\top \boldsymbol{\Sigma}^{-1}\boldsymbol{\mu}\right) d\boldsymbol{x}$$

$$= \log\left(\sqrt{(2\pi)^n \det \boldsymbol{\Sigma}^{-1}} \exp\left(\frac{1}{2}\boldsymbol{\mu}^\top \boldsymbol{\Sigma}^{-1}\boldsymbol{\mu}\right)\right).$$

Plugging this into Eq. (7) and rearranging, one arrives at the well known expression for the multivariate Gaussian density:

$$\mathbb{P}_{\boldsymbol{\theta}}(\boldsymbol{x}) = \exp\left(\langle \boldsymbol{\theta}, \phi(\boldsymbol{x}) \rangle - B(\boldsymbol{\theta})\right)$$

$$= \frac{1}{\sqrt{(2\pi)^n \det \boldsymbol{\Sigma}^{-1}}} \exp\left(-\frac{1}{2}\boldsymbol{x}^\top \boldsymbol{\Sigma}^{-1}\boldsymbol{x} + \boldsymbol{x}^\top \boldsymbol{\Sigma}^{-1}\boldsymbol{\mu} - \frac{1}{2}\boldsymbol{\mu}^\top \boldsymbol{\Sigma}^{-1}\boldsymbol{\mu}\right)$$

$$= \frac{1}{\sqrt{(2\pi)^n \det \boldsymbol{\Sigma}^{-1}}} \exp\left(-\frac{1}{2}(\boldsymbol{x} - \boldsymbol{\mu})^\top \boldsymbol{\Sigma}^{-1}(\boldsymbol{x} - \boldsymbol{\mu})\right).$$

Based on this equivalence, the joint density (2) of an LDS can also be rewritten in terms of exponential families (7)

$$\mathbb{P}_{\boldsymbol{A},\boldsymbol{\Sigma}}(\boldsymbol{x}_{1:T}) = \mathbb{P}_{\boldsymbol{\Sigma}}(\boldsymbol{x}_1) \prod_{t=1}^{T-1} \mathbb{P}_{\boldsymbol{A}_t,\boldsymbol{\Sigma}}(\boldsymbol{x}_{t+1}|\boldsymbol{x}_t)$$

$$= \mathbb{P}_{\boldsymbol{\theta}_1}(\boldsymbol{x}_1) \prod_{t=1}^{T-1} \mathbb{P}_{\boldsymbol{\theta}_{t+1}}(\boldsymbol{x}_{t+1}|\boldsymbol{x}_t)$$

$$= \exp\left(\sum_{t=1}^{T} \langle \boldsymbol{\theta}_t, \phi_t(\boldsymbol{x}_t, \boldsymbol{x}_{t-1}) \rangle - B(\boldsymbol{\theta}_t, \boldsymbol{x}_{t-1})\right)$$

---

[3] The log partition function is usually denoted by $A(\boldsymbol{\theta})$. Since the symbol A is already reserved for transition matrices, we denote the log partition function with B instead.

where we used the exponential family form of each $x_t | x_{t-1} \sim \mathcal{N}(A_{t-1} x_{t-1}, \Sigma)$ and hence, the parameters and sufficient statistic are

$$\theta_t = \begin{pmatrix} -\frac{1}{2} \operatorname{vec}(\Sigma^{-1}) \\ \operatorname{vec}(\Sigma^{-1} A_{t-1}) \end{pmatrix}, \qquad \phi_t(x_t, x_{t-1}) = \begin{pmatrix} \operatorname{vec}(x_t x_t^\top) \\ \operatorname{vec}(x_t x_{t-1}^\top) \end{pmatrix}.$$

Again, we set $x_0 := 0$ and $A_0 := 0$ to compactify notation. To remove the functional dependence between the local log-partition functions $B(\theta_t, x_{t-1})$ and x_{t-1}, we include the corresponding term $-\frac{1}{2} x_{t-1}^\top A_{t-1}^\top \Sigma^{-1} A_{t-1} x_{t-1}$ directly into the parameters and sufficient statistics:

$$\tilde{\theta}_t = \begin{pmatrix} -\frac{1}{2} \operatorname{vec}(\Sigma^{-1}) \\ \operatorname{vec}(\Sigma^{-1} A_{t-1}) \\ -\frac{1}{2} \operatorname{vec}(A_{t-1}^\top \Sigma^{-1} A_{t-1}) \end{pmatrix}, \qquad \tilde{\phi}_t(x_t, x_{t-1}) = \begin{pmatrix} \operatorname{vec}(x_t x_t^\top) \\ \operatorname{vec}(x_t x_{t-1}^\top) \\ \operatorname{vec}(x_{t-1} x_{t-1}^\top) \end{pmatrix}.$$

Finally, the joint probability of the LDS in exponential family form is

$$\mathbb{P}_{\tilde{\theta}}(x_{1:T}) = \exp\left(\langle \tilde{\theta}, \tilde{\phi}(x_{1:T}) \rangle - B(\tilde{\theta}) \right)$$

where, $\tilde{\theta} = (\tilde{\theta}_1, \tilde{\theta}_2, \ldots, \tilde{\theta}_T)^\top$, $\tilde{\phi} = (\tilde{\phi}_1(x_1, x_0), \tilde{\phi}_2(x_2, x_1), \ldots, \tilde{\phi}_T(x_T, x_{T-1}))^\top$, and $B(\tilde{\theta}) = \sum_{i=1}^{T} B(\tilde{\theta}_t)$ are the corresponding parameter, sufficient statistics and log partition function, respectively.

This representation has several drawbacks when compared to the native representation in terms of transition matrices. An obvious disadvantage is, that multiple copies of Σ^{-1} are encoded into the parameters. Moreover, the transition matrices can only be recovered via inversion of Σ^{-1} and subsequent matrix multiplication with the lower part of θ_t which encodes $\Sigma^{-1} A_{t-1}$. Hence, $\mathcal{O}(n^3)$ flops are required to extract A_{t-1} from θ_t which might be prohibitive in a large system.

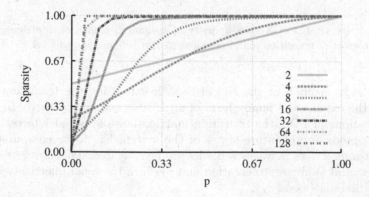

Fig. 1. X-axis: Parameter p of the Bernoulli distribution of the entries of the lower triangular matrix $\tilde{L}$. Y-axis: Average sparsity of $(\tilde{L} + 10 I_n)(\tilde{L} + 10 I_n)^\top$.

3   Reparametrization of LDS

The main goal of this work is a sparse reparametrization of LDS that does not alter the dependences which are encoded in the transition matrices. If l_1-regularization is applied to a transition matrix $\boldsymbol{A}_t$, some of its entries will be pushed to 0, and hence, some flow of information between variables is prohibited. Moreover, if a particular value of $\boldsymbol{A}_t$ does not change much over time, i.e., $[\boldsymbol{A}_t]_{i,j} \approx c$ for all $1 \leq t < T$, l_1-regularization can not exploit this redundancy. Here, we aim at finding an alternative representation that is able to sparsify such redundancies while keeping small interactions between variables intact. For discrete state Markov random fields, this task has already been solved by STRF. The core of STRF is a spatio-temporal reparametrization of the exponential family

$$\theta_t(\boldsymbol{\Delta}) = \sum_{i=1}^{t} \frac{1}{t-i+1}\boldsymbol{\Delta}_i$$

with l_1 and l_2 regularization of the $\boldsymbol{\Delta}_i$.

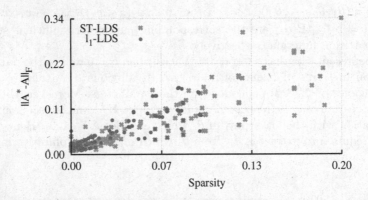

Fig. 2. Sparsity vs. Error. X-axis: Sparsity of estimated transition matrices. Y-axis: Estimation error of transition matrices, measured in Frobenius norm $\|\boldsymbol{A}^* - \boldsymbol{A}\|_F$.

As already mention at the end of Sect. 2, extracting the transition matrices from the exponential family form of an LDS is rather expensive. In practical applications of LDS, the transition matrices are of special interest. Either because a prediction of future states of the system has to be computed, or if particular interactions between variables are investigated. Therefore, we dismiss the exponential family representation and perform the reparametrization w.r.t. the transition matrices.

$$\boldsymbol{A}_t(\boldsymbol{\Delta}) = \sum_{i=1}^{t} \frac{1}{t-i+1}\boldsymbol{\Delta}_i$$

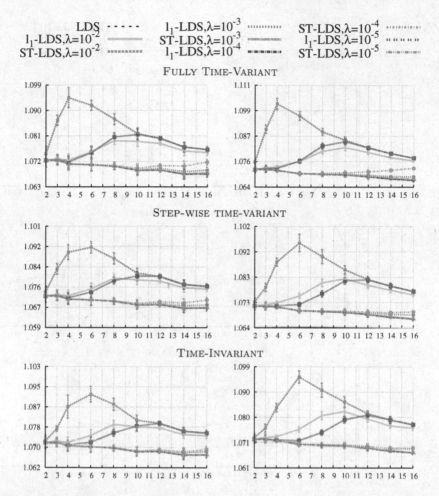

Fig. 3. X-axis: number of variables n. Y-axis: Normalized negative log-likelihood ℓ/n. Left: $T = 4$; Right: $T = 8$. Lower is better.

Analogous to (6), this results in the objective function

$$\ell^{\text{ST-LDS}}(\boldsymbol{\Delta}, \boldsymbol{\Sigma}^{-1}, \mathcal{D}) = \ell(\boldsymbol{A}(\boldsymbol{\Delta}), \boldsymbol{\Sigma}^{-1}, \mathcal{D}) + \lambda \sum_{t=1}^{T-1} \|\boldsymbol{\Delta}_t\|_1 \tag{8}$$

with $\boldsymbol{\Delta} = (\boldsymbol{\Delta}_1, \boldsymbol{\Delta}_2, \ldots, \boldsymbol{\Delta}_{T-1})$. Notice that we perform only l_1-regularization of $\boldsymbol{\Delta}$, since the results in [12] suggest that the impact of l_2-regularization on the sparse reparametrization is neglectable. In addition, $\boldsymbol{\Sigma}^{-1}$ is treated as a constant as explained in Remark 1.

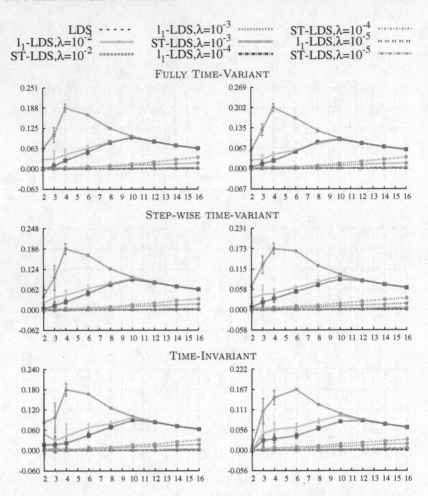

Fig. 4. X-axis: number of variables n. Y-axis: Sparsity of estimated transition matrices. Left: $T = 4$; Right: $T = 8$. Higher is better.

The partial derivatives of ℓ w.r.t. $\boldsymbol{\Delta}_t$ are required to apply the proximal algorithm from Sect. 2.2. We apply the matrix chain rule (see, e.g., [9]) to get

$$\frac{\partial \ell}{\partial \operatorname{vec}(\boldsymbol{\Delta}_t)^{\top}} = \left(\frac{\partial \ell}{\partial \operatorname{vec}(\boldsymbol{A}(\boldsymbol{\Delta}))^{\top}} \right) \left(\frac{\partial \operatorname{vec}(\boldsymbol{A}(\boldsymbol{\Delta}))}{\partial \operatorname{vec}(\boldsymbol{\Delta}_t)^{\top}} \right)$$

with

$$\frac{\partial [\boldsymbol{A}_{t'}(\boldsymbol{\Delta})]_{l,r}}{\partial [\boldsymbol{\Delta}_t]_{i,j}} = \begin{cases} \frac{1}{t'-t+1}, & t' \geq t . \wedge i = l \wedge j = r \\ 0, & \text{else} \end{cases}$$

The block Lipschitz constant $U_t = \sqrt{\sum_{t'=t}^{T-1} \left(n/(t'-t+1)\right)^2}$ of $\boldsymbol{A}(\boldsymbol{\Delta})$ w.r.t. $\boldsymbol{\Delta}_t$ is derived as described in Sect. 2.2, i.e.,

$$U_t = \left\| \left(\frac{\partial}{\partial \operatorname{vec}(\boldsymbol{\Delta}_t)^\top} \right) \left(\frac{\partial \ell}{\partial \operatorname{vec}(\boldsymbol{\Delta}_t)^\top} \right) \right\|_F.$$

Now, since $f = \ell \circ \boldsymbol{A}$ is the composition of two Lipschitz continuous functions, $U_t L_t$ is the t-th block Lipschitz constant of $f(\boldsymbol{\Delta}) = \ell(\boldsymbol{A}(\boldsymbol{\Delta}), \boldsymbol{\Sigma}^{-1}, \mathcal{D})$.

4 Experiments

Experiments are conducted in order to investigate and compare the (i) loss, (ii) sparsity and (iii) estimated transition matrices of the following methods:

– Plain time-variant LDS as defined in (1) with objective function (4)
– l_1-LDS with objective function (6)
– ST-LDS with objective function (8)

Here, *sparsity* is defined as the fraction of zero-entries in a parameter $\boldsymbol{\theta} \in \mathbb{R}^d$, i.e., sparsity$(\boldsymbol{\theta}) = \frac{1}{d} \sum_{i=1}^{d} \mathbb{1}(\boldsymbol{\theta}_i = 0)$. The indicator function $\mathbb{1}(\text{expr})$ evaluates to 1 iff expr is true.

The synthetic data for the experimental evaluation is generated by the following stochastic process:

1. Fix the number of variables n, time-steps T and samples N.
2. Generate a random inverse covariance matrix $\boldsymbol{\Sigma}^{-1}$. This is done by generating a lower triangular binary matrix $\tilde{\boldsymbol{L}}$ where each entry is draw independently from a Bernoulli distribution with parameter p. The sign of each non-zero off-diagonal entry is determined by drawing from another Bernoulli with parameter $1/2$. Then, the $n \times n$ up-scaled identity matrix $10\boldsymbol{I}_n$ is added to $\tilde{\boldsymbol{L}}$ and the result is multiplied by its own transpose, i.e., $\tilde{\boldsymbol{\Sigma}}^{-1} = (\tilde{\boldsymbol{L}} + 10\boldsymbol{I}_n)(\tilde{\boldsymbol{L}} + 10\boldsymbol{I}_n)^\top$. The implied $\tilde{\boldsymbol{\Sigma}}$ is normalized in order to have unit variances. Figure 1 shows the sparsity of the final inverse covariance matrix $\boldsymbol{\Sigma}^{-1}$ as a function of n and p.
3. Generate $T-1$ random transition matrices $\boldsymbol{A}_1, \boldsymbol{A}_2, \ldots, \boldsymbol{A}_{t-1}$. The entries are drawn independently from a uniform distribution over $[-\omega, \omega]$:
   (a) For $[\boldsymbol{A}_t]_{i,j}$ and all $1 \leq t < T$. (FULLY TIME-VARIANT)
   (b) For $[\boldsymbol{A}_1]_{i,j}$ and $[\boldsymbol{A}_{T/2}]_{i,j}$ and then copied to all $[\boldsymbol{A}_t]_{i,j}$ with $1 < t < T/2$ and $T/2 < t < T$, respectively. (STEP-WISE TIME-VARIANT)
   (c) For $[\boldsymbol{A}_1]_{i,j}$ and then copied to all $[\boldsymbol{A}_t]_{i,j}$ for all $1 < t < T$. (TIME-INVARIANT)
4. For $i = 1$ to N
   (a) Draw $\boldsymbol{x}_1^i$ from $\mathcal{N}(0, \boldsymbol{\Sigma})$.
   (b) For $t = 2$ to T
      i. Draw $\boldsymbol{\varepsilon}_t$ from $\mathcal{N}(0, \boldsymbol{\Sigma})$.
      ii. Compute $\boldsymbol{x}_t^i = \boldsymbol{A}_{t-1}\boldsymbol{x}_{t-1}^i + \boldsymbol{\varepsilon}_t$.

This procedure is applied for $n \in \{2, 4, 6, 8, 10, 12, 14, 16\}$, $T \in \{4, 8\}$ and $N = 10000$. Random covariance matrices are generated with $p = 1/4$ and random transition matrices are generated with $\omega = 1/n$. For each combination of n and T, 10 datasets are sampled, which makes a total of 1.6×10^6 data points. The evaluation of regularized methods l_1-LDS and ST-LDS is carried out with $\lambda \in \{10^{-2}, 10^{-3}, 10^{-4}, 10^{-5}\}$. All models are estimated by the proximal algorithm, described in Sect. 2.2. In case of an unregularized objective, the proximal algorithm reverts to block-wise gradient descent.

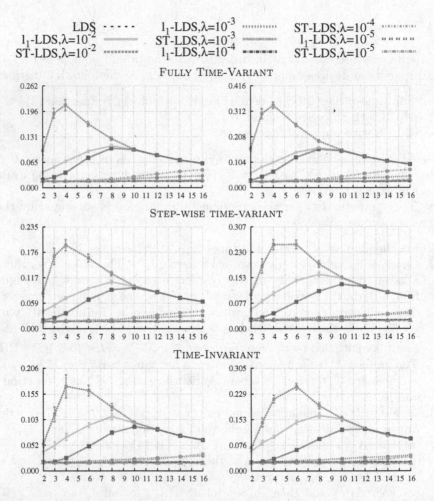

Fig. 5. X-axis: number of variables n. Y-axis: Estimation error of transition matrices, measured in Frobenius norm $\|\boldsymbol{A}^* - \boldsymbol{A}\|_F$. Left: $T = 4$; Right: $T = 8$. Lower is better.

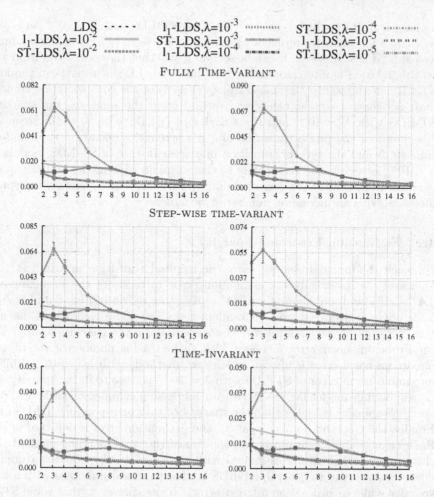

Fig. 6. X-axis: number of variables n. Y-axis: Estimation error of transition matrices, measured in maximum norm $\|A^* - A\|_\infty$. Left: $T = 4$; Right: $T = 8$. Lower is better.

4.1 Likelihood and Sparsity

Results for the average negative log-likelihood (4) and sparsity of the corresponding transition matrices are depicted in Figs. 3 and 4, where each point is averaged over 10 random data sets. Comparing results among different model sizes requires normalization of the loss function values by the corresponding number of variables, hence, Fig. 3 shows ℓ/n. Plots on the left contain results for $T = 4$ time-steps and plots on the right results for $T = 8$ time-steps, respectively. In all cases, a larger value of λ corresponds to more sparsity and a larger loss. Note, however, that the regularization parameter has a different impact on l_1-LDS and ST-LDS models, i.e., sparsity and loss of ST-LDS models with $\lambda = 10^k$ are in the range of l_1-LDS models with $\lambda = 10^{k-1}$. The results suggest the existence of a phase transition which is clearly visible for ST-LDS with $\lambda = 10^{-2}$ and l_1-LDS $\lambda = 10^{-3}$: for small (in terms of n) models, the loss of l_1-LDS is larger than the loss of ST-LDS. However, it can be seen in Fig. 3 that there exists n_0 from which on this relation is interchanged, i.e., the loss of l_1-LDS is lower than the loss of ST-LDS. Remarkably, the sparsity plots (Fig. 4) of the corresponding transition matrices show a similar behavior. Starting from the same n_0, the sparsity of ST-LDS with $\lambda = 10^{-2}$ and the sparsity of l_1-LDS with $\lambda = 10^{-3}$ converge. The point of the phase transition and it's strength depend on the number of time-steps and the type of transition matrices. In case of ST-LDS models with $\lambda = 10^{-5}$ however, the loss is close to that of plain LDS model and the sparsity is larger than that of the corresponding l_1-LDS models. Moreover, the sparsity increases with an increasing number of variables.

4.2   Estimation Error and Sparsity

For each random dataset, we store the original transition matrices A^*. This allows us, to investigate the estimation error in terms of the Frobenius norm $\|A^* - A\|_F$ and maximum norm $\|A^* - A\|_\infty$, as shown in Figs. 5 and 6. Again, each point is averaged over 10 random data sets. The ranking of the methods in terms of estimation error is coherent with the sparsity results. While the Frobenius-norm-error increases with an increasing number of variables, the maximum-norm-error is almost zero for all methods with $\lambda \leq 10^{-3}$. While the maximum-norm-error of ST-LDS with $\lambda = 10^{-4}$ is close to 0, the sparsity of the corresponding model increases with an increasing number of variables. Moreover, it's sparsity is higher than the sparsity of the corresponding l_1-LDS model. Finally, the trade-off between sparsity and estimation error is depicted in Fig. 2. Each error-sparsity pair represents one run of the corresponding method. Transition matrices which are estimated with the plain LDS model are completely dense in any case. In general, ST-LDS is able to produce models with a higher sparsity while incorporating a larger error. Notice, however, that some ST-LDS models achieve about twice the sparsity as l_1-LDS models but with the same (rather low) estimation error.

5   Conclusion

In this article, we investigated a combined reparametrization and regularization approach which is designed to detect redundancies in the dynamics of linear dynamical systems. Based on ordinary linear dynamical systems, the new model, called ST-LDS, was derived and a proximal parameter optimization procedure was presented. Expensive line-search techniques or similar step-size adaption techniques were avoided by deriving the block Lipschitz constants of the corresponding objective function w.r.t. the new reparametrization. Differences to l_1-regularization-based approaches were discussed and an evaluation on synthetic data was carried out. The results show, that with an increasing size of an ST-LDS, the estimation error is close to that of an ordinary LDS while achieving more sparsity than l_1-regularization-based models. An investigation of spatio-temporal regression models with non-Gaussian noise is an appealing direction for future research, since many real world phenomena might be explained better by other probability distributions.

Acknowledgement. This work has been supported by Deutsche Forschungsgemeinschaft (DFG) within the Collaborative Research Center SFB 876 "Providing Information by Resource-Constrained Data Analysis", project A1.

References

1. Bolte, J., Sabach, S., Teboulle, M.: Proximal alternating linearized minimization for nonconvex and nonsmooth problems. Math. Program. **146**(1–2), 459–494 (2014)
2. Cai, T., Liu, W., Luo, X.: A constrained ℓ_1 minimization approach to sparse precision matrix estimation. J. Am. Stat. Assoc. **106**(494), 594–607 (2011)
3. Dagum, P., Galper, A., Horvitz, E.: Dynamic network models for forecasting. In: Proceedings of the 8th Annual Conference on Uncertainty in Artificial Intelligence, pp. 41–48 (1992)
4. Fearnhead, P.: Exact Bayesian curve fitting and signal segmentation. IEEE Trans. Signal Process. **53**(6), 2160–2166 (2005)
5. Friedman, J., Hastie, T., Tibshirani, R.: Sparse inverse covariance estimation with the graphical lasso. Biostatistics **9**(3), 432–441 (2008)
6. Han, F., Liu, H.: Transition matrix estimation in high dimensional time series. In: Proceedings of the 30th International Conference on Machine Learning, pp. 172–180 (2013)
7. Kolar, M., Song, L., Ahmed, A., Xing, E.P.: Estimating time-varying networks. Ann. Appl. Stat. **4**(1), 94–123 (2010)
8. Lauritzen, S.L.: Graphical Models. Oxford University Press, Oxford (1996)
9. Magnus, J.R., Neudecker, H.: Matrix Differential Calculus with Applications in Statistics and Econometrics, 2nd edn. Wiley, Chichester (1999)
10. Rodrigues de Morais, S., Aussem, A.: A novel scalable and data efficient feature subset selection algorithm. In: Daelemans, W., Goethals, B., Morik, K. (eds.) ECML PKDD 2008, Part II. LNCS (LNAI), vol. 5212, pp. 298–312. Springer, Heidelberg (2008)
11. Parikh, N., Boyd, S.: Proximal algorithms. Found. Trends Opt. **1**(3), 127–239 (2014)
12. Piatkowski, N., Lee, S., Morik, K.: Spatio-temporal random fields: compressible representation and distributed estimation. Mach. Learn. **93**(1), 115–139 (2013)
13. Ravikumar, P., Wainwright, M.J., Lafferty, J.D.: High-dimensional ising model selection using ℓ_1-regularized logistic regression. Ann. Appl. Stat. **38**(3), 1287–1319 (2010)
14. Song, L., Kolar, M., Xing, E.P.: Time-varying dynamic Bayesian networks. Adv. Neural Inf. Process. Syst. **22**, 1732–1740 (2009)
15. Tibshirani, R., Saunders, M., Rosset, S., Zhu, J., Knight, K.: Sparsity and smoothness via the fused lasso. J. Royal Stat. Soc. Ser. B **67**(1), 91–108 (2005)
16. Trabelsi, G., Leray, P., Ben Ayed, M., Alimi, A.M.: Dynamic MMHC: a local search algorithm for dynamic bayesian network structure learning. In: Tucker, A., Höppner, F., Siebes, A., Swift, S. (eds.) IDA 2013. LNCS, vol. 8207, pp. 392–403. Springer, Heidelberg (2013)

17. Wainwright, M.J., Jordan, M.I.: Graphical models, exponential families, and variational inference. Found. Trends Mach. Learn. 1(1–2), 1–305 (2008)
18. Wong, E., Awate, S., Fletcher, T.: Adaptive sparsity in gaussian graphical models. JMLR W&CP 28, 311–319 (2013)
19. Xuan, X., Murphy, K.: Modeling changing dependency structure in multivariate time series. In: Proceedings of the 24th International Conference on Machine Learning, pp. 1055–1062. ACM (2007)
20. Zhou, S., Lafferty, J.D., Wasserman, L.A.: Time varying undirected graphs. Mach. Learn. 80(2–3), 295–319 (2010)
21. Zhou, S., Rütimann, P., Xu, M., Bühlmann, P.: High-dimensional covariance estimation based on gaussian graphical models. J. Mach. Learn. Res. 12, 2975–3026 (2011)

Detection of Local Intensity Changes in Grayscale Images with Robust Methods for Time-Series Analysis

Sermad Abbas$^{(\boxtimes)}$, Roland Fried, and Ursula Gather

Department of Statistics, TU Dortmund University, Dortmund, Germany
{sermad.abbas,ursula.gather}@tu-dortmund.de,
fried@statistik.tu-dortmund.de

Abstract. The purpose of this paper is to automatically detect local intensity changes in time series of grayscale images. For each pixel coordinate, a time series of grayscale values is extracted. An intensity change causes a jump in the level of the time series and affects several adjacent pixel coordinates at almost the same points in time. We use two-sample tests in moving windows to identify these jumps. The resulting candidate pixels are aggregated to segments using their estimated jump time and coordinates. As an application we consider data from the plasmon assisted microscopy of nanosize objects to identify specific particles in a sample fluid. Tests based on the one-sample Hodges-Lehmann estimator, the two-sample t-test or the two-sample Wilcoxon rank-sum test achieve high detection rates and a rather precise estimation of the change time.

Keywords: Change points · Jump detection · Spatio-temporal analysis · Image sequences

1 Introduction

In this work we deal with the automatic detection of local intensity changes in time series of digital grayscale images. A single grayscale image can be represented by a matrix with one entry for each pixel coordinate. Each entry stands for the brightness (intensity) of the corresponding pixel coordinate. The intensity ranges from black to white on a grayscale. The observed intensity represents a true intensity signal which is overlaid by noise and possibly outliers.

A local intensity change occurs if the true signals of several neighboring pixel coordinates change abruptly at the same point in time. We assume that these changes are permanent. We aim at their detection and estimation of their time of occurrence. We extract the grayscale values for each pixel coordinate over time. An intensity change induces a jump in the time series for the related pixel coordinates. We use two-sample tests in moving windows for the detection of structural breaks in time series to identify these coordinates. Afterwards, we aggregate them to coordinate sets using temporal and spatial criteria. The convex hulls of these sets represent potential locations of intensity changes.

© Springer International Publishing Switzerland 2016
S. Michaelis et al. (Eds.): Morik Festschrift, LNAI 9580, pp. 251–271, 2016.
DOI: 10.1007/978-3-319-41706-6_13

As a special application, we use data from the PAMONO (Plasmon Assisted Microscopy of Nano-Size Objects) biosensor [28], provided to us by the Leibniz-Institut für Analytische Wissenschaften – ISAS – e.V. in Dortmund. The PAMONO biosensor is a device to detect specific particles in a sample fluid by indirectly proving their existence. A primary objective is the detection of viruses, see e.g. [22,23,28].

The surface of the biosensor is a gold layer which is placed on a glass prism. A sample fluid is pumped through a flow cell atop the surface. Particles in the fluid can adhere on the top side of the layer. A laser beam is directed at the bottom side of the surface. The reflection is recorded by a CCD camera. This delivers a time series of grayscale images. An adhesion on the sensor surface causes, due to some physical effects, a permanent intensity increase in the image sequence for the affected pixel coordinates. In this application we only have to deal with positive jumps in the time series. The objective is to detect the particles on the sensor surface automatically, which is made difficult by artifacts on the sensor surface and noise in the grayscale images.

This work is organized as follows: In Sect. 2 we describe the underlying model for the grayscale images and the time series formally. The method to detect local intensity changes in time series of grayscale images is presented in Sect. 3. The application on the PAMONO data is the main focus of Sect. 4, where we also describe the data sets in more detail. Finally, in Sect. 5 we summarize and discuss the results.

2 Model Assumptions

In this section, we describe our model for the grayscale images and the pixel time series. We refer to the PAMONO data to make the terms better comprehensible.

Let $(\mathcal{Y}_t : t = 1, \ldots, N)$ be a time series of digital grayscale images with resolution $K \times L$, where $K, L \in \mathbb{N}$. The image at time $t \in \mathbb{N}$ can be interpreted as an $L \times K$ matrix $\mathcal{Y}_t = \left(Y_t^{(i,j)} : i = 1, \ldots, L, \ j = 1, \ldots, K \right)$. The random variable $Y_t^{(i,j)}$ describes the intensity at the pixel coordinate which corresponds to the matrix entry in row i and column j. In an ℓ-bit digital image, $\ell \in \mathbb{N}$, $Y_t^{(i,j)}$ has a discrete distribution and takes integer values in $\{0, 1, \ldots, 2^\ell - 1\}$, with the brightness of a pixel in the image increasing with the associated intensity [9]. For the analysis of the digital images we transform the intensities so that $Y_t^{(i,j)} \in \left\{ 0, \frac{1}{2^\ell}, \frac{2}{2^\ell}, \ldots, \frac{2^\ell - 1}{2^\ell} \right\}$.

During the generation of a digital image from a real-life object several environmental and technical aspects influence the result. The observed digital image thus corresponds to the true image which is overlaid by random noise [9].

In the following, we assume that the observed intensity at the pixel coordinates can be described by an additive relationship between a background signal, a target signal, noise and outliers. For the PAMONO data the background signal is the part of the observed intensity which comes from the sensor surface. The target signal would then be the part which is induced by the particles. If no particle is on the surface, it is zero. When a particle adheres on the gold layer the target signal increases to a positive value. The noise and outliers arise from the process of image acquisition.

For the intensity at coordinate (i, j), let $b^{(i,j)}$ describe the true background signal, which is assumed to be constant over time. The time-dependent true target signal is given by $v_t^{(i,j)}$. The random noise is denoted by $\varepsilon_t^{(i,j)}$. Furthermore, we consider an outlier process $\eta_t^{(i,j)}$. Under the assumption that the relationship between these components is additive, the intensity can be modelled as

$$Y_t^{(i,j)} = b^{(i,j)} + v_t^{(i,j)} + \varepsilon_t^{(i,j)} + \eta_t^{(i,j)} . \tag{1}$$

The intensities for one coordinate (i, j) form a time series $\left(Y_t^{(i,j)} : t = 1, \ldots, N\right)$.

Let $t_*^{(i,j)}$ be the change time of the target signal at coordinate (i, j). For a change of magnitude $\Delta_{t_*}^{(i,j)} \in \mathbb{R}$, the target signal is

$$v_t^{(i,j)} = \begin{cases} \Delta_{t_*}^{(i,j)}, & \text{if the target signal changes at time } t_*^{(i,j)} \leq t \\ 0, & \text{otherwise.} \end{cases}$$

Thus, a change in the target signal leads to an intensity change which corresponds to a jump in the observed time series $\left(Y_t^{(i,j)} : t = 1, \ldots, N\right)$.

For simplicity we assume the noise variables $\varepsilon_1^{(i,j)}, \ldots, \varepsilon_N^{(i,j)}$ to be independent, identically and symmetrically distributed over time and between the coordinates. The expectation is $\mathrm{E}\left(\varepsilon_t^{(i,j)}\right) = 0$ and the variance is $\mathrm{Var}\left(\varepsilon_t^{(i,j)}\right) = \sigma^2$. The outlier components $\eta_1^{(i,j)}, \ldots, \eta_N^{(i,j)}$ are zero most of the time but sporadically take large absolute values.

Our objective is to detect local intensity changes in a time series of digital images caused by a specific event. Such relevant changes are assumed to consist of simultaneous shifts of the intensity values of adjacent pixel coordinates. In the PAMONO context, the relevant changes are the ones induced by particle adhesions on the sensor surface. There can be irrelevant changes, e.g. caused by noise or artifacts [22].

To analyze the images, for each coordinate the time series of observed intensities is extracted. This leads to a set of $L \times K$ pixel time series for which we assume that they can be modelled by (1). By combining the background signal $b^{(i,j)}$ and the target signal $v_t^{(i,j)}$ additively to a true underlying signal

$$\mu_t^{(i,j)} = b^{(i,j)} + v_t^{(i,j)},$$

model (1) is seen to be an additive components model as in [7]:

$$Y_t^{(i,j)} = \mu_t^{(i,j)} + \varepsilon_t^{(i,j)} + \eta_t^{(i,j)} .$$

3   Methods

In this section, we present a way to detect local intensity changes in a time series of grayscale images. A general approach is described by Timm et al. [23]. It starts with a preprocessing of the grayscale images. This is followed by an identification step for determining candidate coordinates. In an aggregation step they are combined to sets of coordinates representing the candidates for the local intensity changes.

We use these steps as a guideline for our method. Additionally, we estimate the time of occurrence for each of the potential intensity changes. The following subsections describe a concrete proposal for this analysis process.

3.1   Preprocessing

Even if the background signal is assumed to be constant over time, spatially there can be heavy variations. Furthermore, it can be much stronger than the target signal which makes it difficult to identify relevant local intensity changes. In the PAMONO data, the background signal is induced by the gold layer which is uneven [28], i.e. the structure is not the same for different pixel coordinates. Moreover, the target signal is small in comparison to the background [22].

The background signal does not contain relevant information on the target signal and will be removed. We use a constant background removal [25]. The arithmetic mean of the first $B < N$ images of the sequence is calculated for each coordinate. This gives us a reference image which is subtracted from all images thereafter.

If the grayscale images have a high resolution, a large amount of time series needs to be analyzed. Following [25], we use thresholds to remove time series for which we do not expect that a relevant intensity jump occurs. If we want to detect positive intensity changes, we choose an appropriate threshold so that all time series which never exceed it are removed. Analogously, for negative intensity changes a threshold can be defined so that time series which never fall below it are removed from further analysis. The threshold has to be chosen by the operator. Time series which are almost or exactly constant will also be removed, since they can be seen as irrelevant for our analysis. For this, we estimate the variability of each time series by the median absolute deviation from the median (MAD) [14]. If it is zero, the time series is removed. The MAD is a robust alternative to the empirical standard deviation. For a time series which is constant most of the time, the MAD has the advantage that it can still be zero while the standard deviation would be strictly positive.

3.2   Identification of Candidate Pixels

After extracting the time series $(Y_t^{(i,j)} : i = 1, \ldots, L, \, j = 1, \ldots, K, \, t = 1, \ldots, N)$ from an image sequence we check each coordinate individually for the presence of a jump in the true signal μ_t. For better readability the coordinate index (i, j) is dropped in the following.

Preliminaries. To detect jumps in the true signal μ_t of the time series, we use two-sample tests for the location problem and apply them in a moving time window of width $n = 2 \cdot k$. The subwindow width $k \in \mathbb{N}$ is a tuning parameter and has to be chosen by the operator.

For each point in time $t = k, \ldots, N - k$ we test for a jump between t and $t+1$ by splitting the whole window of width n

$$\boldsymbol{Y}_t = (Y_{t-k+1}, \ldots, Y_t, Y_{t+1}, \ldots, Y_{t+k})'$$

into two subwindows

$$\boldsymbol{Y}_{t-} = \left(Y_{t,1}^-, \ldots, Y_{t,k}^-\right)' \text{ and } \boldsymbol{Y}_{t+} = \left(Y_{t,1}^+, \ldots, Y_{t,k}^+\right)'$$

each of width k, where

$$Y_{t,u}^- = Y_{t-k+u} \text{ and } Y_{t,v}^+ = Y_{t+v}, \ u, \ v = 1, \ldots, k \ .$$

We assume that the signal μ_t is locally constant so that

$$\mu_{t-k+u} = \mu_{t-} \text{ and } \mu_{t+v} = \mu_{t+}, \ u, \ v = 1, \ldots, k \ .$$

A jump of magnitude $\Delta_t \in \mathbb{R}$ in the true signal at time t then corresponds to a shift in location with

$$\mu_{t+} = \mu_{t-} + \Delta_t \ .$$

We call $\boldsymbol{Y}_{t-}$ the reference window and $\boldsymbol{Y}_{t+}$ the test window.

Under the previous assumptions, $Y_{t,1}^-, \ldots, Y_{t,k}^-$ as well as $Y_{t,1}^+, \ldots, Y_{t,k}^+$ are independent and identically distributed. In addition, $Y_{t,1}^-, \ldots, Y_{t,k}^-, Y_{t,1}^+, \ldots, Y_{t,k}^+$ are independent.

The null hypothesis $H_{0,t} : \Delta_t = 0$ to be tested at time t is that there is no jump in the signal of the two subwindows.

In contrast to global methods, the application of moving windows has the advantage that relevant signal changes can be detected even if we have a slow trend in the data. Furthermore, we avoid the fitting of a global parametric model [5].

The subwindow width k has to be chosen under consideration of the application [5]. If k is too small, outliers can have a great influence on the results. If it is too large, the assumption of a locally constant signal is possibly not justified and the detection of a location shift can be delayed.

Selected Tests for the Two-Sample Location Problem. In the following, we present some selected tests for the two-sample location problem. The idea is to compare the test window with the reference window by estimating the location difference and standardizing it with a scale estimator [6].

A popular test which uses this principle is the classical two-sample t-test. The magnitude of the location change is estimated by the difference of the arithmetic means $\overline{Y}_{t-}$ and $\overline{Y}_{t+}$ of the reference and the test window:

$$\hat{\Delta}_t^{(0)} = \overline{Y}_{t+} - \overline{Y}_{t-} \ .$$

It is standardized by the pooled empirical standard deviation

$$\hat{S}^{(0)} = \sqrt{\frac{1}{n-2} \left(\sum_{u=1}^{k} \left(Y_{t,u}^- - \overline{Y}_{t-} \right)^2 + \sum_{v=1}^{k} \left(Y_{t,v}^+ - \overline{Y}_{t+} \right)^2 \right)} \ .$$

The t-test has the disadvantage of not being robust against outliers. Therefore, in [6] the arithmetic mean and the pooled standard deviation are replaced by more robust estimators.

A robust alternative to the arithmetic mean is the sample median. The difference of the sample medians $\widetilde{Y}_{t-}$ and $\widetilde{Y}_+$ for the reference and the test window,

$$\hat{\Delta}_t^{(1)} = \widetilde{Y}_{t+} - \widetilde{Y}_{t-},$$

can be used as an estimator for the location difference. In [6] the median of the absolute deviation from the medians of both subwindows is the applied scale estimator:

$$\hat{S}^{(1)} = 2 \cdot \mathrm{med} \left\{ |Y_{t,1}^- - \widetilde{Y}_{t-}|, \ldots, |Y_{t,k}^- - \widetilde{Y}_{t-}|, |Y_{t,1}^+ - \widetilde{Y}_{t+}|, \ldots, |Y_{t,k}^+ - \widetilde{Y}_{t+}| \right\} \ .$$

We will call this test median-differences test (MD-test).

Another robust alternative to the arithmetic mean is the one-sample Hodges-Lehmann estimator [10]. For both subwindows it is defined as

$$\widehat{Y}_{t-} = \mathrm{med} \left\{ \frac{Y_{t,u}^- + Y_{t,v}^-}{2}, \ 1 \leq u < v \leq k \right\} \ \text{and}$$

$$\widehat{Y}_{t+} = \mathrm{med} \left\{ \frac{Y_{t,u}^+ + Y_{t,v}^+}{2}, \ 1 \leq u < v \leq k \right\} \ .$$

The location difference can be estimated by

$$\hat{\Delta}_t^{(2)} = \widehat{Y}_{t+} - \widehat{Y}_{t-} \ .$$

Additionally, we will use the two-sample Hodges-Lehmann estimator for the location difference [10], which is given by

$$\hat{\Delta}_t^{(3)} = \mathrm{med} \left\{ Y_{t,u}^+ - Y_{t,v}^-, \ u, \ v = 1, \ldots, k \right\} \ .$$

The estimators $\hat{\Delta}_t^{(2)}$ and $\hat{\Delta}_t^{(3)}$ will be standardized by the scale estimator [6]

$$\hat{S}^{(2)} = \mathrm{med} \left\{ |X_{t,u} - X_{t,v}| : 1 \leq u < v \leq n \right\},$$

where $(X_{t,1}, \ldots, X_{t,n})' = \left(Y_{t,1}^- - \widetilde{Y}_{t-}, \ldots, Y_{t,k}^- - \widetilde{Y}_{t-}, Y_{t,1}^+ - \widetilde{Y}_{t+}, \ldots, Y_{t,k}^+ - \widetilde{Y}_{t+} \right)'.$

We will call the test based on the one-sample Hodges-Lehmann-estimator HL1-test and the one using the two-sample Hodges-Lehmann estimator HL2-test.

The distribution of the t-test statistic in small samples is known under the null hypothesis if the data come from a normal distribution. In contrast, the small sample distributions under the null hypothesis for the remaining test statistics are unknown. In [6] the permutation principle is used to construct distribution-free tests. When the tests are applied in moving windows, computation of the permutation distribution for each window causes high computational efforts even in case of small window widths like $k = 10$. Following [16] we approximate the distribution under the null hypothesis via simulation under the assumption that the data stem from a normal distribution. In 50000 replications we generate $n = 2k$ random observations from the standard normal distribution and calculate the value of the test statistic for each sample. We only have to compute the distributions for the standard normal case because the test statistics are invariant with respect to linear transformations of the data. In [1] the noise structure of the PAMONO data is analyzed descriptively. There, the normality assumption is made plausible for the data at hand. Nevertheless, the distributional assumption is a disadvantage.

Alternative distribution-free approaches to test for a location difference between two samples are linear rank tests. Here, we will use the two-sample Wilcoxon rank-sum test. Let $R_{t,1}^+, \ldots, R_{t,k}^+$ be the ranks of $Y_{t,1}^+, \ldots, Y_{t,k}^+$ in the whole window. The test statistic is given by the sum of the ranks.

Significance Level. The significance level $\alpha \in (0, 1)$ is another important tuning parameter [16]. We use $N - n + 1$ tests to check for signal jumps in a time series for a fixed k. Thus, the number of tests increases with a decreasing window width. As we now have a multiple testing problem and to make different window widths better comparable we adjust the significance level. Two possibilities for this are the Bonferroni and the Bonferroni-Holm method [11]. With the Bonferroni method, a preselected global significance level $\alpha_g \in (0, 1)$ is divided by the number of tests which will be carried out. The Bonferroni-Holm method is a sequential procedure in which the p-values $p_1, \ldots, p_{N-n+1}$ of the individual tests are sorted in ascending order $p_{(1)} \leq \cdots \leq p_{(N-n+1)}$. The p-value $p_{(u)}$ is then compared with the adjusted significance level $\frac{\alpha_g}{N-n+u}$, $u \in \{1, \ldots, N - n + 1\}$, starting with $p_{(1)}$, until the first time the p-value is not smaller than it. The hypotheses for which the p-value is smaller are rejected.

The Bonferroni-Holm method can lead to a higher power of the resulting test procedure than the Bonferroni method. In the following, the Bonferroni-Holm method will be abbreviated as Holm method.

3.3   Estimation of Jump Points in Time Series

If the hypothesis $H_{0,t}$ for a jump between t and $t + 1$ is rejected, it is likely that the null hypothesis will be rejected for some of the preceding and following times as well. This is because each window $Y_{t-k+1}, \ldots, Y_{t+k-1}$ contains at least one observation from before or after the jump. Therefore, a single jump may cause several rejections at close-by time points. Besides, a time series can contain multiple jumps at different times. In the PAMONO data this can happen if several particles are overlapping on the surface. That is why we want to estimate the true jump times out of the set of all potential jump times. Let

$$I = I_1 = \{t \in \{k, \dots, N-k\} : H_{0,t} \text{ is rejected}\}$$

be the candidate set of times at which $H_{0,t}$ is rejected by a two-sample test in a moving window. Some ways to estimate the true jump time out of the candidates are described in [16]. The number of estimated jump times is denoted by $\hat{m}$.

Method of Wu and Chu. The method of Wu and Chu [27] is based on the idea that all rejections in the neighborhood of a time t belong to the same jump. Let I_w be the candidate set in step w of the method. Initially, $w = 1$ and $I_1 \neq \emptyset$:

1. The jump time t_w is estimated by choosing the candidate time with the minimal p-value, $\hat{t}_w$.
2. The candidate set is reduced by removing a neighborhood of $\hat{t}_w$:

$$I_{w+1} := I_w \setminus \{\hat{t}_w - c \cdot k, \hat{t}_w - c \cdot k + 1, \dots, \hat{t}_w + c \cdot k\}, \ c \in \{1, 2\} \ .$$

3. If $I_{w+1} = \emptyset$ stop, else set $w := w + 1$ and go back to 1.

The parameter c controls the number of points in time which are considered to belong to the same jump as $\hat{t}_w$. If the time with the minimal p-value is not unique, we take the earliest out of the time sequence in ascending order.

Method of Qiu and Yandell. The method of Qiu and Yandell [18] is another way to estimate the true jump time. It constructs tie sets which include candidate times in an interval of a predefined width $c \cdot k$, $c \in \{1, 2\}$.

Let $t_1 < \dots < t_{|I|}$ be the ordered candidate times in I. Starting with t_1 and ending with $t_{|I|}$, consecutive candidate times are assigned to the same tie set in the following way:

Let $t_{r_{1,w}}$ be the first and $t_{r_{2,w}}$ be the last index in the w-th tie set, $w = 1, \dots, \hat{m}$. A tie set is then defined as follows: If there are two indexes $1 \leq r_{1,w} \leq r_{2,w} \leq |I|$ with

$$t_{r_{1,w}} - t_{r_{1,w}-1} > c \cdot k$$
$$t_{r_{2,w}+1} - t_{r_{2,w}} > c \cdot k$$
$$t_{v+1} - t_v \leq c \cdot k \text{ for } v \in \{r_{1,w}, r_{1,w} + 1, \dots, r_{2,w} - 1\},$$

then

$$B_{r_{1,w}, r_{2,w}, w} = \{t_{r_{1,w}}, t_{r_{1,w}} + 1, \dots, t_{r_{2,w}}\}$$

is the w-th tie set. Furthermore, $r_{1,1} = t_1$ and $r_{2,\hat{m}} = t_{|I|}$. Thus, each tie set contains the candidate times for which it is assumed that they belong to the same jump. The estimated jump time is the candidate time with minimal p-value in each tie set. In case that the minimal p-value in a tie set is not unique, we take the arithmetic mean of the possible times within the tie set. This should not be problematic because all candidate times in a tie set are supposed to belong to the same jump.

It is possible that the two jump-time estimators lead to a different number of estimated jumps.

3.4 Aggregation of Candidate Pixels

Each pixel coordinate with at least one detected jump in its time series is a candidate coordinate. We aggregate these to coordinate sets so that adjacent coordinates which could belong to the same intensity change will be assigned to the same set. These are represented by their convex hull and can be used to separate relevant from irrelevant changes.

For the coordinate (i,j) let $\hat{m}_{i,j}$ and $\hat{t}^{(i,j)} = \left\{ \hat{t}_1^{(i,j)}, \ldots, \hat{t}_{\hat{m}_{i,j}}^{(i,j)} \right\}$ be the number and the set of estimated jump times. By using only the information if a coordinate is a candidate coordinate, the results can be represented by a binary matrix $\mathcal{X} = \left(x^{(i,j)} : i = 1, \ldots, L, \ j = 1, \ldots, K \right)$, where

$$x^{(i,j)} = \begin{cases} 1, & \text{if at least one jump is detected at coordinate } (i,j) \\ 0, & \text{otherwise.} \end{cases}$$

In the following sections, we will make use of this matrix.

Median Filter. It is likely that many candidate coordinates are induced by wrong detections due to the type I error of the tests or misleading structures in the time series. As a local intensity change influences neighboring pixel coordinates it can be assumed that candidate coordinates with no or only a few neighbors do not belong to the event of interest. Therefore, they will be removed by applying a spatial median filter [9] with a square window of width $2 \cdot r + 1$ to $\mathcal{X}$. The parameter r has to be chosen by the operator.

Temporal Segmentation. The temporal segmentation is based on two ideas.

1. We expect that the presence of many similar estimated jump times indicates that a change occurred somewhere in the image sequence.
2. Coordinates with similar estimated jump times could belong to the same change.

For an intensity change at time t_* we expect that the estimated jump times for the influenced coordinates will vary around t_*. For this reason we combine estimated jump times which differ by at most $\gamma_1 \in \mathbb{N}$ time units. We construct overlapping time windows of width γ_1:

$$t_\kappa = \{\kappa, \kappa + 1, \ldots, \kappa + \gamma_1 - 1\}, \ \kappa = 1, \ldots, N - \gamma_1 + 1 \ .$$

The time window t_κ contains all points in time from κ to $\kappa + \gamma_1 - 1$.

For each of these time windows we count the number of estimated jumps which occur within the time window. If the frequency is at least $\gamma_2 \in \mathbb{N}$, the window is retained. This gives us the $n_t \in \mathbb{N}$, $n_t \leq N - \gamma_1 + 1$, time windows

$$t_{\kappa_1}, \ldots, t_{\kappa_{n_t}}, \; 1 \leq \kappa_1 < \ldots < \kappa_{n_t} \leq N - \gamma_1 + 1 \; .$$

It is possible that two successive time windows t_{κ_v} and $t_{\kappa_{v+1}}$ with $\kappa_{v+1} - \kappa_v = 1$ reach or exceed the threshold γ_2. This could be induced by the same jump, as the time windows are overlapping. Hence, we unite consecutive time windows to new sets. For $v = 1, \ldots, n_t - 1$, if $\kappa_{v+1} - \kappa_v = 1$, then t_{κ_v} and $t_{\kappa_{v+1}}$ belong to the same set. If $\kappa_{v+1} - \kappa_v > 1$, a new set is started which again is the union of all subsequent time windows. This leads to the $n_\tau \in \mathbb{N}$ sets $\tau_1, \ldots, \tau_{n_\tau}$. The resulting temporal segments are given by

$$\tau_w = \{(i,j) : \text{there is one } \hat{t} \in \hat{t}^{(i,j)} \text{ with } \hat{t} \in \tau_w\}, \; w = 1, \ldots, n_\tau \; .$$

The tuning parameters γ_1 and γ_2 have to be chosen by the operator. It is important to choose γ_1 not too large because this can lead to many points in time which are assigned to the same temporal segment even if the distance between them is large. If γ_1 is chosen too small, it could happen that no time window reaches or exceeds the threshold γ_2.

Spatial Segmentation. In the next step, we combine coordinates with a similar estimated jump time which are in the same region of the image. The resulting coordinate sets will be called spatial segments and represent the potential local intensity changes. We create the $n_\rho \in \mathbb{N}$ spatial segments $\rho_1, \ldots, \rho_{n_\rho}$ by uniting neighboring coordinates within the same temporal segment.

The segment ρ_w, $w = 1, \ldots, n_\rho$, is constructed in a way that, if $(i,j) \in \rho_w$ then there exists another coordinate $(i',j') \in \rho_w$ with $|i - i'| = 1$ or $|j - j'| = 1$. This means that we remove each coordinate which has no adjacent coordinate. It is unlikely that such coordinates belong to a relevant local intensity change because each local intensity change affects more than one coordinate at the same time.

The spatial segments are represented by their convex hull. In this work we use the area $\gamma_3 \in \mathbb{R}^+$ of the convex hull as a simple criterion for the exclusion of very small segments. This again is a tuning parameter which has to be chosen by the operator. More sophisticated criteria are possible [22].

3.5   Estimation of the Change Time

In the last step, the time of a local intensity change is estimated.

Let ρ be a spatial segment. The estimator for the corresponding change time is based on the time at position $\left\lfloor \frac{|\rho|+1}{2} \right\rfloor$ in the sequence of all estimated jump times in ρ in ascending order. A relevant intensity change at time t_* leads to an intensity jump at time t_*. In a simplified way, the two-sample tests detect the jump at time $t_* - 1$. Therefore, the value 1 is added to the index from above.

4   Application

In this section we use the method of Sect. 3 on the PAMONO data as an example for its application. For the analysis we use the statistical software R [19] version 3.0.1-gcc48-base. We use the R package BatchExperiments [2] to carry out the computations on the Linux HPC cluster LiDO in Dortmund. The graphics are created with the R packages ggplot2 [26] and tikzDevice [21]. The tables are generated using the xtable package [4]. Several other R packages are used for some of the computations, namely AnnotationDbi [17], png [24], sp [3] and e1071 [15].

4.1   The PAMONO Data

We use two synthetical data sets which are provided to us by the Leibniz-Institut für Analytische Wissenschaften – ISAS – e.V. The particles of interest are polysterene particles with a diameter of 200 nm. Both data sets differ in the way the background signal was generated and in the number of particles on the surface.

To generate the data sets, the sensor is applied to a sample fluid which contains the particles. The resulting grayscale images are analyzed with a detection method. From all detected intensity changes, experts try to identify those which are caused by a particle. Polygons around these local intensity changes are drawn by hand. The corresponding signals inside these polygons are extracted and added to the background signals at randomly chosen positions and times. Each data set consists of 400 16-bit grayscale images with a resolution of 1280×256 pixels.

To create the background signal of the first data set, the CCD camera takes a single image of the reflection when no particles are present so that the background signal is constant over time. The background is altered for each point in time according to a Poisson noise model [20]. This is a common noise model for CCD sensors [13]. The data set contains 100 particles and will be referred to as BGPoisson.

The background of the second data set is generated by recording the sensor surface over a specific amount of time while the sample fluid does not contain any particles. The idea is to capture the background signal with all types of disturbances except for the particles. This data set will be called BGReal and has 300 particles.

For each of the two background generation methods we have a training and a test data set. The test data sets differ from their corresponding training data sets in the position and the adhesion time of the particles on the surface.

An intensity time series can contain multiple jumps when different particles overlap. We do not have any information on the expected jump height, but we know that it gets smaller the farther the corresponding coordinate is away from the centre of a local intensity change. Values for the area of a local intensity change are unknown, too.

Moreover, we do not know exactly which coordinates are part of an adhesion. We only have knowledge of the polygon vertices which surround the particles.

4.2   Reference Method

To evaluate the performance of the detection method proposed in Sect. 3 we compare the results with a reference method. It is based on the ideas presented in [22, 23]. Its results are provided to us by the Lehrstuhl für Graphische Systeme of the Department of Computer Science at TU Dortmund University. We only describe the basic idea behind the algorithm in short.

In the preprocessing step a constant background is removed. Afterwards, a Haar wavelet is used to reduce the noise in the pixel time series. Candidate coordinates are identified by comparing the time series with a pattern function in a moving time window. If the sum of squared deviations between the observed values and the pattern is smaller than a predefined threshold, the corresponding coordinate is a candidate coordinate. The aggregation of these candidate coordinates is achieved with a marching squares algorithm [8].

4.3   Simulation Study for the Two-Sample Tests

A comparison of the two-sample tests regarding their power and accuracy of the jump time estimation on the data sets is not easily possible because we do not know exactly which time series belong to an adhesion and which do not. Therefore, a simulation study is conducted in [1] to analyze the ability of the tests in moving time windows to detect jumps in time series with a high accuracy for the true jump time. Different situations with and without jumps, outliers and linear trends are studied.

For each simulation 1000 time series are generated from a standard normal distribution of length 200. At most one jump is included in each time series. Thus, the simulated time series give a simplified picture of the time series in the PAMONO data sets. The evaluated criteria are the percentage of detected jumps to estimate the power of the tests as well as the accuracy of the estimated jump time. As only one jump occurs in the simulated time series, the time index with the minimal p-value is used as an estimator for the jump time. If the minimal p-value is not unique, the arithmetic mean of all indexes with the minimal p-value is used. The tests are applied with the subwindow widths $k = 10, 20, 30, 40, 50$. In the following we summarize the main results of the study.

In the jump situations without outliers, the t-test delivers the best results for both criteria and all subwindow widths. The Wilcoxon test, the HL1- and the HL2-test are suitable alternatives. Additionally, they are more robust and provide a high power and accuracy even if outliers are present in the time series. Concerning the Wilcoxon test, the subwindow width should not be too small in case of outliers. The power of the t-test can be zero when outliers are present in a jump situation. This is also the case for the Wilcoxon test with the small subwindow width $k = 10$. For the HL1- and the HL2-test the loss in power for $k = 10$ is bounded. The MD-test also has good robustness properties, but a considerably smaller power and accuracy than the other tests under normality.

The subwindow width has a large influence on the results. A long subwindow increases the power of the test. But in trend situations the assumption of a locally linear signal is not justified so that this could lead to many wrong detections due to the location difference between the subwindows. Regarding the accuracy, a shorter subwindow seems to be preferable. Furthermore, this helps to resist slow monotonic trends. However, the test decision can be more easily influenced by outliers then.

4.4   Application of Detection Method on the PAMONO Data

In this section, we apply the detection method of Sect. 3 to the PAMONO data. The training data sets are used to find good parameter combinations while the test data serve solely for their evaluation. We choose the following presetting:

All tests from Sect. 3.2 are used. The subwindow widths will be chosen as $k = 30,\ 50$. We use $\alpha_g = 0.2$ as the global significance level. The spatial median filter will be applied with parameter $r = 1$. We take this value to remove single candidate coordinates which do not have many neighbors, but we do not want to reduce the size of the areas which actually belong to an adhesion too much. One-sided tests are applied to detect positive jumps.

The remaining parameters, namely the window width for the temporal segmentation γ_1, the minimal number of detected jumps per window γ_2, the minimal area of the output segments γ_3, the correction method for the significance level and the jump-time estimator will be varied to find a good parameter setting.

For the jump-time estimation we use the methods of Wu and Chu as well as Qiu and Yandell with the width factor $c = 1,\ 2$. We use the following abbreviations: WC1 for the method of Wu and Chu with $c = 1$, WC2 for Wu and Chu with $c = 2$, QY1 for Qiu and Yandell with $c = 1$ and QY2 for Qiu and Yandell with $c = 2$.

Preliminary studies show that the accuracy of the adhesion time estimation depends strongly on the relation between γ_1 and γ_2 with the accuracy getting high if γ_2 is substantially larger than γ_1. Therefore, we choose γ_1 and γ_2 so that $\gamma_2 \geq 2 \cdot \gamma_1$.

To correct the significance level we use the method of Bonferroni and the method of Holm.

In total, we evaluate each combination of the following settings:

– tests: HL1-test, HL2-test, MD-test, t-test, Wilcoxon test
– subwindow widths: $k = 30,\ 50$
– $\gamma_1 = 1,\ 2,\ 3$
– $\gamma_2 \in \{\lambda \in \{1, \ldots, 20\} :\ \lambda \geq 2 \cdot \gamma_1\}$
– $\gamma_3 = 4,\ 4.5,\ 5$
– jump-time estimation: WC1, WC2, QY1, QY2
– global significance level: $\alpha_g = 0.2$
– correction of the significance level: Bonferroni, Holm

It has to be emphasized that this does not necessarily lead to the best possible combinations. Our main intention is to show that the approach presented in Sect. 3 is a reasonable way to handle these data.

Evaluation. For the evaluation, we use three different measures. First, we count the number of output segments. It should be as close as possible to the true number of particles on the sensor surface. In addition, the method should detect as many particles as possible. So we will use the percentage of output segments which belong to a particle as a criterion for the detection quality. As a third measure we calculate the mean absolute difference between the true and the estimated adhesion time. This will be called mean absolute temporal deviation (MATD) in the following.

To find a good parameter combination for each two-sample test, we first choose the one with the highest detection rate. It is likely that this results in more than one possibility. In this case, we use the combination with the minimal MATD. If this is also not unique, we just take the first parameter combination in the list because all these combinations do not differ in the detection rate and the accuracy. This situation occurs when we evaluate the results for the parameter settings on BGReal.

When assigning an output segment to a particle, we have to deal with the difficulty that we do not know exactly which coordinates are affected by particles. We only have information on the polygon vertices which surround the relevant local intensity changes. To find out if an output segment actually belongs to a particle, we check if the segment is completely inside one of the polygons. We assign such segments to a particle by determining all polygons in which it lies completely. If this is the case for more than one polygon, we choose the one with the minimal MATD.

Preprocessing of the Data Sets. The constant background removal is used based on the first 20 images. Subsequently, to save time for the following computations, we use a simple threshold approach to remove time series with low intensities. The maximum for each time series is calculated. We use the arithmetic mean of all maxima as the threshold. If all values of a time series are below the threshold, it is removed. This approach does not necessarily lead to an optimal solution and is questionable when almost all time series have similar maximal intensities. Although we do not know the exact number of time series which belong to a particle and therefore contain a jump, the amount is small compared to the whole number of time series. A more suitable approach could be the inclusion of the threshold as another parameter in the optimization.

After the preprocessing we have 152656 time series for the training data set of BGPoisson and 152494 for the corresponding test data set. For BGReal we have 152311 time series in the training data set and 152174 in the test data set. Each time series is of length 380.

Application on BGPoisson. We use the aforementioned parameter settings on the training data set to find a suitable parameter combination for each of the different two-sample tests.

First, we will briefly comment on the influence of the subwindow width, the correction for the significance level and the jump-time estimator on the percentage of detected particles. Table 1 shows the maximal detection rates for the two-sample tests when one of the parameters has a fixed value. The results indicate that large subwindow widths increase the detection rate. This gets particularly clear for the MD-test where the rate increases from 52 % to 96 %. Furthermore, there does not seem to be a big difference between the two correction methods for the significance level and the jump-time estimators concerning the detection rate. In addition, it can be seen that for the given parameters no two-sample test leads to a detection of all particles.

Table 1. Particle detection rates in the training data set of BGPoisson for the different two-sample tests. Each row shows the maximal detection rate.

Parameter	Value	HL1	HL2	MD	t	Wilcoxon
k	30	0.93	0.89	0.52	0.92	0.92
	50	0.98	0.97	0.96	0.98	0.97
Correction	Bonferroni	0.97	0.97	0.95	0.98	0.97
	Holm	0.98	0.97	0.96	0.98	0.97
Jump time	QY1	0.97	0.96	0.96	0.98	0.97
	QY2	0.98	0.96	0.95	0.98	0.97
	WC1	0.96	0.97	0.95	0.98	0.97
	WC2	0.96	0.95	0.95	0.98	0.97

Table 2 shows the best parameter combinations with respect to the detection rate and the MATD. For all tests, the larger subwindow width $k = 50$ is chosen. Moreover, with $\gamma_3 = 4$, we use the smallest minimal area for the output segments. The values of k, γ_1, γ_2 and γ_3 for the t-test and the Wilcoxon test are quite similar, while the correction of the significance level and the jump-time estimators are different. We make a similar observation for the HL1- and the HL2-test.

In Table 3 the results for the two-sample tests are compared with the reference method on the training and the test data set. Not shown is the number of false detections which is zero for each method. On the training data set, the detection rates for all methods are quite similar. The main differences occur in the number of output segments and the MATD. The HL2-test and the reference method detect several particles multiple times. The number of output segments for the reference method is considerably larger than the 100 existing particles in the data set. Except for the HL2-test, all two-sample tests result in a smaller MATD than the reference method, with the t-test and the Wilcoxon test having the smallest MATD, followed by the HL1-test. The observations on the test data correspond to those on the training data. Again, we have no false detections by any of the methods.

Table 2. Parameter values for each test which lead to the highest particle detection rate and the smallest mean absolute deviation between the estimated and the true change time on the training data set of BGPoisson.

Parameter	HL1	HL2	MD	t	Wilcoxon
k	50	50	50	50	50
Correction	Holm	Holm	Holm	Holm	Bonferroni
γ_1	1	1	3	2	2
γ_2	2	4	12	8	7
γ_3	4	4	4	4	4
Jump time	QY2	WC1	QY1	WC1	QY1

Table 3. Evaluation of the parameter values on the training and test data sets of BGPoisson for the two-sample tests and the reference method.

Data set	Measure	HL1	HL2	MD	t	Wilcoxon	Reference
Training data	Output segments	98	101	96	98	97	149
	Detection rate	0.98	0.97	0.96	0.98	0.97	0.98
	MATD	0.94	7.69	1.39	0.34	0.30	4.83
Test data	Output segments	98	100	92	97	98	143
	Detection rate	0.98	0.97	0.91	0.97	0.98	0.95
	MATD	0.83	7.11	1.25	0.33	0.32	5.15

Application on BGReal. Now we compare the detection methods on the data set BGReal. There are two major challenges with which we have to deal here: First, we have 300 particles which are on the sensor surface in the same time period as in BGPoisson. This makes it more likely that several particles adhere on the surface at similar times and overlap. In this case, the method has to distinguish between at least two different particles. Another difficulty arises because of the noise structure which is now real noise and not artificially generated.

Table 4 shows the maximal detection rates if the subwindow width, the correction method for the significance level or the jump-time estimator are fixed. The detection rates are generally lower than on BGPoisson. Again, the subwindow width $k = 50$ leads to higher detection rates than $k = 30$. Furthermore, at least for the HL1-, the HL2- and the MD-test, the jump-time estimator seems to have a larger influence on the detection rate than on BGPoisson.

In Table 5 the parameter combinations which lead to the highest detection rate and the smallest MATD are shown. The ratio between γ_2 and γ_1 is much higher than for BGPoisson. In contrast to the parameter combinations on BGPoisson, only the method of Qiu and Yandell is selected for the jump-time estimation.

Table 4. Particle detection rates in the training data set of BGReal for the different two-sample tests. Each row shows the maximal detection rate.

Parameter	Value	HL1	HL2	MD	t	Wilcoxon
k	30	0.66	0.58	0.28	0.68	0.65
	50	0.88	0.88	0.74	0.89	0.89
Correction	Bonferroni	0.88	0.88	0.74	0.89	0.89
	Holm	0.88	0.87	0.73	0.89	0.89
Jump time	QY1	0.88	0.87	0.74	0.89	0.89
	QY2	0.88	0.88	0.74	0.89	0.89
	WC1	0.84	0.82	0.70	0.89	0.88
	WC2	0.86	0.84	0.70	0.89	0.88

Again, the values for γ_1, γ_2 and γ_3 are similar for the t-test and the Wilcoxon test. This is also the case for the HL1- and the HL2-test.

Table 6 gives the evaluation results on the training and the test data set. On the training data, only the reference method detects more than 90 % of the particles. Except for the MD-test, the rates for the two-sample tests are just slightly smaller than this. The reference method has a much higher MATD than the test-based methods. The MATD value for the HL2-test is now similar to those for the other two-sample tests. Except for the MD-test, each method leads to one false detection on the training data and to multiple detections of some particles, but its detection rate is much smaller than for the remaining methods. On the test data, no method leads to a false detection. The detection rates are comparable with those achieved on the training data set. All MATD values increase compared to the results on the training data set.

Table 5. Parameter values for each test which lead to the highest particle detection rate and the smallest mean absolute deviation between the estimated and the true change time on the training data set of BGReal.

Parameter	HL1	HL2	MD	t	Wilcoxon
k	50	50	50	50	50
Correction	Holm	Bonferroni	Bonferroni	Holm	Bonferroni
γ_1	1	1	2	2	3
γ_2	12	10	13	18	18
γ_3	4	4	4	4	4
Jump time	QY2	QY2	QY1	QY2	QY2

Table 6. Evaluation of the parameter values on the training and test data sets of BGReal for the two-sample tests and the reference method.

Data set	Measure	HL1	HL2	MD	t	Wilcoxon	Reference
Training data	Output segments	270	269	224	270	270	383
	Detection rate	0.88	0.88	0.74	0.89	0.89	0.92
	MATD	1.88	2.04	2.28	1.39	1.53	8.59
Test data	Output segments	275	260	239	260	263	379
	Detection rate	0.90	0.85	0.78	0.84	0.87	0.92
	MATD	2.63	3.06	2.31	2.35	2.48	9.15

Discussion of the Results. The results show that the test-based method for the detection of local intensity changes from Sect. 3 is able to identify the majority of the particles on the sensor surface. It is possible to achieve detection rates comparable to those of the reference method. Furthermore, the results indicate that a much higher accuracy of the change time estimation than for the reference method can be attained.

Among the considered tests, the MD-test leads to the lowest detection rates, especially on the data set BGReal. This could be due to its poor power in the jump detection. The t-test, the Wilcoxon test and the HL1-test are comparable regarding the accuracy and the detection rates. The HL2-test leads to similar detection rates, but has a higher MATD on BGPoisson. Therefore, the t-test, the Wilcoxon test and the HL1-test seem to be reasonable choices regarding the criteria detection rate and accuracy. On the test data set of BGReal the HL1-test delivered the highest accuracy which could be interpreted in the way that it is more suitable in real-noise situations.

The big difference between the results of the HL1- and the HL2-test for the MATD is surprising as the one-sample and the two-sample Hodges-Lehmann estimators behave, at least asymptotically, similar in case of symmetric distributions. For asymmetric distributions the two-sample Hodges-Lehmann estimator has better asymptotic properties [12]. The jump-time estimator could be the reason for this: For the HL2-test we use the method of Wu and Chu. The time with the minimal p-value is not unique in many cases. Choosing the smallest point in time of those with the minimal p-value can lead to a large estimation error if many points in time lead to the same minimal p-value. This problem does not occur for the t-test where we also use the method of Wu and Chu. For the t-test the minimal p-value should be reached when the reference window contains all non-shifted observations and the test window contains all shifted ones. When the window is moved, the test statistic will be directly influenced by the shifted observation in the reference window. On the contrary, the test statistic of the robust HL2-test will be little influenced by some shifted observations and therefore several subsequent windows can deliver the same p-value.

In general, the two-sample tests lead to a much more accurate estimation of the adhesion time than the reference method. An explanation could be that we use the information on the estimated jump time explicitly in our aggregation step. Furthermore, the reference method detects several particles more than once. Each output segment delivers an estimator for the adhesion time. It is not clear whether all of the corresponding adhesion time estimations are the same for the corresponding particle. Thus, a possible variability could lead to a more imprecise estimation. Moreover, the parameters of the reference method are optimized to get a high detection rate.

Generally, it seems that larger subwindow widths are to be preferred to get a high detection rate because they lead to a high power and thus more candidate coordinates can be identified. A problem with a large fixed subwindow width is that for several observations at the beginning and the end of the time series no two-sample test can be performed. Besides, it gets difficult to distinguish between multiple jumps which appear next to each other. Additionally, in other applications, trends could cause problems for large window widths because they can be confused with a jump.

The relationship between γ_1 and γ_2 seems to get more important when the number of particles in the sample is high. Many particles make it likely that there are multiple jumps at one point in time. Because of possible wrong detections it is necessary that the threshold γ_2 for the minimal number of jumps per time window of width γ_1 is high. The windows should not be too large as it is possible that too many consecutive time windows are unified in the temporal segmentation.

The jump-time estimator only seems to be important when there are many particles in the sample. An explanation could be that a higher amount of particles leads to more time series with multiple jumps as it is more likely that the particles are overlapping. In this case the method of Qiu and Yandell seems to be a good choice as the jump-time candidates are grouped before estimating the jump time.

The adjustment of the significance level does not seem to have much influence on the results. But it has to be noted that we only used one value for the global significance level, so that further improvements might be possible.

5   Conclusions

The identification of local intensity changes in time series of grayscale images based on statistical two-sample tests in moving time windows seems to be a promising approach. A local intensity change causes a permanent alteration of the grayscale values at the related pixel coordinates. We extract a time series of grayscale values for each coordinate. If it belongs to a local intensity change, a jump at the change time is induced in the signal. We identify these jumps by applying two-sample tests for the location problem in moving windows to the time series. Each of the resulting candidate pixels is assigned to a coordinate set depending on the estimated jump time and its coordinates.

For the exemplary application of this approach we use data from the PAMONO (Plasmon assisted Microscopy of Nano-Size Objects) biosensor [28]. This device can be used to detect specific particles, e.g. viruses, in a sample fluid.

Our method achieves high detection rates and a precise estimation of the change time. A test based on the one-sample Hodges-Lehmann estimator, the classical two-sample t-test and the two-sample Wilcoxon rank-sum test deliver the best results regarding these two criteria. Apart from the two-sample test itself, another important parameter is the window width which should not be too small to achieve a high detection rate.

In the evaluation the parameters of our method were optimized in a rather simple way by trying out several combinations, because the aim was only to investigate the general usefulness of the method. A more structured approach could lead to even better values for the detection rate and the accuracy.

The given data do not contain any type of trend. In trend situations the application of two-sample tests for the location problem could be problematic because a trend can be confused with a location shift when the subwindow width or the trend is large. In this case, other test procedures which are capable of dealing with trends or a preliminary trend removal could be preferable.

Taking all output segments of the detection procedure as the relevant local intensity changes can cause many false detections. Artifacts can lead to time series which resemble those of the relevant particles. Additionally, they can have a large spatial extent, so that the median filtering will not remove them. Thus, the output segments can contain particles as well as artifacts. Hence, they only should be seen as candidates and further criteria might be necessary to distinguish between the particles of interest and other structures. A possibility for future research could be the combination of the proposed method with a classification based on polygonal form factors similar to [22].

Acknowledgments. The work on this paper has been supported by Deutsche Forschungsgemeinschaft (DFG) within the Collaborative Research Center SFB 876 "Providing Information by Resource-Constrained Analysis", project C3. The computations were performed on the LiDO HPC cluster at TU Dortmund University. We thank the Leibniz-Institut für Analytische Wissenschaften – ISAS – e.V. in Dortmund for providing the PAMONO data to us and Dominic Siedhoff and Pascal Libuschewski for helpful advice on how to work with the PAMONO data and for providing the results for the reference method.

References

1. Abbas, S.: Detektion von Nanoobjekten in Graustufenbildern und Bildsequenzen mittels robuster Zeitreihenmethoden zur Strukturbrucherkennung. Master's thesis, TU Dortmund University, Dortmund (2013)
2. Bischl, B., Lang, M., Mersmann, O.: BatchExperiments: Statistical Experiments on Batch Computing Clusters. R package version 1.0-0968 (2013)
3. Bivand, R.S., Pebesma, E., Gomez-Rubio, V.: Applied Spatial Data Analysis with R, 2nd edn. Springer, New York (2013)
4. Dahl, D.B.: xtable: Export tables to LaTeX or HTML. R package version 1.7-3 (2014)
5. Fried, R.: On the Robust detection of edges in time series filtering. Comput. Stat. Data Anal. **52**(2), 1063–1074 (2007)

6. Fried, R., Dehling, H.: Robust nonparametric tests for the two-sample location problem. Stat. Methods Appl. **20**(4), 409–422 (2011)
7. Fried, R., Gather, U.: On rank tests for shift detection in time series. Comput. Stat. Data Anal. **52**(1), 221–233 (2007)
8. Gomes, A.J.P., Voiculescu, I., Jorge, J., Wyvill, B., Galbraith, C.: Implicit Curves and Surfaces: Mathematics, Data Structures and Algorithms. Springer, Dordrecht (2009)
9. Gonzalez, R.C., Woods, R.E.: Digital Image Processing, 3rd edn. Prentice Hall, Upper Saddle River (2008)
10. Hodges, J.L., Lehmann, E.L.: Estimates of location based on rank tests. Ann. Math. Stat. **34**(2), 598–611 (1963)
11. Holm, S.: A simple sequentially rejective multiple test procedure. Scand. J. Stat. **6**(2), 65–70 (1979)
12. Høyland, A.: Robustness of the Hodges-Lehmann estimates for shift. Ann. Math. Statist. **36**(1), 174–197 (1965)
13. Jähne, B.: Digitale Bildverarbeitung, 7th edn. Springer, Berlin (2012)
14. Maronna, R.A., Martin, R.D., Yohai, V.J.: Robust Statistics: Theory and methods. Wiley Series in Probability and Statistics. Wiley, Chichester (2006)
15. Meyer, D., Dimitriadou, E., Hornik, K., Weingessel, A., Leisch, F.: e1071: Misc Functions of the Department of Statistics (e1071), TU Wien. R package version 1.6-3 (2014)
16. Morell, O.: On nonparametric methods for robust jump-preserving smoothing and trend detection. Ph.D. thesis, TU Dortmund University, Dortmund (2012)
17. Pages, H., Carlson, M., Falcon, S., Li, N.: AnnotationDbi: Annotation Database Interface. R package version 1.26.0
18. Qiu, P., Yandell, B.: A local polynomial jump-detection algorithm in nonparametric regression. Technometrics **40**(2), 141–152 (1998)
19. R Core Team: R: A Language and Environment for Statistical Computing. R Foundation for Statistical Computing, Vienna (2013)
20. Rangayyan, R.M.: Biomedical Image Analysis. CRC Press, Boca Raton (2005)
21. Sharpsteen, C., Bracken, C.: tikzDevice: R Graphics Output in LaTeX Format. R package version 0.7.0 (2013)
22. Siedhoff, D., Weichert, F., Libuschewski, P., Timm, C.: Detection and classification of nano-objects in biosensor data. In: International Conference on Systems Biology - Microscopic Image Analysis with Application in Biology (MIAAB 2011), pp. 1–6 (2011)
23. Timm, C., Libuschewski, P., Siedhoff, D., Weichert, F., Müller, H., Marwedel, P.: Improving nanoobject detection in optical biosensor data. In: Proceedings of the 5th International Symposium on Bio- and Medical Information and Cybernetics (BMIC 2011), pp. 236–240 (2011)
24. Urbanek, S.: png: Read and Write PNG Images. R package version 0.1-7 (2013)
25. Weichert, F., Gaspar, M., Timm, C., Zybin, A., Gurevich, E.L., Engel, M., Müller, H., Marwedel, P.: Signal analysis and classification for surface plasmon assisted microscopy of nanoobjects. Sens. Actuators B **151**, 281–290 (2010)
26. Wickham, H.: ggplot2: Elegant Graphics for Data Analysis. Springer, New York (2009)
27. Wu, J.S., Chu, C.K.: Kernel-type estimators of jump points and values of a regression function. Ann. Stat. **21**(3), 1545–1566 (1993)
28. Zybin, A., Kuritsyn, Y.A., Gurevich, E.L., Temchura, V.V., Überla, K., Niemax, K.: Real-time detection of single immobilized nanoparticles by surface plasmon resonance Imaging. Plasmonics **5**(1), 31–35 (2010)

SCHEP — A Geometric Quality Measure for Regression Rule Sets, Gauging Ranking Consistency Throughout the Real-Valued Target Space

Wouter Duivesteijn[1(✉)] and Marvin Meeng[2]

[1] Department of Engineering Mathematics, University of Bristol, Bristol, UK
w.duivesteijn@bristol.ac.uk
[2] Leiden Institute of Advanced Computer Science, Leiden University,
Leiden, the Netherlands
m.meeng@liacs.leidenuniv.nl

Abstract. As is well known since Fürnkranz and Flach's 2005 ROC 'n' Rule Learning paper [6], rule learning can benefit from result evaluation based on ROC analysis. More specifically, given a (set of) rule(s), the Area Under the ROC Curve (AUC) can be interpreted as the probability that the (best) rule(s) will rank a positive example before a negative example. This interpretation is well-defined (and stimulates the intuition!) for the situation where the rule (set) concerns a classification problem. For a regression problem, however, the concepts of "positive example" and "negative example" become ill-defined, hindering both ROC analysis and AUC interpretation. We argue that for a regression problem, an interesting property to gauge is the probability that the (best) rule(s) will rank an example with a high target value before an example with a low target value. Moreover, it will do so consistently for all possible thresholds separating the target values into the high and the low. For each such threshold, one can retrieve an old-fashioned binary-target ROC curve for a given rule set. Aggregating all such ROC curves, we introduce SCHEP: the Surface of the Convex-Hull-Enclosing Polygon. This is a geometric quality measure, gauging how consistently a given rule (set) performs the aforementioned separation when the threshold is varied through the target space.

1 Preliminaries and Related Work

Formality first: let us start by introducing some notation, which is necessary to properly discuss the related work. Throughout this paper, we assume a dataset Ω to be a bag of N *records* $x \in \Omega$ of the form $x = (a_1, \ldots, a_k, t)$, where k is a positive integer. We call $a_1, \ldots, a_k$ the *attributes* of x, and t the *target* of x. Whenever necessary, we will distinguish the ith record from other records by annotating it and its elements with a superscript i, i.e., $x^i = (a_1^i, \ldots, a_k^i, t^i)$. We suppose that the attributes are taken from an unspecified domain $\mathscr{A}$, and assume that the target is taken from $\mathbb{R}$, following an unspecified probability density

© Springer International Publishing Switzerland 2016
S. Michaelis et al. (Eds.): Morik Festschrift, LNAI 9580, pp. 272–285, 2016.
DOI: 10.1007/978-3-319-41706-6_14

function $f(y), y \in \mathbb{R}$, of which our $t^1, \ldots, t^N$ form a random sample. Generally, the parameters and the shape of the pdf are unknown. For convenience, we assume that Ω is ordered such that the target values are nondecreasing, i.e., $\forall_{1 \le i \le j \le N} \ t^i \le t^j$.

For notational simplicity, we will let $u_1, \ldots, u_{N-1}$ denote the midpoints of the intervals between t-values:

$$\forall_{i=1}^{N-1} \quad u_i = \frac{1}{2}\left(t^i + t^{i+1}\right)$$

1.1 ROC Space

ROC curves have a rich history in signal detection theory. Traditionally they were used to visualize the tradeoff between hit rates and false alarm rates of classifiers [3,26]. The importance of the area under the ROC curve was recognized [2,10], and its relation to the Wilcoxon test of ranks explored [10]. ROC analysis has found its way in the wider scientific community, through recognition in a paper concerning the behavior of diagnostic systems which appeared in Science [25], and an even more general paper regarding math-based aids for decision making in medicine and industry which appeared in Scientific American [26].

The value of ROC curves for algorithm comparison was first recognized in the late eighties [23], leading to adoption of ROC analysis by the machine learning community. Extensive notes have been written on the application of traditional ROC methodology in machine learning and data mining; see for instance [4,5,9,22]. We will discuss the general gist of the ROC curve in this section; for a more comprehensive overview of its applications in data mining, see [22].

Traditional ROC analysis is done with respect to a binary response variable, i.e., there is a *binary target*. In this case, we denote the desirable value for the target by τ. We then distinguish two bags of records: the *positives* are those records having target value τ (we denote the bag by Pos), and the *negatives* are those records having a target value other than τ (denoted Neg).

Definition 1. *The* positives *are formed by the records whose target has value* τ. *Conversely, the* negatives *are formed by the records whose target has a value other than* τ:

$$Pos = \left\{x^i \in \Omega \,\middle|\, t^i = \tau\right\} \qquad Neg = \left\{x^i \in \Omega \,\middle|\, t^i \ne \tau\right\}$$

This definition also allows for the more general case of a nominal target, in a one-versus-all setting: the one desirable target value τ is contrasted against all other, undesirable target values.

In data mining, ROC is usually employed to analyze *rules*. Sometimes these rules come in the form of classifier decision bounds, sometimes in the form of local patterns in the dataset. Regardless of the particular data mining task at hand, a rule generally strives to separate (all or some) positives from (all or some) negatives. For the purposes of this paper, we disregard the form in which a rule is designed, and concentrate only on which records belong to the rule:

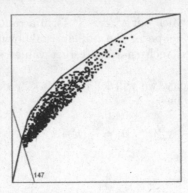

Fig. 1. Rules in ROC space, with convex hull (upper envelope only; the rest isn't interesting)

Definition 2. *A rule R can be any subbag of the dataset at hand: $R \subseteq \Omega$. The records $x \in R$ are said to be* covered *by the rule.*

The quality of a rule $R \subseteq \Omega$ is gauged by considering two quantities: the fraction of positives that are covered by the rule, and the fraction of negatives that are covered by the rule:

Definition 3. *Suppose a rule R. The* True Positive Rate (TPR) *is the proportion of positives covered by R. The* False Positive Rate *is the proportion of negatives covered by R:*

$$TPR(R) = \frac{|R \cap Pos|}{|Pos|} \qquad FPR(R) = \frac{|R \cap Neg|}{|Neg|}$$

The TPR is also known as the *sensitivity* of a rule, while the FPR is equivalent to one minus the *specificity* of a rule.

The ROC methodology analyzes found rules in *ROC space*: the two-dimensional unit square with the FPR on the horizontal axis and the TPR on the vertical axis. Figure 1 displays an example. The *perfect rule* $R = $ Pos would be in the top left corner; its coordinates in ROC space are $(0,1)$. The *empty rule* $R = \varnothing$ can be found in the bottom left corner, with coordinates $(0,0)$, and the *full rule* $R = \Omega$ can be found in the top right corner, with coordinates $(1,1)$.

1.2   Pattern Mining

Pattern mining [8,19] is the broad subfield of data mining where only a part of the data is described at a time, ignoring the coherence of the remainder. One class of pattern mining problems is *theory mining* [18], whose goal is finding subsets S of the dataset Ω that are interesting somehow:

$$S \subseteq \Omega \quad \Rightarrow \quad \text{interesting}$$

Typically, not just any subset of the data is sought after: only those subsets that can be formulated using a predefined *description language* $\mathscr{L}$ are allowed. A canonical choice for the description language is conjunctions of conditions on attributes of the dataset. If, for example, the records in our dataset describe people, then we can find results of the following form:

$$\text{Age} \geq 30 \wedge \text{Smoker} = \text{yes} \quad \Rightarrow \quad \text{interesting}$$

Allowing only results that can be expressed in terms of attributes of the data, rather than allowing just any subset, ensures that the results are relatively easy to interpret for a domain expert: the results arrive at his doorstep in terms of quantities with which he should be familiar. A subset of the dataset that can be expressed in this way is called a *subgroup*.

In the best-known form of theory mining, *frequent itemset mining* [1], the interestingness of a pattern is gauged in an unsupervised manner. Here, the goal is to find patterns that occur unusually frequently in the dataset:

$$\text{Age} \geq 30 \wedge \text{Smoker} = \text{yes} \quad \Rightarrow \quad \text{(high frequency)}$$

The most extensively-studied form of *supervised* theory mining is known as *Subgroup Discovery* (SD) [12], where one (typically binary) attribute of the dataset is singled out as the *target*. The goal is to find subgroups for which the distribution of this target is unusual: if the target describes whether the person develops lung cancer or not, we find subgroups of the following form:

$$\text{Smoker} = \text{yes} \quad \Rightarrow \text{lung cancer} = \text{yes}$$
$$\text{Age} < 30 \quad \Rightarrow \text{lung cancer} = \text{no}$$

The target in SD is commonly assumed to be nominal [12]. However, its initial formulation [13,27] allowed for a more generic target concept, including but not limited to a numeric target. This paper is concerned with this particular setting: Subgroup Discovery with a single target taken from a continuous domain.

1.3   ROC Analysis, and ROC 'n' Rule Learning [6]

Notwithstanding a certain focal difference, the definitions of a rule R in Sect. 1.1 and a subgroup S in Sect. 1.2 allow us to identify the two concepts with one another. Hence, we can analyze both subgroups and rules in ROC space. From now on, we will freely associate subgroups and rules, and to celebrate that confusion, we will from now on write about rule sets denoted by a calligraphic S.

Given the domain at hand, we may assign different costs to the two types of imperfections we can have for a rule R: false positives (i.e., negatives covered by R) and false negatives (i.e., positives not covered by R).

Definition 4. *Suppose a rule set $\mathscr{S}$, and a cost assignment to false positives and false negatives. A rule $R \in \mathscr{S}$ is called* optimal *for this cost assignment within $\mathscr{S}$, if its total cost equals the minimum total cost for any rule in $\mathscr{S}$.*

Clearly many rules in Fig. 1 are not optimal for any cost assignment to the two types of imperfections: for any two rules R_1, R_2, if $\mathrm{TPR}(R_1) \geq \mathrm{TPR}(R_2)$ and $\mathrm{FPR}(R_1) \leq \mathrm{FPR}(R_2)$, and at least one of these inequalities is strict, then R_2 cannot be optimal. Within a rule set $\mathscr{S}$, a rule is optimal for some cost assignment if and only if it lies on the upper envelope of the convex hull of $\mathscr{S}$ in ROC space [22]. We will refer to this ROC convex hull as *ROCCH* and conflate the upper envelope and the convex hull in the remainder of this paper.

For any rule set $\mathscr{S}$, we can use its *Area Under the Convex Hull* (AUCH) to determine how well $\mathscr{S}$ *captures* the target. We assume that both the empty and the full rule are part of every rule set, since these rules can always be 'found'. Hence the AUCH has minimal value 0.5. If the perfect rule is an element of $\mathscr{S}$, then its AUCH equals 1. However, often the perfect rule cannot be found, usually due to the attribute space being too large to fully explore (for instance when many attributes are taken from a continuous domain). In this case, a good rule set contains both relatively small rules that contain almost no negatives, and relatively large rules that contain almost all positives. The best of these rules will lie on the ROCCH. Hence the AUCH is a measure for the degree to which $\mathscr{S}$ captures the target.

The AUCH measures only one particular desirable property of a rule set, and disregards many other desirable properties. For instance, it only looks at the best rules in the set, and does not take into account how much filler the rule set contains: if we add many bad-scoring rules to a rule set its AUCH will not decrease, even though an end-user will prefer the rule set without these bad-scoring rules. Solving this problem [7,16] is beyond the scope of this paper.

1.4   Keep on ROCing in the Real World?

Like many real-world measurements, the target that we try to capture in this paper stems from a continuous domain. Hence, we are looking for a set of rules in a regression setting, or *regression rule set* for the sake of brevity. This breaks down the building blocks of ROC space as introduced in Sect. 1.1: the positives and negatives become ill-defined, which breaks down ROC space and any ensuing analysis. In this Real World, can we still Keep on ROCing?

Some attention has been spent on ROC analysis for the case where the target is nominal, but multi-class rather than binary. When the entire target space is to be managed, the resulting performance can no longer be summarized in two dimensions; higher-dimensional polytopes are needed. Extension of ROC convex hull analysis to multiple classes and multi-dimensional convex hulls has been discussed by Srinivasan [24]. A short paper outlining issues and opportunities was written by Lane [15]. Appropriate attention has also been given to multi-class AUC analysis [9,21]. However, none of these methodologies can be straightforwardly generalized from a multi-class target to a continuous target.

Currently, perhaps the most comprehensive work on ROC curves is the book by Krzanowski and Hand [14], titled "ROC Curves for Continuous Data". Although the word "Continuous" features in the title, this refers merely to the attributes in the data, and not to the target. In the eighth chapter, "Beyond the basics", the authors do discuss the extension of ROC analysis to a multi-class nominal target (cf. Sect. 8.4), but that is as far as the book goes.

Until recently, no canonical adaptation of ROC analysis for regression existed, though many have tried. In 2013, however, José Hernández-Orallo published a paper [11] introducing Regression ROC space (RROC), and showing a direct relation between the Area Over the RROC curve and the population error variance of the regression model. The paper goes to great lengths to derive variants for the regression setting of well-known notions in traditional ROC analysis, including the convex hull properties discussed in Sect. 1.3. Hence, the paper answers affirmatively to the question posed in the first paragraph of this section. We think, however, that we can say more interesting things about a regression rule set, by actively exploiting the insufficient information that is by definition available to us when studying a finite sample of a target from a real-valued domain.

2 SCHEP

As we wrote in Sect. 1, we assume that the target is taken from $\mathbb{R}$, and follows an unspecified probability density function $f(y), y \in \mathbb{R}$. The existence of such a function implies that the probability distribution from which the target is taken is nondegenerate. For many real-life measurements this assumption and the implied properties should hold; we will discuss in Sect. 2.2 that the assumption isn't always met in practice, and how to adapt the forthcoming methods for this occurrence. For the time being, however, we concern ourselves with a target representing a continuous random variable, having an absolutely continuous distribution.

The dataset we consider consists of N records. Our target has a nondegenerate probability distribution, whose form is generally unknown. Since we only have an N-sized sample, there is by definition not enough information about the target space to capture the distribution; the problem is underdetermined. Hence, we propose to use all the information about the target space that we have (Sects. 2.1 and 2.2), and exploit the lack of remaining information to serve our purpose (Sect. 2.3).

2.1 Decomposing the Regression Problem

Given a continuous target, there is usually not one desirable target value τ. Instead, we usually strive for a more generic concept of extremity: having a relatively high or low value for the target. Regardless of this precise concept, the definitions of the bags of positives and negatives, as introduced in Sect. 1.1, become fuzzy. Hence, all ensuing ROC methodology is not readily applicable for the case where the target is continuous. However, a simple thresholding approach can solve that problem. To facilitate this thresholding, we use the following notation:

Definition 5. *Let a dataset Ω and a $y \in \mathbb{R}$ be given. The y-binarized-target version of Ω, denoted D_y, is the dataset obtained from Ω by replacing each t^i by* $\mathbb{1}\{t^i > y\}$.

Here, $\mathbb{1}\{.\}$ denotes the indicator function, which means that the continuous target value t^i is replaced by 1 if it is larger than y, and by 0 if it is not. Hence, D_y divides the target values in positives and negatives, by comparing them to the threshold y.

The intuition runs as follows: suppose we are interested in finding rules discriminating the top-ν from the bottom-$(N - \nu)$ target values. We can do this by defining a new binary target t'. Let $t' = 1$ for a record if its original target is among the top-ν, and $t' = 0$ otherwise. For this new binary target t', we can use binary ROC analysis to determine how well a rule set captures its distribution. In other words, we determine the probability that a random record with a target value in the top-ν is ranked before a random record with a target value in the bottom-$(N - \nu)$.

If we use any point in the target space as a threshold to define a binary target as described in the previous paragraph, we gauge how well a rule set discriminates the top-ν from the bottom-$(N - \nu)$ target values for some choice of ν. Hence, if we let our threshold slide through the target space, we obtain the discriminatory capabilities of a rule set *for all such separations*. Aggregating these performances, we can measure the consistency with which a rule set captures a continuous target.

Sliding the threshold y through the entirety of its domain, $\mathbb{R}$, makes little sense in our setting. After all, we only have information about the target distribution at a sample of N fixed points in target space. Hence, we cannot make reasonable assessments outside of the sample range: the most sensible solution is to disregard values of y such that $y < t^1$ or $y > t^N$.

Within the sample range, our information on the target distribution only changes at the target values present in our dataset. Suppose a value $x \in \mathbb{R}$ that lies strictly between two sample target values: $t^i < x < t^{i+1}$. For a very small amount $\varepsilon > 0$, the information we can infer from dataset D_x is exactly the same as the information we can infer from dataset $D_{x+\varepsilon}$. Intuitively, between two adjacent target values, we only need to consider a binarized-target version of the dataset once, since no more information is available. Since the choice is arbitrary, for the sake of clarity we will consider only the target interval midpoints $u_1, \ldots, u_{N-1}$. In Sect. 2.3, we analyze the ROC convex hulls of precisely these $N - 1$ y-binarized-target versions of the dataset, in order to define a consistency measure for a regression rule set:

Definition 6. *Let a rule set $\mathscr{S}$ and a dataset Ω with continuous target be given. The Collection of Binarized ROCCHs (CBR) is defined as:*

$$\mathrm{CBR}\,(\mathscr{S}, \Omega) = \rho\left(\{\mathrm{ROCCH}\,(\mathscr{S}, D_{u_i})\}_{i=1}^{N-1}\right) \tag{1}$$

Here, the operator $\mathrm{ROCCH}\,(\mathscr{S}, D_y)$ returns the ROC convex hull of the rule set $\mathscr{S}$ on the y-binarized-target version of Ω. The *aggregation operator* ρ remains undefined for now; we will instantiate it in Sect. 2.3.

2.2 Dealing with Duplicates

By imposing a probability density on the target domain (cf. Sect. 1), we have implicitly assumed that all target values are distinct. After all, if we take a finite random sample from a nondegenerate distribution on $\mathbb{R}$, the probability that we draw two or more equal values is 0. However, in many real-life datasets this assumption does not hold. Consider for instance the CMC dataset, available from the UCI Machine Learning Repository [17]. The main task in this dataset is to predict the contraceptive method choice made by women, but one could modify the task to become predicting the age instead. Theoretically, the lifespan of any living creature can be measured at a level that is finegrained enough to make it continuous for all practical purposes. However, since the recorded age of the 1 473 women in the dataset is represented by an integer number of years, we find only 34 different values for the target. This truncating violates the assumption we made, but the age is still better represented as a numeric target rather than nominal. Hence we adapt the CBR definitions to allow recurring target values.

As we have discussed in the previous sections, the information we have about the continuous target distribution remains constant, except for the changes we observe at the target values that are present in our dataset. Hence, we are only interested in the $\mathfrak{N}$ *distinct* target values in the dataset, which we denote by $t^1, \ldots, t^{\mathfrak{N}}$. For convenience, we assume that they are ordered ascendingly, i.e., $\forall_{1 \leq i < j \leq \mathfrak{N}} \; t^i < t^j$. We let $u_1, \ldots, u_{\mathfrak{N}-1}$ denote the corresponding midpoints:

$$\bigvee_{i=1}^{\mathfrak{N}-1} u_i = \frac{1}{2}\left(t^i + t^{i+1}\right)$$

So, in practice, the formula in Eq. (1) becomes:

$$\mathrm{CBR}\left(\mathscr{S}, \Omega\right) = \rho\left(\{\mathrm{ROCCH}\left(\mathscr{S}, D_{u_i}\right)\}_{i=1}^{\mathfrak{N}-1}\right) \tag{2}$$

In this form, the CBR can be determined in practice for every dataset with a numeric target, even if its distribution is degenerate. We have performed experiments on five UCI [17] datasets, extracting a regression rule set by running Subgroup Discovery with Absolute Z-score [20] as quality measure. Details on the datasets can be found in Table 1. The found (non-aggregated) CBRs can be found in Figs. 2, 3, 4, 5 and 6; different colors correspond to different ROCCHs.

Table 1. Dataset characteristics

Dataset	N	$\mathfrak{N}$	k	Target
Automobile	205	185	26	Price
Boston housing	506	229	13	MEDV
Contraceptive Method Choice	1473	34	9	Wifes_age
Year Prediction MSD	515345	90	90	Year
Zoo	101	6	17	Legs

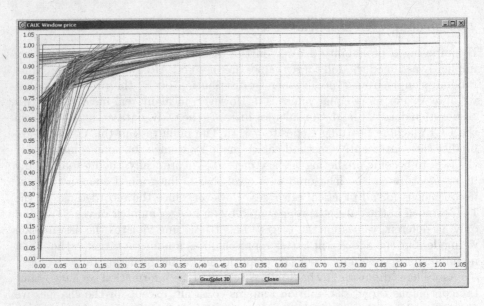

Fig. 2. CBR for a rule set found on the Automobile dataset (Color figure online)

2.3   SCHEP

Now that we can practically create the Collection of Binarized ROCCHs for any dataset, it's time to actually *do* something with it, by instantiating the aggregation operator ρ. We will do so in a very intuitive way, by analyzing what the CBR components set out to do, when given a dataset Ω and rule set $\mathscr{S}$.

Every convex hull in the CBR is associated with one particular threshold value u_i. This threshold converts the continuous-target dataset Ω into its u_i-binarized-target version D_{u_i}. For this dataset, the CH contains information about how well the optimal rules within $\mathscr{S}$ manage to rank positives before negatives. Hence each convex hull represents the performance of selected rules in $\mathscr{S}$ in separating high from low target values, for a particular instantiation of 'high' and 'low'.

The CBR contains lots of information, and there are several meaningful things we could do with it to extract interesting properties of our regression rule set. If we were to compute the AUCH of each ROCCH and aggregate the scores (for instance by simply averaging them), then we obtain an AUC-style measure for regression. This Continuous AUCH could be interpreted as the average probability that a random record with high target value is ranked before a random record with low target value, where the average is taken over the instantiations of 'high' and 'low' present in our target range. This would be a continuous AUC-style measure that is substantially different from the one introduced by Hernández-Orallo [11], which might be an interesting research avenue to explore.

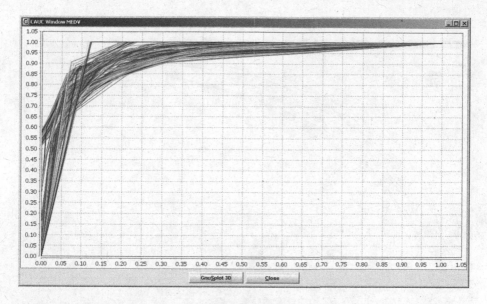

Fig. 3. CBR for a rule set found on the Boston housing dataset (Color figure online)

However, there's more to explore. Each convex hull tells us something about the performance of the optimal rules for a particular threshold value. Notice that the set of rules contributing to this performance may vary with the threshold! A rule may be very informative (and hence lie on the convex hull) for making the separation for a relatively high threshold value, but it is likely to be very irrelevant for making the separation for a relatively low threshold value. Other rules in the rule set may then step up to the plate, and take their place on the convex hull for the lower threshold value. This leads to the observation that the *Collection* of Binarized ROCCHs, the CBR, contains information of how *consistent* a rule set is in finding similarly-well-performing rules for varying thresholds. Hence, we introduce SCHEP:

Definition 7. *The* Surface of the Convex-Hull Enclosing Polygon *(SCHEP) is the instantiation of Eq. (2), where the aggregation operator ρ measures the surface of the smallest polygon enclosing all ROCCHs.*

The two extreme cases are: (min) if all ROCCHs overlap, then SCHEP assumes value 0; (max) if one ROCCH reaches ROC heaven, and another ROCCH is formed by no rules at all, then SCHEP assumes value 0.5. Lower values correspond to more consistent rule sets. In some sense, SCHEP gauges how diverse a rule set $\mathscr{S}$ is. SCHEP assumes lower values if $\mathscr{S}$ contains elements that manage to play to their strengths at different parts of the target domain. No single rule can make every possible separation of high and low target values on its own: instead one needs enough rules that specialize in enough different parts of the target domain to reach the best results. Thus formulated, SCHEP is a metaphor for the ideal research group.

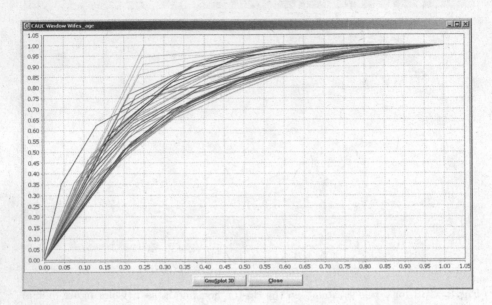

Fig. 4. CBR for a rule set found on the Contraceptive Method Choice dataset (Color figure online)

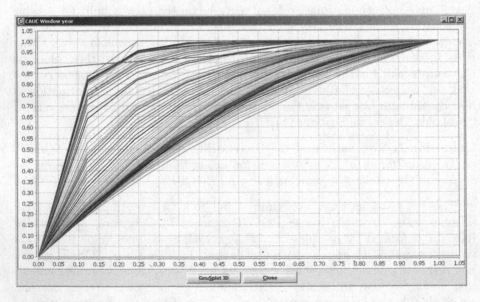

Fig. 5. CBR for a rule set found on the Year Prediction MSD dataset (Color figure online)

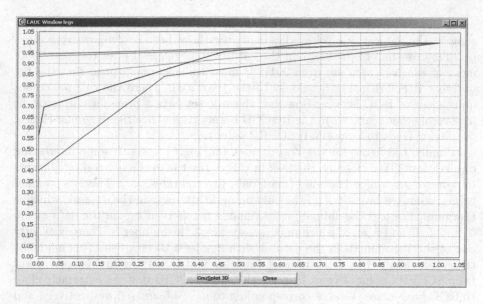

Fig. 6. CBR for a rule set found on the Zoo dataset (Color figure online)

2.4   Concluding Empirical Observations

If we consider the CBRs on display in Figs. 2, 3, 4, 5 and 6, we can make some empirical observations that illustrate where SCHEP is a fruitful addition to existing ROC technology. Traditionally, the rule set found on the Automobile dataset (cf. Fig. 2) would probably be considered best; all ROCCHs are relatively close to ROC heaven, and a decent amount are extremely close. However, the consistency is rather less ideal. There is a striking difference between the ROCCHs for cars priced at $22,500$ and beyond (where the AUCH is extremely close to one), and the cars that are cheaper than that (where the AUCH is substantially lower). This difference is clearly visible in Fig. 2 as a nearly completely white triangle with coordinates $(0,0.75)$, $(0,0.9)$, $(0.07,0.091)$; several ROCCHs soar above a much larger set of ROCCHs that do not come close to the same performance. As a consequence, SCHEP will assign a higher value to this rule set than to the one found on the Boston housing data (cf. Fig. 3), which doesn't come nearly as close to ROC heaven but is much more consistent throughout the target domain.

The rule set found on the Contraceptive Method Choice dataset (cf. Fig. 4) will not have a very high value for a traditional AUC. Its consistency, however, is on roughly the same level as the Automobile rule set, and the SCHEP values will be close. The rule set found on the YearPredictionMSD (cf. Fig. 5), on the other hand, will have a SCHEP value close to 0.5. This is largely because this dataset has a very skewed target distribution. It consists of over half a million songs from the Million Song Dataset, and the task is to predict the year in which the song was released, based on timbre features. The problematic point is that from 1922, when the oldest song in the dataset was released, until 2011, when the youngest song in the dataset was released, songwriters have become progressively more prolific. Over half of the songs in the dataset were released after the year 2000, most of which in turn are from 2011. It must come as no surprise that the big outlying ROCCH in Fig. 5 belongs to the year 2011, where the rule set comes quite close to ROC heaven. However, the rule set pays a price for this good fit in several other years, when the ROCCH barely rises above the diagonal. Hence, overfitting leads to large (bad) SCHEP values.

Finally, the figure for the Zoo dataset (cf. Fig. 6) leads to some interesting observations. Traditionally, the task in this dataset is to classify zoo inhabitants into one of seven broad families (mammals, invertebrates, ...). This is a nominal target, so instead we focus on the only numeric attribute in the dataset, which is the number of legs the animal has. Available numbers of legs in the dataset are 0, 2, 4, 5, 6, and 8, all of which make sense except the number 5. There is exactly one of the 101 animals in the dataset with five legs: the starfish. This might be tricky to classify. However, it turns out that that is not the problematic point from SCHEP's point of view. Instead, the two ROCCHs that get extremely close to ROC heaven are the ones corresponding to $u_i = 4.5$ and 5.5, respectively, and the red ROCCH (which comes also pretty close) belongs to $u_i = 7$. Instead, by far the most problematic separation for the found rule set to make is the one between animals with less or more than three legs; SCHEP has no problems with the starfish.

References

1. Agrawal, R., Mannila, H., Srikant, R., Toivonen, H., Verkamo, A.I.: Fast discovery of association rules. Adv. Knowl. Disc. Data Min. **12**, 307–328 (1996)
2. Bradley, A.P.: The use of the area under the ROC curve in the evaluation of machine learning algorithms. Pattern Recogn. **30**, 1145–1159 (1997)
3. Egan, J.P.: Signal Detection Theory and ROC Analysis. Series in Cognition and Perception. Academic Press, New York (1975)
4. Fawcett, T.: An introduction to ROC analysis. Pattern Recogn. Lett. **27**, 861–874 (2006)
5. Flach, P.A., Hernández-Orallo, J., Ferri Ramirez, C.: A coherent interpretation of AUC as a measure of aggregated classification performance. In: Proceedings of the ICML, pp. 657–664 (2011)
6. Fürnkranz, J., Flach, P.A.: ROC 'n' rule learning - towards a better understanding of covering algorithms. Mach. Learn. **58**(1), 39–77 (2005)
7. Grosskreutz, H., Paurat, D.: Fast and memory-efficient discovery of the top-k relevant subgroups in a reduced candidate space. In: Gunopulos, D., Hofmann, T., Malerba, D., Vazirgiannis, M. (eds.) ECML PKDD 2011, Part I. LNCS, vol. 6911, pp. 533–548. Springer, Heidelberg (2011)
8. Hand, D., Adams, N., Bolton, R. (eds.): Pattern Detection and Discovery. LNCS, vol. 2447. Springer, Heidelberg (2002)
9. Hand, D.J., Till, R.J.: A simple generalization of the area under the ROC curve to multiple class classification problems. Mach. Learn. **45**(2), 171–186 (2001)
10. Hanley, J.A., McNeil, B.J.: The meaning and use of the area under an ROC curve. Radiology **143**, 29–36 (1982)

11. Hernández-Orallo, J.: ROC curves for regression. Pattern Recogn. **46**(12), 3395–3411 (2013)
12. Herrera, F., Carmona, C.J., González, P., del Jesus, M.J.: An overview on subgroup discovery: foundations and applications. Knowl. Inf. Syst. **29**(3), 495–525 (2011)
13. Klösgen, W.: Explora: a multipattern and multistrategy discovery assistant. In: Fayyad, U.M., Piatetski-Shapiro, G., Smyth, P., Uthurusamy, R. (eds.) Advances in Knowledge Discovery and Data Mining, pp. 249–271. MIT Press, Cambridge (1996)
14. Krzanowski, W.J., Hand, D.J.: ROC Curves for Continuous Data. Chapman and Hall, London (2009)
15. Lane, T.: Extensions of ROC analysis to multi-class domains. In: Proceedings of the ICML 2000 Workshop on Cost-Sensitive Learning (2000)
16. van Leeuwen, M., Knobbe, A.J.: Diverse subgroup set discovery. Data Min. Knowl. Disc. **25**(2), 208–242 (2012)
17. Lichman, M.: UCI Machine Learning Repository. School of Information and Computer Science, University of California, Irvine (2013). http://archive.ics.uci.edu/ml
18. Mannila, H., Toivonen, H.: Levelwise search and borders of theories in knowledge discovery. Data Min. Knowl. Disc. **1**(3), 241–258 (1997)
19. Morik, K., Boulicaut, J.F., Siebes, A. (eds.): Local Pattern Detection. Springer, New York (2005)
20. Pieters, B.F.I., Knobbe, A., Džeroski, S.: Subgroup discovery in ranked data, with an application to gene set enrichment. In: Proceedings of the Preference Learning workshop (PL 2010) at ECML PKDD (2010)
21. Provost, F., Domingos, P.: Well-trained PETs: improving probability estimation trees. CeDER Working Paper #IS-00-04, Stern School of Business, New York University (2001)
22. Provost, F.J., Fawcett, T.: Robust classification for imprecise environments. Mach. Learn. **42**(3), 203–231 (2001)
23. Spackman, K.A.: Signal detection theory: valuable tools for evaluating inductive learning. In: Proceedings of the International Workshop on Machine Learning, pp. 160–163 (1989)
24. Srinivasan, A.: Note on the location of optimal classifiers in n-dimensional ROC space. Technical report PRG-TR-2-99, Oxford University Computing Laboratory, Oxford, England (1999)
25. Swets, J.: Measuring the accuracy of diagnostic systems. Science **240**, 1285–1293 (1988)
26. Swets, J.A., Dawes, R.M., Monahan, J.: Better decisions through science. Sci. Am. **283**, 82–87 (2000)
27. Wrobel, S.: An algorithm for multi-relational discovery of subgroups. In: Proceedings of the PKDD, pp. 78–87 (1997)

Bayesian Ordinal Aggregation of Peer Assessments: A Case Study on KDD 2015

Thorsten Joachims[⊠] and Karthik Raman

Cornell University, Ithaca, NY 14853, USA
{tj,karthik}@cs.cornell.edu

Abstract. Peer assessment is the most common approach to evaluating scientific work, and it is also gaining popularity for scaling evaluation of student work in large and distributed classes. The key idea is that each peer reviewer or grader rates a relatively small subset of the items, and that some method of manual, semi-automatic, or fully-automatic aggregation of all assessments defines the eventual rating of all items – the grade in peer grading, or whether to accept or reject a scientific manuscript. In this paper, we explore in how far a Bayesian Ordinal Peer Assessment (BOPA) method can provide additional decision support when making acceptance/rejection decisions for a scientific conference. Using data from the 2015 ACM Conference on Knowledge Discovery and Data Mining (KDD), where this system was deployed, we discuss the potential merit of the BOPA approach compared to conventional decision support offered by the Microsoft Conference Management System (CMT).

Keywords: Peer review · Peer grading · Ordinal feedback · Rank aggregation

1 Introduction

Scientific conferences and large university courses both share the problem of evaluating large sets of items (e.g. scientific papers, project reports), where the quality of each item is difficult to evaluate automatically. A common approach is to use peer reviewing, where each reviewer assesses the quality of a small subset of the items. In such assessments, reviewers are typically asked to assign numeric scores regarding aspects and overall quality of the item, justifying each score with a written explanation. While this approach scales well with the number of items and allows complex criteria under which to evaluate quality, the key problem lies in aggregating the scores of a large number of reviewers into a coherent assessment of the items.

For scientific conferences, the final assessment comes down to the decision of whether to accept or reject a paper. The most widely used approach for aggregating reviewer scores into an acceptance decisions relies on a hierarchy of reviewers, meta-reviewers, and program chairs. This is also the approach taken

© Springer International Publishing Switzerland 2016
S. Michaelis et al. (Eds.): Morik Festschrift, LNAI 9580, pp. 286–299, 2016.
DOI: 10.1007/978-3-319-41706-6_15

at the 2015 ACM Conference on Knowledge Discovery and Data Mining (KDD), which will serve as a case study in this paper. Each of the reviewers assessed a small subset of all submissions, providing an average of 3.9 reviews per paper. Based on these reviews, meta-reviewers were then asked to make acceptance recommendations for their subset of papers. The program chairs made the final acceptance decisions based on the meta-reviewers' recommendations, oversaw the process, and intervened in the reviewing process where necessary.

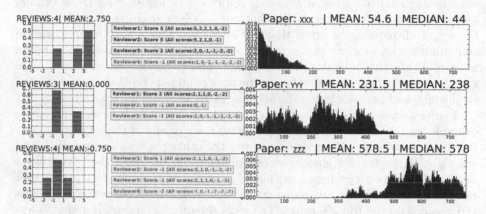

Fig. 1. Information provided to the meta-reviewers and program chairs for three example papers. In the posterior marginal rank distributions at the right of each panel, the x-axis shows the rank of the paper and the y-axis shows the probability of the paper placing at this rank. The plots also show posterior mean and median of the marginal distribution.

Under this decision making approach, both the meta-reviewers and the program chairs are faced with the problem of interpreting the numeric scores given by the reviewers. In particular, some reviewers may be more liberal in their use of "strong accept" (score +5) than others, and reviewers may disagree in their use of the numeric scale more generally. Such biases make it problematic to simply average numeric scores across a small number of reviewers, and using such average scores as a sorting criterion when displaying papers in an online interface may consciously or subconsciously impact the decision process in an unfair way.

In order to overcome this bias, our aim at KDD 2015 was to provide meta-reviewers and program chairs (which we jointly refer to as "decision makers") with more information that helps interpret reviewer scores. In particular, the aims were the following:

Mitigate Reviewer Bias. We would like to present decision makers with information that identifies whether a reviewer is more liberal or strict, and an aggregation of the reviewer scores that is unaffected (or at least less affected) by different reviewer rating scales.

Communicate Uncertainty. Averaging scores provides a point estimate of paper quality, but does not communicate the uncertainty of this estimate. To communicate uncertainty more effectively, we aim to provide decision makers with a full posterior distribution of the paper's predicted quality.

To address the problem of mitigating reviewer bias in using the rating scale, we explore an alternative method for interpreting reviewer scores [14,15]. Instead of interpreting a reviewer's assessment on an absolute scale, we merely derive an ordering from it. Using a Bayesian approach to aggregating these ordinal assessments, we infer posterior distributions of where each paper ranks among the set of all papers. We argue that the latter provides a very natural way to communicate uncertainty of the quality estimate on an intuitively meaningful scale. Overall, this provides meta-reviewers and program chairs with a more global assessment of each paper (w.r.t. the pool of all papers) that is not distorted by mismatched monotonic transformations of the assessment scale.

Figure 1 shows the information provided to the meta-reviewers for three example papers. The left of each panel shows a histogram of the reviewer scores, the middle shows how the reviewers scored the other papers they reviewed, and the right shows the marginal posterior distribution of where the paper ranks among all papers according to the model explored in this paper. The posterior rank distribution of the first paper shows that virtually all its probability mass is contained on the top 200 ranks. The second paper has a posterior that is less peaked and communicates that the model is very uncertain about where it ranks. For the third paper, the model is confident that the paper ranks below the top 300 submissions.

In the following, we outline our approach to inferring these posterior rank distributions from ordinal reviewer assessment. We first formalize the learning problem and then adapt a Bayesian aggregation model that was originally developed for peer grading [14]. We then perform a retrospective analysis of how well the inferred posterior distributions of this model reflect the outcome of the reviewing process, and how presentation biases interact with the predictions of the model.

1.1   Peer Assessment Approaches

In the standard reviewing process of computer science conferences, we are faced with the following *peer assessment* problem. Given is a set of $|D|$ *papers* $D = \{d_1, ..., d_{|D|}\}$ for each of which we need to make a decision y_d whether to accept or reject. The assessment is performed by a set of $|G|$ reviewers $G = \{g_1, ..., g_{|G|}\}$. Each reviewer g receives a subset of papers $D_g \subset D$ to assess. As feedback, each reviewer g provides a score $y_d^{(g)}$ for each of the papers in D_g.

In KDD 2015, there were $|G| = 595$ reviewers and $|D| = 752$ for which we provided decision support analytics. Each reviewer g_i received a subset D_g of average size 4.9. This provided on average 3.9 cardinal assessments for each papers. The assessment scale was "Strong Reject", "Reject", "Weak Reject",

"Weak Accept", "Accept", "Strong Accept". Based on these reviews, 68 meta-reviewers were then asked to make acceptance recommendations for a subset of on average 11.1 papers.

1.2 Cardinal Peer Assessment

The traditional approach of aggregating assessment scores for each paper that is embedded in the CMT Conference Management System is to assign a numeric score $y_d^{(g)}$ to each level of the assessment scale, and then average the numeric scores to get a quality estimate for each paper d

$$\hat{s}_d = \frac{1}{|\{g : d \in D_g\}|} \sum_{g:d\in D_g} y_d^{(g)} \tag{1}$$

We refer to this aggregation method as *score averaging*. This average score can then be used by the meta-reviewers to sort the papers for triage. However, it is also likely to bias how the meta-reviewers perceive the quality of a paper. In particular, it depends on the mapping of assessment levels to scores. Following past years and given the arbitrariness of this mapping, the Program Chair decided to keep the mapping $y_d^{(g)}$ of "Strong Reject"$=-5$, "Reject"$=-2$, "Weak Reject"$=-1$, "Weak Accept"$=1$, "Accept"$=2$, "Strong Accept"$=5$ (Fig. 1).

1.3 Ordinal Peer Assessment

An alternative to assigning scores to levels is to merely interpret these scores in an ordinal way. In particular, we can derive a weak ordering $\sigma^{(g)}$ of the papers in D_g for each g. This avoids mapping the assessment levels to (arbitrary) scores and abstract from different interpretations of the assessment scale by the reviewers. A possible downside is some loss of information, since different assessments may lead to the same ranking. In order to mitigate this information loss and "anchor" the ordinal scale, we add a fictitious "borderline" paper $d_{borderline}$ to each reviewer set D_g, which is given a fictitious rating between "weak reject" and "weak accept" that only this one paper receives. This models that every reviewer has an acceptance threshold by comparing the assigned papers to a fictitious paper that they consider to be right on the acceptance threshold.

Given a collection of rankings from reviewers $\sigma^{(g)}$ for subsets D_g, we aim to estimate an overall ranking of all papers in D. We argue that an overall ranking provides an easy to understand and intuitive way to communicate paper quality, more so than the average of somewhat arbitrary scores as in Score Averaging. Furthermore, in order to achieve our goal of communicating uncertainty, we go beyond a single point estimate of the ranking as in [15] and provide a Bayesian posterior distribution of the rankings.

Table 1. Notation overview and reference.

G, g	Set of all reviewers, Specific reviewer
D, d	Set of all papers, Specific paper
D_g	Set of items graded by reviewer g
$\sigma^{(g)}$	Ranking feedback (with possible ties) from g
η_g	Predicted reliability of reviewer g
$r_d^{(\sigma)}$	Rank of paper d in ordering σ (rank 1 is best)
$d_2 \succ_\sigma d_1$	d_2 is preferred/ranked higher than d_1 (in σ)
$\pi(A)$	Set of all rankings over $A \subseteq D$
$\sigma_1 \sim \sigma_2$	$\exists$ way of resolving ties in σ_2 to obtain σ_1
$\hat{\sigma}$	Estimated ordering of papers
σ^*	(Latent) True ordering of papers

2   Bayesian Ordinal Peer Assessment (BOPA)

The goal in Bayesian Ordinal Peer Assessment (BOPA) is to infer a posterior distribution

$$P(\sigma | \{\sigma^{(g)}; \forall g\}) = \frac{P(\{\sigma^{(g)}; \forall g\} | \sigma) P(\sigma)}{\sum_{\sigma' \in \pi(D)} P(\{\sigma^{(g)}; \forall g\} | \sigma') P(\sigma')}$$

of the true quality ranking of papers σ^* from the set of peer rankings $\sigma^{(g)}$. Following [14], we select the data likelihood $P(\{\sigma^{(g)}; \forall g\} | \sigma)$ and a prior $P(\sigma)$ as follows.

For the prior $P(\sigma)$, we make the natural choice of using the uniform distribution over all rankings, since any other choice would lead to an unfair assessment.

For the data likelihood $P(\{\sigma^{(g)}; \forall g\} | \sigma)$, there is a whole range of possible options. Several extensions of classical models such as the Mallows and Bradley-Terry model are explored in [15]. We focus on the Mallows-based method for its simplicity and good performance in [14,15]. The Mallows-based model defines a distribution over rankings in terms of the Kendall-Tau distance [7] from the true ranking σ^* of assignments.

Definition 1. *The Kendall-τ Distance δ_K between rankings σ_1 and σ_2 is the number of incorrectly ordered pairs between the two rankings and is given by*

$$\delta_K(\sigma_1, \sigma_2) = \sum_{d_1 \succ_{\sigma_1} d_2} \mathbb{I}[[d_2 \succ_{\sigma_2} d_1]]. \tag{2}$$

Given the reviewer orderings $\sigma^{(g)}$, we can define the data likelihood (if the overall ranking was σ) as

$$P(\{\sigma^{(g)}; \forall g\} | \sigma) = \left\{ \prod_{g \in G} \frac{\sum_{\sigma' \sim \sigma^{(g)}} e^{-\delta_K(\sigma, \sigma')}}{Z_M(|D_g|)} \right\}, \tag{3}$$

Algorithm 1. Sampling from Mallows Posterior using Metropolis-Hastings

---

1: **Input:** Grader orderings $\sigma^{(g)}$, Grader reliabilities η_g and MLE ordering $\hat{\sigma}$.
2: Pre-compute $x_{ij} \leftarrow \sum_{g \in G} \eta_g \mathbb{I}[d_i \succ_{\sigma^{(g)}} d_j] - \sum_{g \in G} \eta_g \mathbb{I}[d_j \succ_{\sigma^{(g)}} d_i[$
3: $\sigma_0 \leftarrow \hat{\sigma}$ ▷ Initialize Markov Chain using MLE estimate
4: **for** $t = 1 \ldots T$ **do**
5:     Sample σ' from (**MALLOWS**) jumping distribution: $J_{MAL}(\sigma'|\sigma_{t-1})$
6:     Compute ratio $r_t = \frac{P(\sigma'|\{\sigma^{(g)};\forall g\})}{P(\sigma_{t-1}|\{\sigma^{(g)};\forall g\})}$ using Eq. 7
7:     With probability $\min(r_t, 1)$, $\sigma_t \leftarrow \sigma'$ else $\sigma_t \leftarrow \sigma_{t-1}$
8:     Add σ_t to samples (if burn-in and thinning conditions met)

---

where the normalization constant Z_M is easy to compute as it only depends on the ranking length.

$$Z_M(k) = \prod_{i=1}^{k} \left(1 + e^{-1} + \cdots + e^{-(i-1)}\right) = \prod_{i=1}^{k} \frac{1 - e^{-i}}{1 - c^{-1}} \tag{4}$$

Note that in Eq. 3, ties in the grader rankings are modeled as *indifference* (*i.e.*, agnostic to either ranking), which leads to the summation in the numerator is over all total orderings σ' consistent with the weak ordering $\sigma^{(g)}$.

Under the uniform prior, the posterior distribution of the inferred rankings σ i.e., $P(\sigma|\{\sigma^{(g)}; \forall g\})$ is defined as

$$P(\sigma|\{\sigma^{(g)}; \forall g\}) = \frac{P(\{\sigma^{(g)}; \forall g\}|\sigma)}{\sum_{\sigma' \in \pi(D)} P(\{\sigma^{(g)}; \forall g\}|\sigma')}. \tag{5}$$

With the posterior distribution in hand, we can derive the desired marginal rank distributions of each assignment, or we can predict a single ranking that minimizes posterior expected loss.

However, exact computations with this posterior are infeasible given the combinatorial number of possible orderings of all assignments. To help us ascertain information from the posterior, we will employ MCMC based sampling as previously used for Ordinal Peer Grading of student assessments in [14]. Markov Chain Monte Carlo (or MCMC in short) are a set of techniques for sampling from a distribution by constructing a Markov Chain which converges to the desired distribution asymptotically. Metropolis-Hastings is a specific MCMC algorithm which is particularly common when the underlying distribution is difficult to sample from (as is the case here) especially for multi-variate distributions.

Thus to help us estimate the posterior we will design a Markov Chain whose stationary distribution is the distribution of interest: $P(\sigma|\{\sigma^{(g)}; \forall g\})$. Along with the theoretical guarantees accompanying these methods, an added advantage is the fact that we can control the desired estimation accuracy (by selecting the number of samples).

This results in a simple and efficient algorithm, shown in Algorithm 1. To begin, we pre-compute statistics of the net cumulative weighted total each assignment d_i is ranked above another assignment d_j. We then initialize the

Markov Chain using the MLE estimate of the ordering: $\hat{\sigma}$. While computing the Maximum-Likelihood Estimator (MLE) of Eq. 3 is NP-hard [6], several simple and tractable approximations that are shown to work well in practice are presented in [15].

At each timestep, to propose a new sample σ' given the previous sample σ_{t-1}, we sample from a jumping distribution (Line 5). In particular, we use a Mallows-based jumping distribution:

$$J_{MAL}(\sigma'|\sigma) \propto e^{-\delta_K(\sigma',\sigma)}. \tag{6}$$

This is a simple distribution to sample from and can be done efficiently in $|D|log|D|$ time. Furthermore, as this is a symmetric jumping distribution (*i.e.*, $J_{MAL}(\sigma'|\sigma) = J_{MAL}(\sigma|\sigma')$), the acceptance ratio computation is simplified.

When it comes to computing the (acceptance) ratio r_t (Line 6), we can rely on the pre-computed statistics to do so efficiently. In particular, we can simplify the expression for the ratio to:

$$\frac{P(\sigma_a|\{\sigma^{(g)}; \forall g\})}{P(\sigma_b|\{\sigma^{(g)}; \forall g\})} = \prod_{g \in G} e^{\delta_K(\sigma^{(g)},\sigma_b) - \delta_K(\sigma^{(g)},\sigma_a)}$$

$$= \prod_{i,j} e^{x_{ij}(\mathbb{I}[d_i \succ_{\sigma_a} d_j] - \mathbb{I}[d_i \succ_{\sigma_b} d_j])} \tag{7}$$

This expression is again simple to compute and can be done in time proportional to the number of flipped pairs between σ_a and σ_b, which in the worst case is $O(|D|^2)$. Overall, the algorithm has a worst-case time complexity of $O(T|D|^2)$.

The resulting samples produced by the algorithm can be used to *estimate* the posterior distributions including the marginal posterior of the rank of each assignment *i.e.*, $P(r_d|\{\sigma^{(g)}; \forall g\}$, as well as statistics such as the entropy of the marginal, the posterior mean and median etc.

In order to improve the quality of the resulting estimates, we ensure proper mixing by targeting a moderate acceptance rate and by thinning samples (in our experiments we thin every 10 iterations). Furthermore we draw samples once the chain has started converging *i.e.*, we use a burn-in of around 10,000 iterations. In total we used 50,000 samples drawn from the Markov Chain in this manner.

We also derive a Metropolis-Hastings based extension of the Mallows model with reviewer reliabilities. Following [14, 15], reviewer reliability can be included into the model via

$$P(\{\sigma^{(g)}; \forall g\}|\sigma, \{\eta_g\}) = \left\{ \prod_{g \in G} \frac{\sum_{\sigma' \sim \sigma^{(g)}} e^{-\eta_g \delta_K(\sigma,\sigma')}}{Z_M(\eta_g, |D_g|)} \right\}.$$

In addition to sampling the orderings, we also sample the reliabilities using a Gaussian jumping distribution (also symmetric). However the acceptance ratio computation is now more involved and hence less efficient than that for Algorithm 1, but nonetheless can be computed fairly efficiently. We omit the precise equation and computations for the purpose of brevity.

Software and an online service that implements these methods is available at http://www.peergrading.org/.

2.1 Relation to Existing Rank Aggregation Literature

The ordinal peer assessment problem can be viewed as a specific kind of rank aggregation problem. It is closely related to the ordinal peer grading problem as discussed in [14,15], with only one main difference. In peer grading it is equally important to estimate the rank of an assessment anywhere in the ranking, while for ordinal peer assessment it is more important to get the right order toward the top of the ranking.

More generally, rank aggregation [8] covers a wide class of problems where the goal is the combination of ordinal (ranking) information from multiple different sources. Voting Systems (or Social Choice [1]) are one of the most common applications of rank aggregation techniques. The goal of these systems is to merge the preferences of a set of individuals. Condorcet voting methods such as *Borda count* amongst others [6,10] are commonly used to tackle these problems. Search Result Aggregation (also known as Rank Fusion or Metasearch [2]) is perhaps the most well-known rank-aggregation problem. Given rankings from different sources (typically different algorithms), the goal is to merge them and produce a single output ranking. Extensions of classical techniques such as the Mallows model [11] and Bradley-Terry model [3] have become popular for these problems [4,9] and have been used to improve ranking performance in different settings [12, 13,16]. While our work also extends the classical Mallows model, a key difference is the fact that unlike other rank aggregation problems, a single ordering of assignments does not suffice since it does not communicate uncertainty.

Related to this work are also the recent experiments conducted as part of the reviewing process of the Neural Information Processing (NIPS) conference [5]. Their controlled experiment investigated the variability of the acceptance decisions. Their findings in part motivated our decision to increase the number of reviews per paper.

3 Empirical Analysis

We now analyze the BOPA approach outlined above on the reviewing data of KDD 2015. To give some insights into the data, we first outline the reviewing process.

On February 20, 2015, a total of 819 paper were submitted. Reviewer assignments D_g were made though CMT's built-in optimizer based on reviewer bids. The Program Committee included 595 reviewers that produced a total of 2919 reviews. Reviewers were asked to finish their reviews by March 27, when authors were given the opportunity to write a short response to the reviews. On April 14, Meta-Reviewers were asked to initiate discussion among the reviewers. The decision recommendations by the Meta-Reviewers of whether to accept or reject

a paper were due on May 1. However, many Meta-Reviewers submitted their recommendations late, but eventually everybody delivered well before the author
notification on May 12. In the time from May 1 to May 12 the Program Chairs
reviewed the Meta-Reviewer recommendations and made final accept/reject decisions. In many cases, the Program Chairs initiated additional discussions for
controversial papers or papers where the meta-reviewer was not confident, using
a variety of strategies to resolve remaining issues (e.g., assigning a second meta
reviewer). In the end, 160 papers were accepted.

On April 15, we took a snapshot of all available reviews at that time and
applied the BOPA model outlined in this paper. We only consider the reviewers
answer to the question

"What is your overall recommendation?"

that is answered on the scale given in Sect. 1.2. We then distributed the results
via email to the Meta-Reviewers for all papers assigned to them on April 29. The
delay was due to creating the PDFs summarizing the results. This means that
most Meta-Reviewer decisions were made without access to the BOPA results.
However, for the more controversial papers which Meta-Reviewers tend to make
decisions on last, the Meta-Reviewers had access to the BOPA results. However, since access to BOPA results was outside the CMT system, the summary
statistics and ranking that CMT provides were probably more salient.

The analysis we conduct below is based on a review snapshot from May 4,
when most reviews and meta-reviews were submitted and in their final revision.
It covers all 752 papers for which BOPA analytics were provided to the Meta-
Reviewers.

3.1   Do Aggregated Reviewer Scores Predict the Number of Accepted Papers?

The first aspect we evaluate is in how far BOPA and Score Averaging (with the
numeric scale given in Sect. 1.2) predict how many papers will be accepted. A
natural acceptance threshold for Score Averaging is 0. This would predict that
240 papers[1] will be accepted. This substantially exceeds the actual number of
accepted papers of 160.

For BOPA, it is natural to use the mode of the posterior of the artificial
borderline paper $d_{borderline}$. The mode is located at 202, with 95 % tails spanning
the interval [184, 219]. This is closer to the actual number of accepted papers,
but still significantly high.

Overall, there seems to be a substantial difference in the aggregated opinions
of the reviewers and the final decisions, where papers need to substantially exceed
the aggregate vote threshold of the reviewers in order to be accepted.

---

[1] Papers with average score of exactly 0 were counted as 0.5 each.

3.2 How Different Are the Predictions of BOPA and Score Averaging?

The second question we investigate is whether BOPA and Score Averaging actually make different predictions. If they did not, then any further analysis and comparison would be somewhat pointless.

In order to calibrate their acceptance threshold to the actual acceptance number, we adjust the acceptance threshold of Score Averaging to 0.3. This leads to 161 accepted papers for Score Averaging.

For BOPA, its probabilistic model makes it straightforward to compute the optimal decisions. We compute

$$P(y_d = accept | \{\sigma^{(g)}; \forall g\}) = P(r_d \leq 160 | \{\sigma^{(g)}; \forall g\}) \tag{8}$$

and predict a paper to be accepted, if it has a probability of being among the top 160 papers that is greater than 0.5. This predicts that 164 papers are accepted.

Counting the number of papers where Score Averaging and BOPA make different acceptance decisions leads to 51 papers. This is quite a substantial difference, given 160 accepted papers. As a reference point for the magnitude of this difference, consider score averaging with a different numeric mapping. In particular, instead of using the scale $[-5, -2, -1, 1, 2, 5]$, consider the scale $[-3, -2, -1, 1, 2, 3]$. Score Averaging with this alternative scale differs in only 2 papers from the original scale. This highlight how different BOPA and Score Averaging are in their predictions (and how pointless it was for the Program Chairs to agonize over the selection of the mapping scale).

3.3 How Closely Do Review Aggregation Methods Predict Acceptance Decisions?

As the previous section showed, BOPA and Score Averaging make substantially different predictions. Which of these predictions more accurately reflect the actual accept/reject decisions?

Table 2. Confusion matrices for predicting paper acceptance using BOPA (left) and Score Averaging (right).

BOPA	predict accept	predict reject
true accept	123	37
true reject	41	551

Score Averaging	predict accept	predict reject
true accept	125	35
true reject	36	558

Table 2 shows the confusion matrices for both methods. Overall, BOPA disagrees with the actual decisions on 78 papers and Score Averaging disagrees on 71 papers. The difference between these two disagreement counts is not significant (McNemar's test with 0.95 confidence threshold). These relatively high

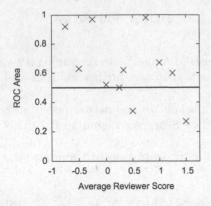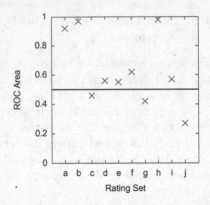

Fig. 2. Area under the ROC Curve (AUC) when ranking papers in the same equivalence class by BOPA's posterior probability of acceptance. On the left, papers with the same reviewer rating average are considered equivalent. On the right, papers with identical sets of ratings are considered equivalent. Only equivalence classes with more than 10 papers are shown.

disagreement rates indicate that many decisions are not clear-cut and that especially the Meta-Reviewers use their own insights and their interpretation of the review text to make the decisions.

The probabilistic nature of the BOPA model makes it possible to verify, if these disagreement rates were expected by the model. In particular, BOPA's predicted error rate can be computed as

$$disagreement = \sum_{d \in D} \min\{P(y_d = accept|\{\sigma^{(g)}; \forall g\}), P(y_d = reject|\{\sigma^{(g)}; \forall g\})\}.$$

(9)

For our data, the disagreement as predicted by BOPA is 65.3, which not far off the actual disagreement of 78. This provides a first indication that BOPA is able to quantify the amount of uncertainty in the aggregated reviewer scores. We will further investigate this in Sect. 3.5.

3.4 Can BOPA Distinguish Paper Quality Between Papers with the Same Reviewer Scores?

The previous section showed that the amount of disagreement of BOPA does not seem to be better than that of Score Averaging. However, there are biases that may have influenced that statistic. First, the Score Average was readily available in CMT for sorting, which may have biased the Meta-Reviewers' perception of the paper's quality. Second, the reviewers acceptance scores are communicated to the authors, but not the BOPA ranks. Thus, going against the cardinal score average requires effort from the Meta-Reviewer to justify that recommendation, which disincentivises the Meta-Reviewer from deviating from the score average.

In order to get results that are unaffected by such biases, we now consider subgroups of papers that have equal bias. First, Fig. 2 (left) shows how BOPA performs for papers that have the same score average. In particular, for each score average value, we rank all papers with that score average by their probability of acceptance $P(y_d = accept|\{\sigma^{(g)}; \forall g\})$ as predicted by BOPA. The left plot of Fig. 2 shows the Area under the ROC Curve (AUC) for all score average values that have at least 10 papers and for which the AUC exists. For most values, the AUC is greater than 0.5, indicating that BOPA sorts the papers better than random. The average AUC over all score averages weighted by the number of papers in the equivalence class is 0.630, which is substantially better than 0.5. The right plot in Fig. 2 shows the equivalent results, where the conditioning is not on the score average, but on a particular set of ratings. The weighted AUC here is 0.627.

This provides evidence that BOPA is indeed able to mitigate the problem of different reviewer scales, since it is able to identify papers that are more likely to be accepted even if they have exactly the same ratings. However, an alternative explanation is that this may also be affected by bias, since Meta-Reviewers were given the BOPA results, even if late in the decision process. To fully resolve this question beyond doubt, a controlled trial may be necessary.

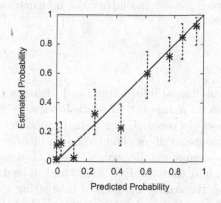

Fig. 3. Calibration of BOPA posterior acceptance probabilities. Binning is done via quantiles so that each bin contains roughly 40 papers. The x-axis shows the averaged posterior acceptance probabilities, and the y-axis the observed fraction of accepted papers per bin (with 95 % binomial confidence intervals).

3.5   How Calibrated Are the BOPA Acceptance Probabilities?

The estimated disagreement rate of BOPA already provided some evidence in Sect. 3.3 that BOPA is able to accurately capture the uncertainty inherent in the review process. We now investigate more closely, if BOPA indeed produces well-calibrated probabilities. In particular, we compute $P(y_d = accept|\{\sigma^{(g)}; \forall g\})$ as in Eq. (8) and ask whether a predicted $P(y_d = accept|\{\sigma^{(g)}; \forall g\})$ of value p indeed means that the paper d has a p-percent probability of being accepted.

Figure 3 shows a calibration plot, where papers are binned by $P(y_d = accept|\{\sigma^{(g)}; \forall g\})$ falling into specific intervals $[p_1, p_2]$. The intervals are selected to include roughly 40 papers each (except the interval closes to 0, which contains 399 papers), and the average value of $P(y_d = accept|\{\sigma^{(g)}; \forall g\})$ for each bin is plotted on the x-axis. The y-axis shows the ratio of accepted papers in each bin with 95 % binomial confidence intervals.

For perfectly calibrated prediction probabilities, all points should lie on the diagonal. Overall, calibration of the BOPA probabilities is remarkably good, especially in the high-probability region. This verifies that BOPA does indeed convey an accurate impression of uncertainty, as was desired in our original goals.

3.6   Anecdotal Qualitative Feedback

As mentioned above, the information as illustrated in Fig. 1 was emailed to all 68 Meta Reviewer. While we did not ask for a response to this email, 14 Meta Reviewer responded to this email. The vast majority of these responses indicated strong support for providing such information, calling it "helpful" and "useful". No response raised any concerns or was negative. Several emails included suggestions for how to better present and layout the information, and how to better integrate it with CMT.

4   Conclusions

We investigated how additional information and aggregation of reviewer information can provide decision support to Meta-Reviewers and Program Chairs for making accept/reject decisions. Using data from KDD 2015, we adapted a Bayesian ordinal rank aggregation method to the problem of estimating posterior rank distributions of submissions. Regarding the goal of providing information about uncertainty, we find that the BOPA method indeed captures accurately calibrated probabilites. Regarding the goal of mitigating mismatching reviewer scales, we find evidence that this is also achieved by BOPA. However, final confirmation about whether Meta-Reviewers and Program Chairs actually make better decisions using the additional information can only be conclusively answered through controlled experiments, which are outside the scope of this study.

Acknowledgments. This research was funded in part by NSF Awards IIS-1217686, IIS-1247637, IIS-1513692, the JTCII Cornell-Technion Research Fund and a Google Ph.D. Fellowship.

References

1. Arrow, K.J.: Social Choice and Individual Values, 2nd edn. Yale University Press, New Haven (1970). http://cowles.econ.yale.edu/P/cm/m12-2/index.htm
2. Aslam, J.A., Montague, M.: Models for metasearch. In: SIGIR, pp. 276–284 (2001). http://doi.acm.org/10.1145/383952.384007

3. Bradley, R.A., Terry, M.E.: Rank analysis of incomplete block designs: I. The method of paired comparisons. Biometrika **39**(3/4), 324–345 (1952). http://www.jstor.org/stable/2334029
4. Chen, X., Bennett, P.N., Collins-Thompson, K., Horvitz, E.: Pairwise ranking aggregation in a crowdsourced setting. In: WSDM, pp. 193–202 (2013). http://doi.acm.org/10.1145/2433396.2433420
5. Cortes, C., Lawrence, N.: The NIPS experiment (2014). http://inverseprobability.com/2014/12/16/the-nips-experiment/
6. Dwork, C., Kumar, R., Naor, M., Sivakumar, D.: Rank aggregation methods for the web. In: WWW, pp. 613–622 (2001). http://doi.acm.org/10.1145/371920.372165
7. Kendall, M.: Rank Correlation Methods. Griffin, London (1948)
8. Liu, T.Y.: Learning to rank for information retrieval. Found. Trends Inf. Retr. **3**(3), 225–331 (2009). http://dx.doi.org/10.1561/1500000016
9. Lu, T., Boutilier, C.: Learning mallows models with pairwise preferences. In: ICML, pp. 145–152, June 2011
10. Lu, T., Boutilier, C.E.: The unavailable candidate model: a decision-theoretic view of social choice. In: EC, pp. 263–274 (2010). http://doi.acm.org/10.1145/1807342.1807385
11. Mallows, C.L.: Non-null ranking models. Biometrika **44**(1/2), 114–130 (1957). http://www.jstor.org/stable/2333244
12. Niu, S., Lan, Y., Guo, J., Cheng, X.: Stochastic rank aggregation. CoRR abs/1309.6852 (2013)
13. Qin, T., Geng, X., Liu, T.Y.: A new probabilistic model for rank aggregation. In: NIPS, pp. 1948–1956 (2010)
14. Raman, K., Joachims, T.: Bayesian ordinal peer grading. In: ACM Conference on Learning at Scale (LS) (2015)
15. Raman, K., Joachims, T.: Methods for ordinal peer grading. In: KDD 2014, NY, USA, pp. 1037–1046 (2014). http://doi.acm.org/10.1145/2623330.2623654
16. Volkovs, M.N., Zemel, R.S.: A flexible generative model for preference aggregation. In: WWW, pp. 479–488 (2012). http://doi.acm.org/10.1145/2187836.2187902

Collaborative Online Learning
of an Action Model

Christophe Rodrigues[1], Henry Soldano[1], Gauvain Bourgne[2],
and Céline Rouveirol[1]([⊠])

[1] L.I.P.N, UMR-CNRS 7030, Univ. Paris-Nord, 93430 Villetaneuse, France
{Christophe.Rodrigues,Henry.Soldano,
Celine.Rouveirol}@lipn.univ-paris13.fr
[2] LIP6, Université Pierre et Marie Curie, Sorbonne Universités, 75005 Paris, France
gauvain.bourgne@lip6.fr

Abstract. A number of recent works have designed algorithms that
allow an agent to revise a relational action model from interactions with
its environment and uses this model for building plans and better exploring its environment. This article addresses Multi Agent Relational Action
Learning: it considers a community of agents, each rationally acting following some relational action model, and assumes that the observed effect
of past actions that led an agent to revise its action model can be communicated to other agents of the community, potentially speeding up the
on-line learning process of agents in the community. We describe and
experiment a framework for collaborative relational action model revision where each agent is autonomous and benefits from past observations
memorized by all agents of the community.

1 Introduction

Adaptive behavior studies how an autonomous agent can modify its own behavior in order to adapt to a complex, changing and possibly unknown environment.
Whereas [8] and more recently [9] focused on adapting and grounding a symbolic representation for an agent from sensing information, we assume here a
fixed vocabulary and address the problem of learning an action model from
interactions with the environment. Considering that an adaptive agent needs
to simultaneously learn from its experience and act to fulfill various goals, an
adaptive system thus needs to integrate some kind of *online* learning together
with action selection mechanisms. Adaptation within relational representations
has been primarily addressed by Relational Reinforcement Learning (RRL) [5]
by extending the classical Reinforcement Learning (RL) problem to handle first
order representations. In the *indirect* or *model-based* Reinforcement Learning
framework [20], the agent explicitly learns such an *action model*, allowing it to
predict the effect of actions, and uses it as an input of a symbolic planner, whose
output is a plan to execute in order to reach its current goal. *Indirect* RL proved
to be very efficient when handling relational – Datalog – representations [4,10].

© Springer International Publishing Switzerland 2016
S. Michaelis et al. (Eds.): Morik Festschrift, LNAI 9580, pp. 300–319, 2016.
DOI: 10.1007/978-3-319-41706-6_16

We have recently proposed a relational revision algorithm, implemented in the IRALe software [17]. IRALe models an agent acting on its environment according to an action model that predicts the effect of the actions the agent can perform when applied in the current state. Adapting the inspiring sloppy modeling paradigm [12] to action model learning, an IRALe agent starts from a (usually empty) action model and performs online revision of this model, here a deterministic conditional STRIPS-like action model. The main features of the learning algorithm in IRALe are that the action model is represented as a relational rule set and that this model is revised, in a bottom-up way, each time some observation contradicts the model, i.e. the model makes an error when predicting the effect of some action in the current state. Among all examples sequentially encountered by the agent, the IRALe agent only memorizes those, that we call *counter-examples*, associated to a prediction error and that have therefore enforced a revision of its model.

In this article, and following previous work [18,19], we study a community of autonomous IRALe agents. Each agent acts in its environment following its current relational action model, and exchanges information with other agents following the general multi agent learning protocol SMILE [2,3]. Intuitively, the SMILE protocol is based on a "consistency maintenance" process: when some new observation contradicts its current model, the agent first revises its current model in order to ensure that the revised model is *consistent* with the observations it has memorized, then the agent communicates this revised model to the other members of the community, and possibly receives past observations they have memorized and that in turn contradict the revised model. After a number of such revision/criticism interactions, resulting in a *global revision* associated to the initial contradiction, the revised model is stated as *globally consistent* with the observations memorized by all the agents.

To adapt the general SMILE framework for learning action models, we have stated a number of restrictions: the target relational action model is supposed to be deterministic, the actions performed by the agents do not interfere, and there is no common goal for which agents would need coordination. As a counterpart, agents are autonomous, do not have any shared memory, thus preserving privacy, and each agent, when revising its model, is ensured to benefit from all past observations memorized by the community. This allows modelling complex learning situations in which agents act in independent parts of a global environment, and still benefit from information exchange.

For instance, consider a classroom of robots, each training on some simple tasks, as stacking colored cubes. Each robot works on its own table and cubes, and has to learn the same, universal, action model. The SMILE framework states how these robots can be each helped by their classmates, by exchanging elements of their past experience on a utility basis: only past observations in an agent memory that contradict at some point the current model of some other agent are transmitted to the latter. Another example would be mobile devices assisting their owner exploring the same unknown country: the underlying world model is the same, but one can reasonably assume that agents are far enough so that the actions of some agents only have local effects that do not affect the results of other agents actions. Therefore, each mobile device can safely build its action model, benefiting from other similar devices' experiences.

We investigate in this article how the revision of the current model of an agent can benefit from interactions between agents, in the context of relational action model learning. In the collaborative relational action model learning proposed here, each agent uses IRALe as a revision mechanism, is equipped with a symbolic planner and tries to form and execute plans in order to reach some random goal. Various agents behaviors have been investigated in the SMILE framework, considering how agents behave with respect to the other members of the community, in particular considering how they take into account the other agents' models they are aware of. We investigate here the *individualistic* variant of SMILE [2], denoted by iSMILE, in which each agent plainly cooperates with the other agents, by communicating counter-examples to their current model, but only modifies its own model when performing a global revision. In other words, agents always prefer the model they are currently revising to other agents' models.

The paper is organized as follows: IRALe is described in Sect. 2, adaptation of iSMILE to relational action model learning is described in Sect. 3. Finally, experiments are detailed and interpreted in Sect. 4.

2   IRALe: Revising a Relational Action Model

In this section, we discuss the relational action model revision algorithm IRALe [17] that each agent uses to revise its relational action model. IRALe learns a STRIPS-like action model as a set of rules from state/action/effect triples. Several rules can be associated to each action, where each rule completely describes the effect of the action in a given context. In this way, the model allows to represent conditional effects. IRALe only memorizes *counter-examples*, namely examples that have raised a prediction error (the observed effect is not the predicted one) at some point during the model construction. IRALe learns deterministic rules, i.e. once the preconditions of some rule are satisfied, the rule always predicts the same effect. The algorithm is primarily intended to learn in a *realizable* case, i.e., when there exists an action model exactly predicting the outcomes of any action in any state. Note, however, that IRALe has been proven to be accurate when learning in the presence of some amount of noise [16].

Related Work. Learning planning operators has been studied intensively, including the problem of learning the effects of actions, in the context of Relational Reinforcement Learning (RLL). The first model that integrated an incremental action model and policy learning is MARLIE [4]. Learning relational action rules has also been studied in the context of inductive logical programming by Otero et al. [14]. In both cases, the model predicts the value of each possible effect literal (positive or negative) separately. Let us also mention the work of Xu and Laird [21] and Mourao et al. [13], that both learn black box models in batch mode. Mourao and colleagues propose an additional step for extracting rules after a black box model has been learned. [22,23] learn models as sets of rules from plan traces, but they do not incrementally revise this model.

Other works [10,15] address stochastic effects. Learning is then performed from scratch and needs prior memorization of the whole set of observations.

2.1   States, Actions, Examples and Rules

States and actions are represented by objects and relations between them. Examples are observations resulting from the agent actions and the agent minimally revises the action model when needed. Relations between objects in a state are described using predicates applied to constants. In the following, objects are denoted by constants and a lower-case character $(a, b, f, \ldots)$. Variables are denoted by an upper-case character $(X, Y, \ldots)$, and may instantiate to any object of the domain. A *term* is here a constant or a variable. Actions and relations between objects are denoted by *predicate* symbols.

Examples are described as conjunctions of ground literals. Following a STRIPS-like notation, state literals that are not affected by the action are not described in the effect part. The examples are denoted by $x.s/x.a/x.e.add, x.e.del$, with $x.s$ a conjunction of literals, $x.a$ a literal of action and, regarding the effect part, $x.e.add$ a conjunction of positive literals and $x.e.del$ a conjunction of negated literals. Some examples may have an empty effect list (i.e., $x.del = x.add = \emptyset$), accounting for illegal action applications in specific contexts [17].

Example 1. Figure 1 displays an example e of the action *move* in a blocks world: $onTable(a), onTable(b), on(c, a)/move(c, b)/on(c, b), \neg on(c, a)$. $\qquad\qquad$ □

IRALe builds an action model, made of a set of rules T, according to a set of observed examples O that have been memorized during the agent history. Each rule r is composed by a precondition $r.p$, an action $r.a$ and an effect $r.e$, and is denoted by $r.p/r.a/r.e$. The precondition $r.p$ is a conjunction of positive literals which have to be satisfied to apply the rule, $r.a$ a is a literal defining the performed action, $r.e$ is composed of two sets of literals: $r.e.add$ is the set of literals getting true when the rule is applied, and $r.e.del$ is the set of literals

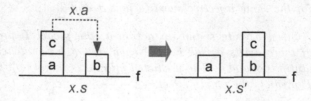

Fig. 1. Example of a *move* action in a simple blocks world

getting false when the rule is applied. According to a rule r, an action $r.a$ has no other effects but those described by $r.e$.

Example 2. In the same blocks world of Example 1, $r = on(X,Z)/move(X,Y)/$ $on(X,Y), \neg on(X,Z)$ is an action rule. When an agent performs the action $move(c,b)$ in the current state represented on the left of Fig. 1, according to this rule, the agent should reach, as its new current state, the state represented on the right of Fig. 1. $\qquad\square$

Note that in an action rule, preconditions and effect may contain (existential) variables that do not appear in the action literal. To complete the model, a default rule is implicitly added to T: for any action a, whenever no rule for a applies, the action is predicted to have no effect, i.e. $e.del = e.add = \emptyset$.

2.2   Rule Covering and Contradiction

Matching operations between rules and examples relies on subsumption under Object Identity, denoted as OI-subsumption [6], which is an intuitive partial order relation when learning action rules for planning [15].

Rule matching definition relies on the definitions of pre-matching $\overset{sa}{\sim}$ and post-matching $\overset{ae}{\sim}$ functions. Pre-matching checks whether a given rule may apply to predict the effect of a given action in a given state, and post-matching checks which rule(s) of the action model may explain the effect observed in the example.

Definition 1. *For any rule r, and example $x = (s/a/e)$, r pre-matches x, $(r \overset{sa}{\sim} (s,a))$ iff there exists two injective substitutions σ and θ such that (i) $(r.a)\sigma = a$, and (ii) $(r.p)\sigma\theta \subseteq s$. r post-matches x $(r \overset{ae}{\sim} (a,e))$ iff there exists two injective substitutions σ and θ such that (i) $(r.a)\sigma = a$ (ii) $(r.e)\sigma\theta = e$.*

The question of whether the action model contradicts or is consistent with an example is addressed through the following definitions. Given an example and a rule pre-matching the example, *covering* checks whether the effect part of the example is accurately explained/predicted by the rule, while rule contradiction appears whenever the rule incorrectly predicts the outcomes of the action.

Definition 2. *For any rule r and example x, r covers x (denoted $r \approx x$) iff r pre-matches x and post-matches x for the same injective substitutions σ and θ. Conversely, x contradicts r (denoted $r \not\approx x$) iff r pre-matches x and does not post-match x for the same injective substitutions σ and θ.*

Definition 3. *Given a state s and an action a, the model T predicts a non empty effect iff there exists a rule $r \in T$ such that r pre-matches (s,a) with injective substitutions σ and θ. The predicted effect e is then $(r.p)\sigma\theta$. If no such rule r exists, T predicts the empty effect.*

The model T needs to be revised whenever the current action model fails to predict the observed effect of some action in the current state. The (state/action/effect) example is then said to *contradict* the model, is stated as a *counter-example* of T, and is then memorized in O. We give hereunder the necessary definitions.

Definition 4. $x = (s, a, e)$ *is a* counter-example *and is said to* contradict *the model T iff the predicted effect of T given (s, a) (see Definition 3) $\neq e$. x may make T incoherent: there is a rule of T pre matching (s, a) and predicting a non empty effect whereas $x.e = \emptyset$. Alternatively, x may make T incomplete, in that case $x.e \neq \emptyset$ and there is no rule pre-matching x.*

Example 3. Consider again the block world situation of Fig. 1. The rule of Example 2 both pre-matches and post-matches the example x of Example 1, with substitutions $\sigma = \{X/c, Y/b\}$ and $\theta = \{Z/a\}$. This means that the rule r correctly predicts the effect $x.e$ of the action $x.a$ in the state $x.s$. Consider now the following variant y of example x: $y = onTable(a), onTable(b), on(c, a)/move(c, b)/ on(c, b)$. r still pre-matches y but does not post-match y as the observed effect $on(c, b)$ is only part of the predicted effect $on(c, b), \neg on(c, a)$.   □

A model T is said *complete* with respect to O whenever no example in O makes T incomplete. T is said *coherent* w.r.t. O whenever no example in O makes T incoherent. In both cases of contradiction, the model T needs to be updated to T' in order to preserve coherence and completeness *w.r.t.* x and the other past counter-examples in O.

Definition 5. *A model T is* consistent *with respect to a set of examples O, denoted* $cons(T, O)$ *whenever no example in O contradicts the model, i.e. whenever T is both complete and coherent w.r.t. O.*

2.3   Online Revision of the Action Model

The interactions between the agent and the environment produce examples, and when an example contradicts the model, the latter has to be revised by modifying or adding one or several rules. When such a new counter-example x_u is encountered, two kinds of modifications may have to be performed, either generalization or specialization (see [17] for details). We focus here on the generalization process, that takes place in order to preserve completeness of the model, whenever no rule of T pre-matches x_u. The rules r of T which are candidates for generalization are such that r, up to the generalization of some constants into variables, post-matches x_u. Preconditions are then generalized with x_u using least general generalization under OI subsumption. If such a generalization does not contradict any example in O (preserving coherence), r is replaced by the new minimally generalized rule. If no consistent generalization exists, x_u becomes a rule and is added as such to T. Finally, x_u, as a counter-example, is stored in O.

Note that only counter-examples are memorized in O_i, i.e., observations that contradicted the current model at some time point. This is sufficient to ensure that learning converges, in the realizable case [16].

3 Action Model Learning in a Community of Agents

In this section, we present a framework for collective incremental action model learning relying on the SMILE framework [2,3]. A community of n agents, or n-MAS[1], is a set of agents $a_1, \ldots, a_n$. Each agent a_i has a model, here a set of action rules T_i, that will be revised during the learning process, and a set of internal counter-examples O_i. The set of all counter-examples stored in the MAS is denoted by O (for all agents $j \in \{1, \ldots, n\}, O = \cup_{j \in \{1,\ldots,n\}} O_j$).

The *a-consistency* and *mas-consistency* properties are defined as follows.

Definition 6.

- *An IRALe agent a_i is a-consistent iff T_i is consistent with respect to O_i (see Definition 5), i.e., the agent model T_i correctly predicts observed effects for all counter-examples in O_i.*
- *An IRALe agent a_i is mas-consistent iff T_i is consistent with respect to O, i.e., to all counter-examples stored by agents of the n-MAS.*

How to derive mas-consistency of an agent from consistency of this agent with respect to other agents memories needs the property of a-consistency to be *compositional*: this means that given two sets of examples O_1 and O_2 and a set of action rules T, $Cons(T, O_1 \cup O_2)$ is true if and only if $Cons(T, O_1)$ and $Cons(T, O_2)$ are both true. Compositionality is realized for IRALe agents described above, provided they act in independent environments – an action realized by an agent does not affect other agents' environments. If M denotes the incremental (local) revision mechanism described in Sect. 2.3, this means that when encountering a contradiction, an agent applies M to recover its a-consistency. M is then called a *a-consistent revision mechanism*. Furthermore, observations coming from other agents may be used as if the agent had observed them. It is therefore straightforward to define a *mas-consistent global revision mechanism M_s*, i.e. a mechanism that guarantees that the agent applying M_s recovers its mas-consistency. Such a global revision mechanism consists in a set of interleaved revisions, using M, and interactions with other agents, until there is no more contradiction to the current revised action model with respect to counter-examples stored by all agents of the community.

3.1 A Global Revision Mechanism

We describe below this global revision mechanism M_s, first by defining an interaction, then by detailing an iteration together with the termination condition of the mechanism.

The mechanism is triggered by an agent a_i upon direct observation of a *contradictory* observation x, denoted as an *internal counter-example*. This counter-example breaks a-consistency, enforcing revision of T_i into T_i' and is stored in O_i. An interaction $I(a_i, a_j)$ between the *learner* agent a_i and another agent a_j, acting as a *critic*, is as follows:

---

[1] MAS stands for Multi Agent System.

1. Agent a_i sends the revision T_i' to a_j;
2. Agent a_j checks the revision T_i'. If T_i' is a-consistent with respect to its set of counter-examples O_j, a_j sends a notification of *acceptance* of T_i' to a_i. Otherwise, a_j sends a counter-example $x' \in O_j$, denoted as an *external counter-example* for a_i, such that x' contradicts T_i'. Then, x' is stored in O_i.

An iteration of M_s is then composed of a local revision performed by the *learner* agent a_i, followed by a sequence of interactions $I(a_i, a_j)$. If an external counter-example x' is transmitted to a_i, this triggers a new iteration, starting with a new revision of the learner to restore its a-consistency. When all critics have sent a notification of acceptance of a proposed revision, a_i notifies the other agents that its mas-consistency is restored. This ensures that, at the end of the revision process, a_i is mas-consistent. We have then the following property [2,3].

Proposition 1. *Let $a_1, \ldots, a_n$ be a MAS in which agent a_i receives an observation x breaking its a-consistency, and M be an a-consistent local revision mechanism. The global revision mechanism M_s, described above, always terminates and is mas-consistent.*

Example 4. Let us assume that we have a community of three agents, a_1, a_2, a_3 with memories O_1, O_2, O_3 and action models T_1, T_2, T_3; each agent works on its own table on which lies various blocks (3 blocks only in each agent's environment). At the current time, agent a_1 has an empty model $T_1 = \emptyset$ together with an empty memory $O_1 = \emptyset$ and its world is in the state described on the left of Fig. 1, while a_2 and a_3 have one example each in their memories (their model by the way is equal to this example). Agent a_1 applies the action $move(c, b)$ and then observes the effects $on(c, b), \neg on(c, a)$. As this contradicts its empty model, it memorizes the resulting example e_1 (see Example 1) in O_1 and also as a rule r_1 in T_1, which is now consistent with O_1. Then, a_1 sends T_1 to agent a_2 which finds in its memory the example $e_2 = onTable(g), onTable(h), on(k, g)/move(k, h)/on(k, h), \neg on(k, g)$.

This is a counter-example of T_1 and a_2 sends it to agent a_1. Agent a_1 then revises T_1, generalizing $r_1 = e_1$ and e_2 in a new rule: $onTable(X), onTable(Y), on(Z, X)/move(Z, Y)/on(Z, Y), on(Z, X)$, where each constant in e_1 has been turned into a variable.

At that point, the revised T_1 is consistent with O_2 and agent a_1 sends T_1 to agent a_3. Agent a_3 finds in turn a counter-example e_3 in its memory, represents an observation in which the block to move was, before the move, on top of a stack of three blocks: $e_3 = on(l, m), on(n, l), onTable(n)/move(l, table)/onTable(l), \neg on(l, m)$. As the observed effects do not post-match r_1's effect, a specific rule (e_3) is learned. The new revised version of T_1 is proposed to agent a_3: $T_1 = \{onTable(X), onTable(Y), on(Z, X)/move(Z, Y)/on(Z, Y), on(Z, X);$ $on(l, m), on(n, l), onTable(n)/move(l, table)/onTable(l), \neg on(l, m)\}$.

It happens that this new revised version is accepted by both a_2 and a_3, and as a consequence T_1 is now consistent with all the agent memories, and agent a_1 is mas-consistent. $\qquad\qquad\qquad\qquad\qquad\qquad\qquad\qquad\qquad\qquad\qquad\qquad\quad\square$

Consider a community of agents, each equipped with such a global revision mechanism. Note that the external counter-examples transmitted by other agents and memorized by the *learner* agent are *redundant* as they are already present in the memory of the other agents that, as critics, transmitted them. As a consequence, the size of $O = \cup O_i$, i.e., the overall number of examples, is smaller than the sum of the sizes of the counter-examples O_i stored in all agents.

Resources needed by the n-MAS both to perform local revisions and to perform the interactions between the agents have to be considered. We consider that the cost of an interaction is bounded by some constant d. The cost of a local revision $c(m)$ depends on the example memory size $m = |O_i|$ of the learner agent. Hereunder, an interaction is stated as *contradictory* when the critic answers by sending an external counter-example.

Proposition 2. *Let d be the cost of an interaction and c be the revision cost function. When an MAS of n agents has received n_e examples, in the worst case:*

1. *The total number of local revisions performed during the history of the MAS is less than $n_e * n$*
2. *The total cost of interactions is less than $n_e \cdot (n+1) \cdot (n-1) \cdot d$*
3. *The total revision cost is less than $n_e \cdot n \cdot c(n_e)$.*

This means that, for a given n_e, the learning cost (considering only contradictory interactions) is linear with the number of agents n.

3.2   iSMILE: A Revision Mechanism for Individualistic Agents

We consider here the case where an agent never modifies its own current hypothesis but for internal or external counter-examples. Such agents are denoted as *individualistic*.

When the learner agent a_i observes an internal counter-example x w.r.t its current hypothesis T_i, applying global revision M_s results in T_i', now consistent with the set of counter-examples $O \cup \{x\}$ stored in the MAS. However, the other agents a_j for $j \neq i$, i.e. the critics, are not guaranteed to be consistent with x. We define hereunder a weaker property, *delayed mas consistency*.

Definition 7. *Let $O^t = \cup O_j^t$ be the information stored in the MAS at time t*

- *An agent a_i is mas-consistent^t iff $Cons(T_i, O^t)$ is true.*
- *A n-MAS $a_1, \ldots, a_n$ is consistent^t iff each agent a_i is mas-consistent^t. It will be said to be* delayed consistent *iff it is consistent^t with $t = min\{t_1, \ldots, t_n\}$ where each t_i is the time of the last revision of agent a_i.*
- *A revision mechanism is* delayed mas-consistent *iff an agent applying at some time t this global revision mechanism maintains the delayed consistency of the MAS.*

The following property is then the basis of individualistic learning [2].

Proposition 3. M_s *is* delayed mas-consistent.

At a given time t, a n-MAS $a_1, \ldots, a_n$ is *delayed consistent*. The set of examples the MAS is consistent with is O^{t_m} where $t_m = min(t_i)$ and t_i is the time of last revision of agent each a_i immediately preceding t. Counter-examples that have been handled during the interval $t - t_m$ have only been seen by subsets of agents of the MAS.

3.3 The Agent Behavior Model

We consider here a community of individualistic agents acting in their own environment. The behavior of an agent i is as follows: at a given moment, the agent has its own current action model T_i and corresponding counter-examples memory O_i. It is also provided with some goal it has to reach, as for instance stacking block b on top of block c. The agent tries to build a plan, using some planning mechanism. If it succeeds in building a plan, this means that its current action model predicts some effect $\hat{e}$ of the first action a of the plan in the current state s. It will then perform this action, observing the effect e. If $e = \hat{e}$, this means that the new current state s' is as intended in the plan execution and the agent will apply the next action of the plan. Otherwise, this prediction error defines a new counter-example x with $x.s = s, x.a = a, x.e = e$, the current action model is revised locally and the new model is transmitted to the other agents, therefore triggering the iSMILE M_s global revision process. If planning fails, random actions are selected and performed (note that illegal actions, i.e., actions that do not produce any observable effect in the current state are not filtered out) and planning is attempted again until a new plan can be tentatively executed.

We denote as the *desynchronization effect,* the expected decrease in average performance (accuracy of the model or number of successful plans built with this model) resulting from the delay between the various revision times t_i (see Sect. 3.2). This desynchronization effect is expected to increase as the number of agents increases. *Partial resynchronization* is an agent behavior whose purpose is to reduce the desynchronization effect. When considering an agent a_i, its *delay* is defined as $t - t_i$ where times t and t_i correspond respectively to the time of the last revision performed in the whole n-MAS and to the time of the last revision performed by agent a_i. If, since time t_i, no other agent has observed a contradictory example enforcing a revision, then the delay is equal to 0. However, when this delay increases, this means that new examples, that have been observed and memorized by other agents, might contradict the agent a_i theory. The idea of Partial resynchronization is to bound for each agent a_i some monotonic function $f(t - t_i)$ of its delay. We thus add the following behavior to agents: whenever an agent a_i detects that $f(t - t_i) > \delta$, a_i starts a global revision of its current model by sending it to the other agents in order to potentially receive contradictory examples. We use in our experiments $f(t - t_i) = n(t) - n(t_i)$ where $n(t)$ is the total number of examples provided by the environment at time t that contradicted the model of some agent, and therefore triggered a revision in the n-MAS.

4 Experiments

We tested our approach on two domains. The first one is a variant of the blocks world domain in which color predicates for blocks are introduced[2]. This domain requires to learn several rules for capturing the impact of blocks color on the effect of the action *move*. In the *colored-blocks* world, when the agent performs the action $move(a, b)$, a is actually moved on top of b only if a and b have the same color. Otherwise, a is not moved and its color shifts to the same color as b. We run experiments for the 7 blocks with 2 colors domain (7b2c) whose target action model needs 7 rules to model the action *move* and whose state space is composed of nearly 5 million states.

The second domain is the Rover domain from the International Planning Competition[3], which is substantially more complex with a larger number of possibly irrelevant action conditions to consider. This domain has been used previously to investigate action learning [13], but in a different experimental protocol in which examples are generated independently and randomly such that 50 % of the examples $s/a/e$. correspond to a non void effect e. As mentioned above, in our experimental setting, examples are observations from the agent trajectory, and the agent revises its current model online after each action whose outcome contradicts the model. In the Rover domain, the probability that acting randomly leads to a void effect is greater than 90 %. In this domain, an agent corresponds to a base monitoring a team of r rovers equipped with c cameras. The rovers navigate on some area, divided in w way-points, of a planet surface and the team has to perform o objectives regarding science gathering operations. The results of the experiments are communicated to the base. A particular Rover domain in our experiments is described as the tuple (r, w, o, c) and is denoted Rover-rwoc.

Main features of the two domains, i.e. maximal arities of actions, number of state and effect predicates, total number of actions and rules in the target model, are reported hereunder:

Domain	Actions		State/Effects		#rules
	#act.	arity	#pred.	arity	
7b2c	1	2	4	2	7
Rover	9	6	27	3	12

An experiment consists of N runs and is performed for communities of 1, 5 and 30 agents in the colored blocks domain and 1, 5, 20 agents in the Rover domain. For each agent, a run is divided into episodes of at most 50 actions each. The agent starts the first episode with an empty model[4] and the current model at the end of an episode is the starting model at the beginning of the next

---

[2] A problem generator for the colored blocks world problem is available at http://lipn. univ-paris13.fr/~rodrigues/colam.

[3] http://ipc.icaps-conference.org/.

[4] Except in the Rover domain where communication rules are assumed to be known by the agent.

episode. During an episode, the agent explores its environment, starting from a random state, and tries to reach a random goal, both provided by some external controller. Predictive accuracy estimates the probability of correct prediction following the current model and is computed on 100 random state/action pairs whose effect is obtained using the correct model. Collaborative learning follows the iSMILE protocol and exploration is performed according to the agent behavior described in Sect. 3.3. Each agent uses FF [7] as a planner. For that purpose, the goal, domain and action model are translated into an equivalent PDDL [11] planning task. The FF planner is then allowed a short time (2 s) to find a plan or state that planning has failed. The whole framework is implemented using PROLOG threads.

The Colored-Blocks Domain. We are interested first in the predictive accuracy of an agent as a function of the total number t of actions it has performed since the start from the empty model. Figure 2(a) displays the averaged accuracies on 100 runs for communities of 1, 5 and 30 agents. In the same figure, we have also reported the predictive accuracy of a single agent using a baseline relational action model learner, referred to as TILDE on the Fig. 2(a). This baseline learner closely follows the method implemented in MARLIE [4], but uses the more stable state of the art batch relational tree learner TILDE [1] which learns a model from scratch given a set of examples rather than the TG incremental relational decision tree learner used by MARLIE. The example set used, as for IRALe, only contains counterexamples. Clearly, this baseline learner starts with very low accuracies when compared to IRALe. This is because the IRALe starts from the empty model, that always predicts an empty effect. As many state/action pairs in the colored blocks world do result in an empty effect, IRALe accuracy starts at a high level. The MARLIE-like baseline learner does not benefit from this bias and needs 400 actions to reach the IRALe accuracy.

To interpret the benefit of collaboration, we first note that there is a strong relation between the accuracy and the total number of revisions the agent has performed, i.e. the number of internal and external counterexamples in his memory. Such a memory size is obtained for far less actions than those performed by an isolated agent (see Table 1). The benefit of collaborative learning then lies in the proportion of external counterexamples in the agent memory. Figure 2(b) displays both the total number of counterexamples and the number of internal counterexamples in the agent memory *versus* the number of actions it has performed.

We are now interested in the communication costs. Figure 3(a) displays the communication cost per agent, i.e. the number of messages exchanged during its trajectory, as a function of the number of actions performed by the agent, for communities of 5 and 30 agents. In a community of 30 agents, the learned model of an agent is accurate (at level 0.99) as soon as the agent performs 40 actions. It has then exchanged in average 200 messages, which is far from the worst case SMILE bound (see proposition 2) of $40 * 31 * 29 = 35960$ messages. In a community of 5 agents, the same accuracy level requires the agent to perform

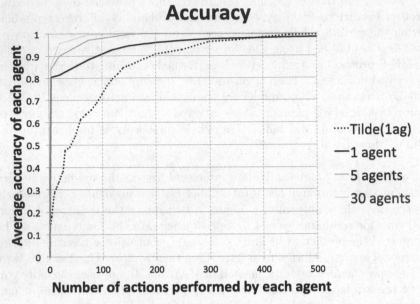

(a) Predictive accuracy of an agent

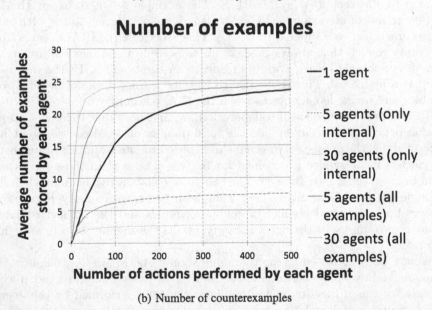

(b) Number of counterexamples

Fig. 2. Predictive accuracy and number of counterexamples (internal only and all counterexamples) vs. Number of actions performed of an agent learning the colored-blocks problem within communities of 1, 5 and 30 agents

Table 1. Number of actions and number of counterexamples (total and internal only) in an agent memory when the agent reaches a fixed accuracy level in 1, 5 and 30 agents communities

#ag.	#actions	Accuracy	#ex.	#intern. ex.
1	250	97.0	21.31	21.31
5	100	96.8	21.32	6.16
30	30	97.2	21.36	2.00

about 100 actions, and to exchange 76 messages. Clearly, the communication cost does not explode when the number of agents increases.

In Fig. 3(b), we report the average number of goals achieved by an agent during a run, as a task oriented measure of learning success. We observe that for all community sizes, there is a critical number of actions an agent has to perform before it makes accurate plans and reaches its random goals. This number is much smaller in the 30-MAS case but a benefit is yet obtained in the 5-MAS case.

We finally investigate in our two application domains the two following settings. The first setting studies *Partial Resynchronization* as an agent behavior (see Sect. 3.3), the second one checks whether *transfer capabilities* are observed within the community, thanks to the high level relational representation of action models. For instance, if such transfer occurs, agents learning in the 7 blocks - 2 colors domain can benefit from external observations transmitted by agents acting in simpler domains such as 3 blocks - 2 colors, for which the target action model is the same. The former setting is denoted as the *Resynchronization* setting, while the latter is referred to as the *Variable* setting.

For the colored blocks domain, and for the *Partial Resynchronization* setting, resynchronization parameter δ is set to 5. In the *Variable* setting, each agent learns in a n blocks 2 colors domain with n randomly drawn in the range [3..7]. For the Rover domain, in the base line experiment denoted *Base*, all agents have the same Rover-2433 domain. In the *Resynchronization*, δ is set to 3, while in the *Variable* setting, each agent has a domain with a parameter tuple (r, w, o, c) randomly drawn in the range ([1–2], [3–4], [2–3], [2–3]).

Curves are provided in Fig. 4 for the colored blocks domain and Fig. 5 for the Rover domain, that report the average accuracy of agents *vs.* the total number of actions performed in the community. This allows comparing the accuracies of 1 and 30 agents communities with respect to the total information available to the community. When comparing agents accuracies in a *per agent* view, as in the previous figures, we focus on time as a resource measured in performed actions per agent, emphasizing that increasing the number of agents gives a clear advantage when considering the total number of goals achieved. The *total number of actions* view emphasizes that there is a price to pay in individual accuracies when the resources (as, for instance, the total energy available) is limited.

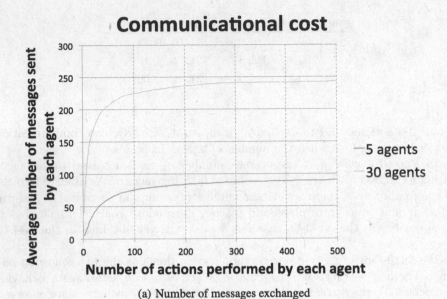

(a) Number of messages exchanged

(b) Number of goals achieved during a run

Fig. 3. Number of messages exchanged and number of goals achieved by an agent learning the colored-blocks problem within communities of 1, 5 and 30 agents

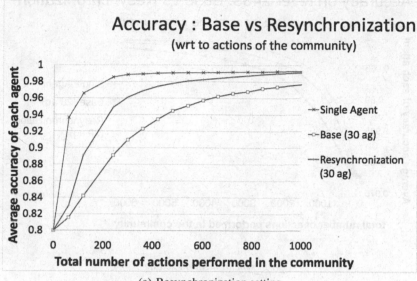

(a) Resynchronization setting

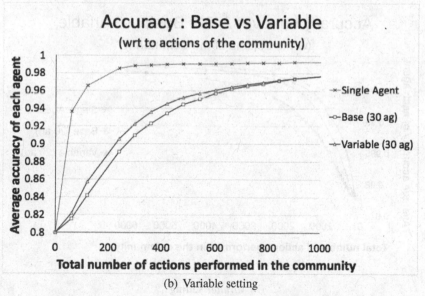

(b) Variable setting

Fig. 4. Average accuracy of agents in the *Resynchronization* setting (left) and in the *Variable setting* (right) versus total number of actions in a 30 agents community learning the 7 blocks 2 colors action model. In both figures, we also display average accuracies of a single agent and a 30 agents communities in the standard setting.

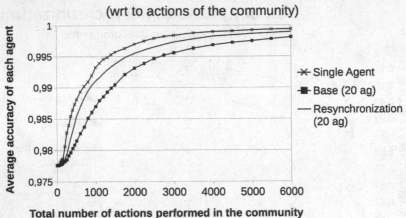

(a) *Resynchronization* setting

(b) *Variable* setting

Fig. 5. Average accuracy of agents in the *Resynchronization* setting (left) and in the *Variable* setting (right) versus the total number of actions in a 20 agents community learning the Rover-2533 action model. In both figures, we also display average accuracies of a single agent and a 20 agents community in the standard setting.

In the colored blocks domain, Fig. 4(a) shows that the *Resynchronization* behavior with $\delta = 5$ strongly reduces the negative effect of delayed consistency. Limiting the resynchronization to $\delta = 5$ does not allow to plainly compensate this negative effect but avoids an excessive price to pay in terms of communication cost between agents (data not shown). In Fig. 4(b), we see the transfer effect: the accuracies are measured in the 7 blocks 2 colors domain, while each agent learns in a n blocks 2 colors domain with n randomly drawn in the range [3..7]. These transfer capabilities rely on the relational representation of the action models.

Figure 5 shows similar trends for the Rover domain. In Fig. 5(a), we compare the average accuracies of a Rover-2433 learning agent belonging to a 20-MAS in the *Base*, *Variable* and *Resynchronization* settings. We observe again that both the *Variable* and *Resynchronization* settings speed up learning of each agent with respect to the *Base* case. In the *Variable* setting, the agent accuracy shows even larger improvement over the *Base* setting in the Rover domain than in the blocks world, which is a strong support for learning relational models in complex domains.

5   Conclusion

In this article, we have modeled and simulated a community of agents who revise their relational action model. Each agent, when revising its current action model, benefits from past observations communicated by other agents on a utility basis: only observations contradicting the current model of the learner agent are transmitted. The experiments give various insights about the benefits of communicating in a utility basis. We observe that when collaborating, the agents do learn faster, i.e. reach an accurate model with much less direct observations than when they are isolated agents, and they learn with a relatively small number of observations communicated by other agents. Moreover, we see on both domains addressed that agents exhibit some transfer capabilities, that follow from their use of first-order explicit domains: counter examples acquired in different (simpler) contexts are useful to other agents and improve convergence to a correct model. Resynchronization, by ensuring that agents regularly submit their models to critics and as a consequence revise their model, also substantially improves the average behaviour of agents in term of accuracy.

Agents are considered as autonomous and separate entities, as for instance robots or mobile devices, with no access to controllers agents or shared memory. We argue that such autonomous entities, still able to communicate with similar entities while preserving their privacy, will play an important role in the future. The framework proposed here is a first step towards more sophisticated situations as the plain multi agent learning case, in which agents interfere as they act in the same environment.

References

1. Blockeel, H., De Raedt, L.: Top-down induction of first-order logical decision trees. Artif. Intell. **101**(1–2), 285–297 (1998)
2. Bourgne, G., Bouthinon, D., El Fallah Seghrouchni, A., Soldano, H.: Collaborative concept learning: non individualistic vs. individualistic agents. In: Proceedings of ICTAI, pp. 549–556 (2009)
3. Bourgne, G., El Fallah-Seghrouchni, A., Soldano, H.: SMILE: sound multi-agent incremental learning. In: Proceedings of AAMAS, p. 38 (2007)
4. Croonenborghs, T., Ramon, J., Blockeel, H., Bruynooghe, M.: Online learning and exploiting relational models in reinforcement learning. In: Proceedings of IJCAI, pp. 726–731 (2007)
5. Dzeroski, S., De Raedt, L., Driessens, K.: Relational reinforcement learning. Mach. Learn. **43**, 7–50 (2001)
6. Esposito, F., Ferilli, S., Fanizzi, N., Basile, T.M.A., Di Mauro, N.: Incremental learning and concept drift in inthelex. Intell. Data Anal. **8**(3), 213–237 (2004)
7. Hoffmann, J.: FF: the fast-forward planning system. AI Mag. **22**, 57–62 (2001)
8. Klingspor, V., Morik, K., Rieger, A.: Learning concepts from sensor data of a mobile robot. Mach. Learn. **23**(2–3), 305–332 (1996)
9. Kulick, J., Toussaint, M., Lang, T., Lopes, M.: Active learning for teaching a robot grounded relational symbols. In: Proceedings of IJCAI (2013)
10. Lang, T., Toussaint, M., Kersting, K.: Exploration in relational domains for model-based reinforcement learning. JMLR **13**, 3725–2768 (2012)
11. McDermott, D.: The 1998 AI planning systems competition. AI Mag. **21**(2), 35–55 (2000)
12. Morik, K.: Sloppy modeling. In: Morik, Katharina (ed.) Knowledge Representation and Organization in Machine Learning. LNCS, vol. 347, pp. 107–134. Springer, Heidelberg (1989)
13. Mourão, K., Zettlemoyer, L.S., Petrick, R.P.A., Steedman, M.: Learning STRIPS operators from noisy and incomplete observations. In: Proceedings of UAI, pp. 614–623 (2012)
14. Otero, R.: Induction of the indirect effects of actions by monotonic methods. In: Kramer, S., Pfahringer, B. (eds.) ILP 2005. LNCS (LNAI), vol. 3625, pp. 279–294. Springer, Heidelberg (2005)
15. Pasula, H.M., Zettlemoyer, L.S., Kaelbling, L.: Learning symbolic models of stochastic domains. JAIR **29**, 309–352 (2007)
16. Rodrigues, C., Gérard, P., Rouveirol, C., Soldano, H.: Incremental learning of relational action rules. In: Proceedings of ICMLA, pp. 451–458. IEEE Press (2010)
17. Rodrigues, C., Gérard, P., Rouveirol, C., Soldano, H.: Active learning of relational action models. In: Muggleton, S.H., Tamaddoni-Nezhad, A., Lisi, F.A. (eds.) ILP 2011. LNCS, vol. 7207, pp. 302–316. Springer, Heidelberg (2012)
18. Rodrigues, C., Soldano, H., Bourgne, G., Rouveirol, C.: A consistency based approach on action model learning in a community of agents. In: Proceedings of AAMAS, pp. 1557–1558 (2014)
19. Rodrigues, C., Soldano, H., Bourgne, G., Rouveirol, C.: Multi agent learning of relational action models. In: Proceedings of ECAI, pp. 1087–1088 (2014)
20. Sutton, R.S.: Dyna, an integrated architecture for learning, planning, and reacting. SIGART Bull. **2**, 160–163 (1991)
21. Xu, J.Z., Laird, J.E.: Instance-based online learning of deterministic relational action models. In: Proceedings of AAAI (2010)

22. Yang, Q., Wu, K., Jiang, Y.: Learning action models from plan examples using weighted MAX-SAT. Artif. Intell. **171**(2–3), 107–143 (2007)
23. Zhuo, H.H., Nguyen, T.A., Kambhampati, S.: Refining incomplete planning domain models through plan traces. In: Proceedings of IJCAI (2013)

Ontology-Based Classification – Application of Machine Learning Concepts Without Learning

Thomas Hoppe[✉]

Datenlabor Berlin, Berlin, Germany
thomas.hoppe@datenlabor.berlin
http://www.datenlabor.berlin/

Abstract. The application of machine learning algorithms on real world problems rarely encounters ideal conditions. Often either the available data are imperfect or insufficient, or the learning situation requires a rather complicated combination of different approaches. In this article I describe an application, which – in an ideal world – would be solvable by a conventional supervised classification algorithm. Unfortunately, the available data are neither reliably classified nor could a manual correct reclassification be derived under restricted available resources. Since we had a large domain thesaurus available, we were able to develop a new approach, skipping the learning step and deriving a classification model directly from this thesaurus. The evaluation showed that for the intended use the quality of the classification model is more than sufficient.

Keywords: Supervised learning · Classification · Ontology · Thesaurus · Validation · Evaluation · Accuracy · Precision · Recall · Bootstrapping

1 Background

In 2011 the German federal states Berlin and Brandenburg came to an agreement to implement a common innovation strategy. Goal of this strategy is the development of so called cross-border innovation clusters. These clusters summarize economical sectors and branches which are of particular interest for the development of the region Berlin-Brandenburg.[1] These clusters describe sectors with a high potential for economical development of the region or a need for action:

- Health
- Energy and Environment Technologies
- Transport, Mobility and Logistics
- ICT, Media and Creative Industries
- Optics (including Microsystems Technologies)
- Tourism
- Food
- Metal
- Plastics/Chemistry

---

[1] http://www.berlin.de/sen/wirtschaft/politik/innovationsstrategie.en.html. Accessed 19 July 2014.

© Springer International Publishing Switzerland 2016
S. Michaelis et al. (Eds.): Morik Festschrift, LNAI 9580, pp. 320–330, 2016.
DOI: 10.1007/978-3-319-41706-6_17

Development of these clusters requires not only the financial support of research institutions and companies in the region, but also the availability of qualified employees. Hence the support of professional training and continued education for these clusters becomes important. This support consists of offering appropriate education offers, supporting an easy access to these offers and supporting the transition from established jobs to new job descriptions.

In order to ease the access to professional training and continued education both federal states have developed and support the "Weiterbildungsdatenbank Berlin-Brandenburg" (WDB-BB; database for continued education) which is operated by EUROPUBLIC GmbH. Purpose of this database is to make education offers in the region Berlin-Brandenburg easy accessible and thus to support the development of the innovation clusters. Currently this database contains about 30.000 education offers.

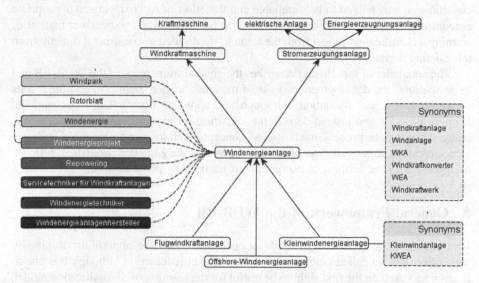

Fig. 1. Language Usage for denoting "Windenergieanlage"

With a semantic, thesaurus-based search engine the Ontonym GmbH has developed the means for making the education offers of the WDB-BB easily searchable by means of modeling the language usage of users searching for continued education [1]. Figure 1 shows an excerpt of this thesaurus for terms used by users searching for continued education offers in the context of renewable energies. Grey scales of the nodes denote different term categories, arrows denote term subsumption and dashed lines general relations between terms.

We could show by direct comparison against the original full text search that this semantic search reduces the search time of a user per query by 10 % on average and that the quality of the search results can be increased by avoiding on average 55 % of inappropriate search results and substituting them with 27-44 % better fitting results, thus reducing the entire amount of search results by 12-27 % [2].

In a precursor project, this thesaurus was extended by cluster-specific and additional related terms for the nine innovation clusters. The thesaurus currently consists of about 11.300 recruitment and continued education specific concepts with roughly 17.200 (partially multilingual) terms for which the semantic search is capable of generating nearly 33.600 alternative writing forms[2].

2   Project Goal

Goal of the project described in this chapter was the development and exploration of a classification approach, which can be used to classify educational offers into the clusters of the innovation strategy. On the first sight this looked like a conventional supervised learning task, but although the education offers where pre-classified, their classification was judged to be unreliable and the effort of reclassifying an appropriate amount of several thousand offers appeared to be prohibitively expensive. Instead of learning a classification model from the scratch, we decided to construct it directly from the existing thesaurus.

The remainder of this chapter describes the general framework of the WDB-BB and its restrictions, the development process of the classification algorithm and the results of its validation and evaluation. Although this approach does not use conventional learning methods and instead derives the classification model from the thesaurus, the entire development process made use of concepts well-known in machine learning. Hence it can be seen as an example which shows that the *good data analytical practice* of applied machine learning is useful even if no training phase is used.

3   General Framework of the WDB-BB

Until spring 2015 WDB-BB gave educational institutions the opportunity to classify their offers into so called cutting-edge fields ("Zukunftsfelder"). Although this classification appeared on the first sight to be useful for the learning of a classification model, the discussion with EUROPUBLIC showed that

- just few institutions classified their offers, hence
- just a minor part of offers were actually classified, and
- the quality of the classifications were often judged as imprecise or incorrect.

Although nearly 30.000 offers are available in the WDB-BB, it turned out during the project that just roughly 2/5 of them where unique and that the others are just repeated editions of the same course, which are described by the same text. Of those unique offers only 3.500 where poorly classified according to the cutting-edge fields. Of course, reclassification of this amount of unique offers would in principle be feasible, but the resources for a correct reclassification of these offers were neither available, neither in form of available budget nor skilled employees.

---

[2] This number still excludes all variants derivable by stemming, which cannot be estimated.

The semantic search implemented for the WDB-BB uses the thesaurus to compute annotations of the textual descriptions of the education offers and for queries posed by users. The thesaurus is used on one hand for the recognition of terms in the text and on the other hand for mapping the recognized terms into the controlled vocabulary defined by the thesaurus. Thus, the semantic search has for every text document, i.e. every educational offer, already an annotation based on a controlled vocabulary available, which annotates recruitment, continued education and cluster-specific concepts used in the offer's textual description.

4   Development of Ontology-Based Classification

Since the manual classification of the educational offers by their authors was found to be impracticable the goal of the project was the development of an automatic classification approach, which either produces highly correct classification or at least produces good suggestions for suitable classes.

Usually the task of developing such a classification system would be realized with a supervised learning approach, which uses pre-classified training examples to derive the classification model. However in this project we were confronted with the restriction, that neither a reliable pre-classified training set exists, nor that a larger training set could be classified manually. So instead of learning the classification model we decided to construct it.

Input for this construction was a set of relevant terms for each cluster and the thesaurus. Already during the precursor project this thesaurus was extended by cluster-relevant terms plus additional sub and super concepts, synonyms and relations between them. Hence, the concepts characterizing a cluster already existed in the thesaurus, but were distributed over different categories, like position and job titles, branches, skills and competencies, technologies, tasks and activities, etc.

In order to construct the classification model we needed to collect all the cluster's characterizing terms and derive some measure of their importance for the cluster. Collection of the characterizing terms was realized by traversal of the thesaurus and collecting for each relevant term all directly related terms. Determination of the weights was realized by counting how frequently the characterizing terms were encountered during the traversal.

4.1   Concept Centrality

The occurrence count of the concepts encountered during the traversal represents some kind of measure of the "concept centrality" with respect to the cluster. Concepts frequently encountered during the traversal can be considered as more central to the cluster than concepts encountered less often, since the former are related to a larger number of other cluster concepts.

Hence, the traversal process derived not only a list of concepts characterizing the respective cluster, but also a numerical value representing the centrality of the concept. We call this list the "centrality set" of the cluster.

Although this concept list and their centrality values do not represent the usual frequency count of terms in some existing document collection, these counts can be interpreted as term frequencies of some virtual document describing the cluster. With this interpretation trick, conventional measures for the comparison of term frequencies between each education offer's annotation and the centrality set of the cluster could be applied.

4.2   Comparison Metrics

For the classification of education offers against the cluster descriptions, we use the simple approach of measuring the similarity of the offer's annotation with respect to the centrality set. We investigated two metrics: Cosine Similarity and Naïve Bayes. For both measures we considered different variations of parameters:

1. Ignoring concepts with small centrality values
2. Different weightings of the title's and textual description's annotations
3. Weighting the term frequency within the offer's annotation

As we found during a first validation, both metrics gave comparable results. Interestingly, neither the variation of the weightings (2) nor the use of the offer's concept frequencies (3) had a significant impact on the classification quality. However, ignoring concepts with small centrality values improved the classification quality.

4.3   Improving the Classification Quality

These findings gave us the crucial clue that the means for improving the classification quality reside in the centrality sets (i.e. the constructed classification models). An inspection of those sets showed that between some clusters a large overlap existed between their centrality sets and that the centrality sets included – because of the rather unintelligent – extraction process a number of very general concepts, not indicative for any cluster.

Manual minimization of the overlap between centrality measures and the removal of general, non-indicative terms led to a significant improvement during the second step of the design phase of the classification approach.

5   Validation and Evaluation of the Approach

As usual in machine learning we used the approach of splitting the data into a training set and a test set. The set of 3.500 poorly classified education offers was split into a set of 2.800 offers used for the development and a set of 700 offers restrained for the final evaluation. Since we didn't used a training phase, we used the training set entirely for validation purposes to check whether the parameter modifications of the comparison metrics described in Sect. 4.2 led to improvements during the classification, and to check whether the minimization of the overlap of the centrality sets described in Sect. 4.3 led to the expected improvements.

5.1 Validation

For the validation of the parameter modifications we used the 2.800 education offers poorly classified in to the cutting-edge fields, which where automatically mapped into the nine innovation cluster. Since this process maps potential faulty classifications from the classes of the cutting-edge fields to the classes of the innovation cluster, this mapping does not improve the quality of the original data. Misclassified offers remain misclassified.

Of course, it is justified to ask, what do we gain with such a mapping, if the data are still error-prone? Are these error-prone data really helpful to validate improvements of the classification approach? The important point here is to note that we do not measure the overall absolute correctness of the classification approach with respect to these data; we just use them to estimate the relative improvement in the reduction of misclassified cases. For this purpose even erroneous classifications can be used as long as they are not modified between modifications of the classification approach.

The developed approach classifies annotations of offers against constructed annotations of clusters. Although the annotation of a cluster was derived from a thesaurus capturing the meaning of cluster's terms partially, the comparison metrics used for determining the fitting clusters are purely syntactical and numerical. Hence, it can neither differentiate the different meanings a term might have in different clusters, nor does it allow disambiguating the term's meaning in the context of the clusters, nor is it possible to make the cluster's annotations completely disjoint. Therefore, misclassifications cannot be avoided completely; instead they can only be reduced as far as possible.

Inspection of the Training Data. Already the inspection of the training data showed, that the cluster's distribution is skewed and that the number of offers per cluster was unevenly distributed (see Table 1). Therefore, we expected to obtain from the clusters "Ernährung", "Kunststoff", "Optik", and "Umwelttechnik" unreliable measurements.

Table 1. Distribution of offers

Cluster	German cluster label	No. of offers
Food	Ernährung	19
Health	Gesundheitswirtschaft	427
ICT, Media & Creative Industries	IKT	1837
Plastics/Chemistry	Kunststoff	8
Metal	Metall	101
Optics (incl. Microsystem Technology)	Optik	28
Tourism	Tourismus	93
Energy & Environment Technologies	Umwelttechnik	46
Transport, Mobility and Logistics	Verkehr	162

Validation of the Best Parameter Settings. We experimented with different parameter settings (see Sect. 4.2) in order to determine the best parameter configurations. Table 2 reports for both investigated metrics accuracy, precision and recall of the best parameter configurations. For both metrics these parameter settings ignored terms of the centrality set with a centrality smaller than 3. While Naïve Bayes gave best results, when the number of term occurrences in the offers was taken into account, the cosine measure worked – as expected – best, when this frequency information was not used.

As can be seen from Table 2 both metrics achieve a comparable overall accuracy, while the precision and recall vary significantly for each cluster. Especially the greyed out clusters "Kunststoff" and "Optik" cannot be correctly estimated, because of the small number of available training examples.

Table 2. Validation of best parameter settings

	Naïve Bayes		Cosine	
	Precision	Recall	Precision	Recall
Ernährung	59,1%	18,3%	43,8%	29,6%
Gesundheitswirtschaft	84,4%	74,3%	93,7%	68,7%
IKT	88,0%	90,6%	86,6%	88,9%
Kunststoff	5,9%	12,5%	6,7%	12,5%
Metall	96,8%	81,3%	86,9%	94,6%
Optik	6,3%	66,7%	7,3%	100,0%
Tourismus	56,4%	75,7%	37,8%	97,1%
Umwelttechnik	56,7%	95,0%	66,7%	92,5%
Verkehr	78,0%	76,7%	79,6%	73,5%
Accuracy		82,2%		81%

5.2 Evaluation

The final evaluation of both metrics and their best parameter settings on the 700 retained test cases[3] showed a similar behavior and interestingly resulted in a higher overall accuracy, as can be seen from Table 3.

The clusters "Kunststoff" and "Optik" gave completely wrong resp. no results, since just few offers resp. no offers where contained in the test set. Because of the low number of test cases for the clusters "Ernährung" and "Tourism" their determined classification quality needs to be interpreted with care.

Statistical Bootstrapping. Obviously, splitting the available cases into a training set and a test set implies that the selected test cases belong to a single randomly chosen distribution. This distribution might or might not be representative for the entire unknown population of all cases. Hence a single evaluation of the test cases accuracy, precision and recall gives only a single randomly chosen view on the overall quality of the approach. What is really needed is a statement about the quality of the approach for a larger number of possible distributions in the form of confidence intervals of those measures.

---

[3] Only 600 of these test cases could be classified into the innovation clusters.

Table 3. Evaluation of both metrics

	No. Of Offers	Naïve Bayes		Cosine	
		Precision	Recall	Precision	Recall
Ernährung	13	50,0%	23,1%	30,0%	23,1%
Gesundheitswirtschaft	238	89,3%	97,9%	93,2%	92,4%
IKT	132	87,6%	85,6%	84,7%	84,1%
Kunststoff	4	0,0%	0,0%	0,0%	0,0%
Metall	65	98,2%	81,5%	90,8%	90,8%
Optik	0	0,0%	0,0%	0,0%	0,0%
Tourismus	14	46,2%	85,7%	43,8%	100,0%
Umwelttechnik	36	85,3%	80,6%	83,9%	72,2%
Verkehr	98	94,1%	80,6%	96,4%	81,6%
Accuracy			87%		85,5%

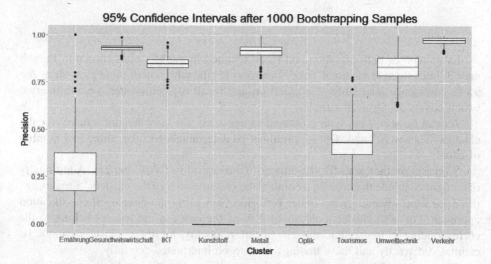

Fig. 2. Confidence Intervals for the Classification's Precision

In order to determine these confidence intervals we used bootstrapping [3, 4] on the classified examples. Since the designed classification approach does not use a learning step and since only the test set was correctly classified, we had to adopt the bootstrapping for the evaluation slightly. Instead of randomly choosing for each bootstrap evaluation the training and test set new, in order to learn the classification models from a large number of potential data sets, we used for the bootstrapping only the test cases. Under the assumption, that their distribution is representative for the entire unknown population, the distribution of the test cases can be considered a "virtual population" from which we drew 1.000 random samples with replacement.

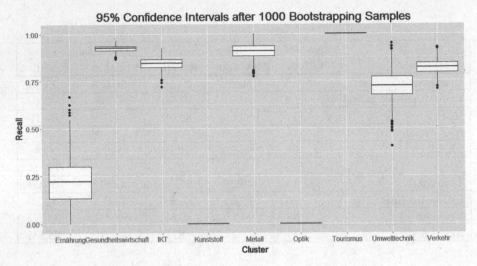

Fig. 3. Confidence Intervals for the Classification's Recall

Evaluation of both precision and recall resulted in the box plots shown in Figs. 2 and 3. In contrast to the usual Tukey box plots [5] the whiskers of these plots show the 95 % confidence interval for the precision and recall of each cluster, i.e. the interval which covers 95 % of all samples.

As can be seen from both plots and as expected, the classification quality for the clusters "Kunststoff" and "Optik" could not be determined, because of the low number of cases.

Surprisingly the recall for the cluster "Tourismus" is 100 % perfect. Additionally one might conclude that tossing a coin for the classification of the clusters "Ernährung" or "Tourismus" would give a better precision than using the developed classification approach. For both clusters only about 2 % of the cases in the test set belong to the corresponding clusters, thus the classification quality for these clusters cannot be estimated correctly and these findings need to be interpreted carefully.

Since only 6 % of the test cases belong to the cluster "Umwelttechnik" it is reasonable to assume that the large variation in its confidence interval could be reduced if additional test cases would be available for this cluster.

The classification quality for the cluster "Umwelttechnik" can be considered good and for the remaining clusters "Gesundheitswirtschaft", "IKT", "Metall" and "Umwelttechnik" as very good.

6 Summary

The preconditions for the application of machine learning approaches are not always fulfilled. In the case of the WDB-BB, we had no reliably classified cases available to realize a learning phase. However, we have a large thesaurus available from which a classification model could be constructed. By extracting for each cluster all concepts related to the class together with a numerical value describing the "concept's centrality" within a cluster and interpreting these values as some term frequency of a "virtual document" describing the cluster, we were able to construct a classification model from the thesaurus instead of learning it.

Already during the validation, we found that the clusters were not evenly distributed and therefore not for all clusters an evaluation would give reliable results. During the validation phase we cleaned up the cluster's centrality sets and determined the parameter configurations which resulted in the highest overall accuracy.

The test of the final configurations of the similarity measures eventually showed that very good classifications results are obtainable with this approach. Although not for all clusters valid results could be achieved during the evaluation, we are confident, that the quality of this approach will hold for these clusters too, if additional cases become available for the evaluation of their classification quality.

6.1 Implementation of the Approach

Purpose of the design of this approach is the automization of the classification of continued education offers. Often such a course does not fall into just one cluster. E.g. a course about healthy nutrition may fall into the cluster "health" as well as into the cluster "food" or a continued education about "secure welding" might fall into the cluster "metal", "energy" or "transportation". Hence it is not so important to determine the best cluster into which a continued education can be classified; instead it is completely sufficient to determine the three best clusters under which an offer should appear.

Moreover, the classification approach should be applied in a context where either the descriptions of the offers are input manually or where they are obtained from a bulk upload. In both cases it is important to know, which automatically obtained classifications are reliable and which are not. The analysis of the confidence intervals has identified the cluster, where we cannot be completely confident about the quality of the approach. Identifying the offers which fall into these clusters and asking either the course supplier or the WDB-BB staff for a manual classification of the corresponding offers, helps to ensure good classification results as well as it allows the purposeful collection of additional evaluation cases for these clusters.

References

1. Hoppe, T.: Modellierung des Sprachraums von Unternehmen – Was man nicht beschreiben kann, das kann man auch nicht finden. In: Humm, B., Reibold, A., Ege, B. (eds.) Corporate Semantic Web – Wie semantische Anwendungen in Unternehmen Nutzen stiften. X.media.press, Springer, Heidelberg (2015)
2. Hoppe, T., Junghans, H.: Messung des Nutzens semantischer Suche. In: Humm, B., Reibold, A., Ege, B. (eds.) Corporate Semantic Web – Wie semantische Anwendungen in Unternehmen

Nutzen Stiften. X.media.press, Springer, Heidelberg (2015)
3. Efron, B.: Bootstrap methods: another look at the jackknife. Ann. Stat. **7**(1), 1–26 (1979)
4. Hesterberg, T., Monaghan, S., Moore, D.S., Clipson, A., Epstein, R.: Bootstrap Methods and Permutation Tests, Companion Chapter 18 to The Practice of Business Statistics. W.H. Freemann and Company, New York (2003). http://bcs.whfreeman.com/pbs/cat_160/PBS18.pdf. Accessed 2 Mar 2015
5. Tukey, J.W.: Exploratory data analysis. Addison-Wesley, Boston (1977)

Deep Distant Supervision: Learning Statistical Relational Models for Weak Supervision in Natural Language Extraction

Sriraam Natarajan[1][✉], Ameet Soni[2], Anurag Wazalwar[1],
Dileep Viswanathan[1], and Kristian Kersting[3]

[1] Indiana University, Bloomington, USA
{natarasr,anurwaza,diviswan}@indiana.edu
[2] Swarthmore College, Swarthmore, USA
soni@cs.swarthmore.edu
[3] TU Dortmund, Dortmund, Germany
kristian.kersting@cs.tu-dortmund.de

Abstract. One of the challenges to information extraction is the requirement of human annotated examples, commonly called gold-standard examples. Many successful approaches alleviate this problem by employing some form of distant supervision i.e., look into knowledge bases such as Freebase as a source of supervision to create more examples. While this is perfectly reasonable, most distant supervision methods rely on a given set of propositions as a source of supervision. We propose a different approach: we infer weakly supervised examples for relations from statistical relational models learned by using knowledge outside the natural language task. We argue that this deep distant supervision creates more robust examples that are particularly useful when learning the entire model (the structure and parameters). We demonstrate on several domains that this form of weak supervision yields superior results when learning structure compared to using distant supervision labels or a smaller set of labels.

1 Introduction

In this chapter, we consider the problem of Information Extraction (IE) from Natural Language (NL) text, where the goal is to learn relationships between attributes of interest; e.g., learn the individuals employed by a particular organization, identifying the winners and losers in a game, etc. There have been two popular forms of supervised learning used for information extraction.

1. The classical machine learning approach. For instance, the NIST Automatic Content Extraction (ACE) RDC 2003 and 2004 corpora, has over 1000 documents that have human-labeled relations leading to over 16,000 relations in the documents [10]. ACE systems use textual features – lexical, syntactic and semantic – to learn mentions of target relations [21,28].

© Springer International Publishing Switzerland 2016
S. Michaelis et al. (Eds.): Morik Festschrift, LNAI 9580, pp. 331–345, 2016.
DOI: 10.1007/978-3-319-41706-6_18

2. *Distant supervision*, a more practical approach, where labels of relations in the text are created by applying a heuristic to a common knowledge base such as Freebase [10, 19, 23].

However, both these methods have inherent issues. Pure supervised approaches are quite limited in scalability due to the requirement of high quality labels, which can be very expensive to obtain for most NL tasks. While distant supervision appears to be a viable alternative, the quality of the generated labels is crucially dependent on the heuristic that is being used to map the relations to the knowledge base. Consequently, there have been several approaches that aim to improve the quality of these labels ranging from multi-instance learning [5, 19, 22] to using patterns that frequently appear in the text [23]. As noted by Riedel et al. [19], the distant supervision assumption can be too strong, particularly when the source used for labeling the examples is external to the learning task at hand.

In our earlier work [13], on which the present chapter is based, we took a drastically different approach that we summarize and explain in this chapter. Our insight is that the labels are typically created by "domain experts" who annotate the labels carefully. These domain experts have some inherent rules in their mind that they use to create examples. For example, when reading a sports column, there is an inherent bias that we expect that "home teams are more likely to win a game" or that "in most games, there is a winner and a loser and they are not the same team". We call this knowledge *world knowledge* as it describes the domain (or the world) and not specific language constructs. We aim to use such knowledge to create examples for learning from NL text. More precisely, our hypothesis – which we verify empirically – is that the use of world knowledge will help in learning from NL text. This is particularly true when there is a need to learn a model without any prior structure since the number of examples needed can be large. These weakly supervised examples can augment the gold-standard examples to improve the quality of the learned models. To this effect, we use the probabilistic logic formalism called *Markov Logic Networks* [4] to perform *weak supervision* [2] to create more examples. Instead of directly obtaining the labels from a different source, we perform inference on *outside knowledge* (i.e., knowledge not explicitly stated in the corpus) to create sets of entities that are "potential" relations. This outside knowledge forms the *context MLN* – CMLN – to reflect that they need not be linguistic models. An example of such knowledge could be that "home teams are more likely to win a game". Note that this approach enables the domain expert to write rules in first-order logic so that the knowledge is not specific to any particular textural wording but is general knowledge about the world (in our example, about the games played). During the information extraction (IE) phase, unlabeled text is then parsed through some entity resolution parser to identify potential entities. These entities are then used as queries to the CMLN which infers the posterior probability of relations between these entities. These inferred relations become the probabilistic examples for IE. This is in contrast to distant supervision where statistical learning is employed at "system build time" to construct a function from training examples.

So far, the major hurdle to learning the full models in IE is the large number of features leading to increased complexity in the search [17]. Most methods use a prior designed graphical model and **only** *learn the parameters*. A key issue with most structure-learning (model-learning) methods is that, when scoring every candidate structure, parameter learning has to be performed in the inner loop. We, on the other hand, employ an algorithm based on *Relational Functional Gradient Boosting* (RFGB) [6,7,12] for learning **the structure**. It must be emphasized clearly that the main contribution of the paper is not the learning algorithm, but instead is presenting a method for generation of weakly supervised examples to augment the gold standard examples. Specifically, we adapt the RFGB algorithm to learn in the presence of probabilistic examples by explicitly optimizing the KL-divergence between the weak labels and the current predicted lables of the algorithm.

To demonstrate the concept, we perform knowledge extraction on the TAC 2015 Cold Start Knowledge Base Population (KBP) task[1]. Given a large, unannotated collection of documents, the KBP task aims to construct a knowledge base of entities (*entity discovery*) and facts about these entities (*slot filling*). Utilizing a subset of target predicates from the slot filling task, we compare our weak supervision approach to using gold-standard examples alone. More precisely, we perform 5-fold cross validation on these tasks and show that the proposed approach outperforms learning only from gold-standard data.

When we bias the learner with examples created from commonsense knowledge, we can distantly learn model structure. Because we have more training examples than the limited supply of gold-standard examples and they are of a higher quality than traditional distant labeling, the proposed approach allows for a better model to be learned. Our algorithm has two phases:

1. *Weak supervision phase*, where the goal is to use commonsense knowledge (CMLN). This CMLN could contain clauses such as "Higher ranked teams are more likely to win". "Home teams are more likely to win," etc. Given this CMLN, parameters (weights) are learned from a knowledge base by looking at the previously completed games. Of course, these weights could also be provided by the domain expert. Once these weights are learned, predictions are made on entities extracted from unlabeled text and these predictions serve as weakly supervised examples for our next phase. This phase can be independent of the linguistic information (once relations are extracted from text) and simply relies on world knowledge.
2. *Information extraction phase*, where the noisy examples are combined with some "gold-standard" examples and a relational model is learned using RFGB on textual features from the gold-standard and the weakly supervised documents. Note that this phase only uses the text information for learning the model. The world knowledge is ignored when learning from linguistic features.

The potential of such advice giving method is not restricted to NL tasks and is more broadly applicable. For instance, this type advice can be used for labeling

---

tasks [24] or to shape rewards in reinforcement learning [3] or to improve the number of examples in a medical task. Such advice can also be used to provide guidance to a learner in unforseen situations [9].

We proceed as follows: after reviewing the related work, we present the two phases of our approach in greater detail. We then present the experimental setup and results on the three different tasks before concluding by outlining future research directions.

2   Related Work

We now present the related work in distant supervision and probabilistic logic models.

Distant Supervision: As mentioned earlier, our approach is quite similar to *distant supervision* [2,10] which generates training examples based on external knowledge bases. The external database provides pairs of related entities; sentences in which any of the related entities are mentioned are considered to contain positive training examples. These examples along with the few annotated examples are provided to the learning algorithm. These approaches assume that the sentences that mention the related entities probably express the given relation. Riedel et al. [19] relax this assumption by introducing a latent variable for each mention pair to indicate whether the relation is mentioned or not. This work was further extended to allow overlapping relations between the same pair of entities (e.g. Founded(Jobs, Apple) and CEO-of(Jobs, Apple)) by using a multi-instance multi-label approach [5,22]. We employ a model based on non-linguistic knowledge to generate the distant supervision examples. Although we rely on world knowledge to obtain the relevant input relations for our CMLN model, one can imagine tasks where such relations are available as inputs or extracted earlier in the pipeline.

Statistical Relational Learning: Most NLP approaches define a set of features by converting structured output such as parse trees, dependency graphs, etc. to a flat feature vector and use propositional methods such as logistic regression. Recently, there has been a focus of employing Statistical Relational models that combine the expressiveness of first-order logic and the ability of probability theory to model uncertainty. Many tasks such as BioNLP [8] and TempEval [25] have been addressed [16,18,27] using SRL models, namely Markov Logic Networks (MLNs) [4]. But these approaches still relied on generating features from structured data. Sorower et al. [20] use a similar concept in spirit where they introduce a mention mode that models the probability of facts mentioned in the text and use a EM algorithm to learn MLN rules to achieve this. We represent the structured data (e.g. parse trees) using first-order logic and use the RFGB algorithm to learn the structure of Relational Dependency Networks (RDN) [14]. Relational Dependency Networks (RDNs) are SRL models that consider a joint distribution as a product of conditional distributions. One of the important advantages of RDNs is that the models are allowed to be cyclic. As shown in the next section, we use MLNs to specify the weakly supervised world knowledge.

3 Structure Learning for Information Extraction Using Weak Supervision

One of the most important challenges facing many natural language tasks is the paucity of "gold standard" examples. Our proposed method, shown in Fig. 1, has two distinct phases: *weak supervision phase* where we create weakly supervised examples based on commonsense knowledge and *information extraction phase* where we learn the structure and parameters of the models that predict relations using textual features. The key idea in the weak supervision phase is to use commonsense knowledge with basic linguistic patterns while "inferring" the labels of the set of interesting entities. On the other hand, the information extraction phase considers extensive linguistic patterns when learning the relations. The intuition is that the labeling phase merely uses the "inductive bias" of the domain expert while the information extraction phase merely analyzes the text. We explain each of these in turn now providing the key intuitions. For more technical details, we refer to our paper [13].

3.1 Weak Supervision Phase

We now explain how our first phase addresses the key challenge of obtaining additional training examples. As mentioned earlier, the key challenge is obtaining annotated examples. To address this problem, we employ a method that is commonly taken by humans. For instance, consider identifying a person's family relationship from news articles. We may have an *inductive bias* towards believing two persons a sentence with the same last name are related. To formalize

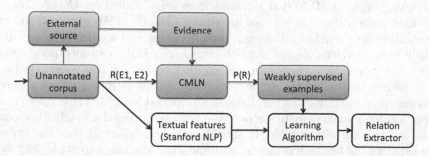

Fig. 1. Flowchart of our method. The top-half represents the weak supervision phase where we generate the examples using the CMLN and facts from an external source. The bottom-half represents the information extraction phase where we learn a SRL model using the weakly supervised and gold standard examples.

this notion, we seek to construct a model that captures this inductive bias along with gold-standard examples.

We consider a formalism called Markov Logic Networks (MLNs) that extend a sub-group of first order logic (with finite domain elements) by allowing for softening of the clauses using numeric weights. We employ MLNs to capture the world knowledge. MLNs [4] are relational undirected models where first-order logic formula correspond to the cliques of a Markov network and formula weights correspond to the clique potentials. A MLN can be instantiated as a Markov network with a node for each ground predicate (atom) and a clique for each ground formula. All groundings of the same formula are assigned the same weight.

Intuitively, a possible world with a formula having a higher weight is more probable than one with a lower weight[2]. There have been several weight learning, rule learning (also called structure learning) and inference algorithms proposed for MLNs.

We use MLNs because they provide an easy way for the domain expert to provide the knowledge as simple first-order logic rules. The weights of these rules can then be learned using training data. Or in other cases (such as our experiments), we can simply set these weights by a rule-of-thumb where highly probably rules have higher weights. In our experiments, we found that the difference between multiple weight settings do not affect the results as long as the ordering between the rules is maintained. This is due to the fact that we potentially obtain the same ordering of the weakly supervised labels – only the *scale* of the probabilities varies. In our experiments, we observed this pattern and hence note that specific weights do not have a high impact on our labels.

Another important aspect of this phase is that we use the scalable Tuffy system [15] to perform inference. One of the key attractions of Tuffy is that it can scale to millions of documents. The inference algorithm implemented inside Tuffy appears to be robust and hence serves as an ideal package to be used for capturing the world knowledge.

Our proposed approach for weak supervision is presented in Fig. 2. The first step is to design an MLN that captures knowledge, called an *CMLN*. For the KBP task, some rules that we used are shown in Table 1. For example, the first rule identifies any number following a person's name and separated by a comma is likely to be the person's age (e.g., "Sharon, 42"). Rules can incorporate more textual features, such as the fourth rule and fifth rule which state the appearance of the lemma mother or father between two persons is indicative of a parent relationship (e.g., "Malia's father, Barack, introduced her..."). These rules can be simply written by the domain's expert and softened using the weights presented in the left column of the table. It should be noted, however, that the type of knowledge we choose to leverage is not limited to domain expertise. Any type of knowledge (e.g., commonsense or linguistic patterns) can be utilized for weak supervision, as demonstrated by our sample rules in Table 1. The relative merits

---

[2] The ratio is actually a log-odds of their weights. We refer to the book for more details [4].

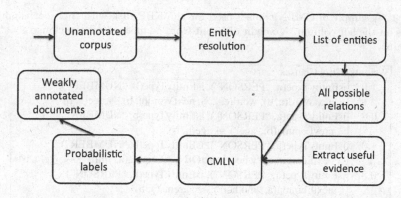

Fig. 2. Steps involved in creation of weakly supervised examples.

between the different rules can be judged reasonably even though a domain's expert may not fully understand the exact impact of the weights. It is quite natural in several tasks, as we show empirically, to set "reasonable" weights.

The next step is to create weakly supervised learning examples. We use *Stanford NLP* toolkit to perform entity resolution to identify the potential persons, organizations, countries, etc. in each document. The target predicates of interest for the KBP task were binary slots – that is, a relation with two entities as arguments. We use the CMLN to obtain the posterior probability on the relations being true between entities mentioned in the same sentence – for example, a parent and child pair. Note that to perform inference, evidence is required. Hence, we constructed a basic set of facts for each word in the document, including entity type, lemmas, and neighboring words.

Recall that the results of inference are the posterior probabilities of the relations being true between the entities extracted from the same sentence and they are used for annotations. One simple annotation scheme is using the MAP estimate (i.e., forcing the soft predictions to a hard true or false). An alternative would be to use a method that directly learns from probabilistic labels. Choosing the MAP would make a strong commitment about several examples on the borderline. Since our world knowledge is independent of the text, it may be the case that for some examples perfect labeling is not easy. In such cases, using a softer labeling method might be more beneficial. The second phase of our approach – structure learning using RFGB – is capable of handling either approach. In our experiments, we chose to maintain probabilistic labels to reflect the confidence in a weakly supervised prediction. Now these weakly supervised examples are ready for our next step – *information extraction*.

3.2 Learning for Information Extraction

Once the weakly supervised examples are created, the next step is inducing the relations. We employ the procedure from Fig. 3. We run both the gold standard and weakly supervised annotated documents through Stanford NLP toolkit to

Table 1. A sample of CMLN clauses used for the KBP task with their corresponding weights on the left column. A weight different from ∞ means that the clause is a "soft" constraint.

Weights	MLN Clauses
1.0	isEntityType(a, "PERSON"), isEntityType(b, "NUMBER"), nextWord(a, e), word(e, ","), nextWord(e, b) → age(a, b)
0.6	isEntityType(a, "PERSON"), isEntityType(b, "NUMBER"), prevLemma(b, "age") → age(a, b)
1.0	isEntityType(a, "PERSON"), isEntityType(b, "NUMBER"), nextLemma(a, "who"), nextOfNextLemma(a, "turn") → age(a, b)
0.8	isEntityType(a,"PERSON"), isEntityType(b, "PERSON") nextLemma(a, "mother") → parents(a, b)
0.8	isEntityType(a, "PERSON"), isEntityType(b, "PERSON") nextLemma(a, "father") → parents(a, b)
0.6	isEntityType(a, "PERSON"), isEntityType(b, "PERSON") lemmaBetweenEntities(a, b, "husband") → spouse(a, b)
1.0	isEntityType(a, "PERSON"), isEntityType(b, "NATIONALITY") prevWord(a, b) → origin(a,b)
1.0	isEntityType(a, "PERSON"), isEntityType(b, "COUNTRY") prevOfPrevLemma(b,"citizen"), prevLemma(b,"of") → origin(a,b)

create linguistic features. Once these features are created, we run the RFGB algorithm [12]. This allows us to create a joint model between the target relations. We now briefly describe the adaptation of RFGB to this task.

Our prior work – triggered by the intuition that finding many rough rules of thumb can be faster and more accurate than finding a single, highly accurate local model – turned the problem of learning relational models into a series of relational function approximation problems using functional gradient-based boosting [11]. The key idea is to represent each relation's conditional distribution as a sum of regression models grown incrementally.

Using functional-gradient boosting for learning the structure has several advantages. First, being a non-parametric approach the number of parameters

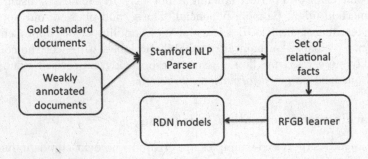

Fig. 3. Steps involved in learning using probabilistic examples.

grows with every boosting iteration. Due to the incremental updates of the structure and greedy tree-learning, predicates are introduced only as needed; as a result the potentially large search space is not explicitly considered. Second, such an approach can take advantage of off-the-shelf regression-tree learners. Moreover, advances made in the tree learners, such as being able to handle continuous features, can be utilized easily. Third, the use of functional-gradient boosting makes it possible to learn the structure and parameters simultaneously, which is an attractive feature as structure learning in SRL models is computationally quite expensive. Finally, given the success of ensemble methods in machine learning [1,26], it can be expected and demonstrated extensively that this approach is superior in predictive performance compared to the other structure learning methods.

Similar to previous approaches in functional-gradient boosting, we use the sigmoid function ($\frac{e^x}{e^x+1}$) to represent the probability distribution of each example, x_i:

$$P(x_i = true|\mathbf{Ne}(x_i)) = \frac{e^{\psi(x_i;\mathbf{Ne}(x_i))}}{e^{\psi(x_i;\mathbf{Ne}(x_i))} + 1} \tag{1}$$

$$\log P(x_i = true|\mathbf{Ne}(x_i)) = \psi(x_i;\mathbf{Ne}(x_i)) - log\left(e^{\psi(x_i;\mathbf{Ne}(x_i))} + 1\right)$$

where $\mathbf{Ne}(x_i)$ corresponds to the neighbors of x_i that influence x_i. In directed graphs, $\mathbf{Ne}(x_i)$ is the parents of the variable, whereas for a Markov network it is the Markov blanket. For our approach, we define the joint probability distribution as a product of conditional distributions. In our experiments, when we perform joint relation extraction, we simply learn the distribution of every variable assuming the others are observed and repeat the process for all the relations.

In standard graphical models literature, this is called the pseudo-log-likelihood (PLL) and is defined as:

$$PLL(\mathbf{x}) \equiv \log P(\mathbf{X} = \mathbf{x}) = \log \prod_{x_i \in \mathbf{x}} P(x_i|\mathbf{Ne}(x_i)) = \sum_{x_i \in \mathbf{x}} \log P(x_i|\mathbf{Ne}(x_i))$$

Functional-gradient boosting first computes the functional gradients ($\frac{\partial}{\partial \psi(x)}$) of the score that we wish to maximize. In most previous work, our goal was to learn a model that maximizes the PLL of the examples in the training data. Hence, we calculated the functional gradient of PLL for every example.

$$\frac{\partial \log P(\mathbf{X} = \mathbf{x})}{\partial \psi(x_i;\mathbf{Ne}(x_i))} = \frac{\partial \log P(x_i|\mathbf{Ne}(x_i))}{\partial \psi(x_i;\mathbf{Ne}(x_i))}$$

$$= \frac{\partial \left(\psi(x_i;\mathbf{Ne}(x_i)) - log\left(e^{\psi(x_i;\mathbf{Ne}(x_i))} + 1\right)\right)}{\partial \psi(x_i;\mathbf{Ne}(x_i))}$$

$$= I(x_i = true) - \frac{1}{e^{\psi(x_i;\mathbf{Ne}(x_i))} + 1} \frac{\partial \left(e^{\psi(x_i;\mathbf{Ne}(x_i))} + 1\right)}{\partial \psi(x_i;\mathbf{Ne}(x_i))}$$

$$= I(x_i = true) - \frac{e^{\psi(x_i; \mathbf{Ne}(x_i)))}}{e^{\psi(x_i; \mathbf{Ne}(x_i))} + 1}$$

$$= I(x_i = true) - P(x_i = true; \mathbf{Ne}(x_i)) = \Delta(x_i) \qquad (2)$$

where I is the indicator function, which returns 1 if x_i is a positive example here, otherwise returns 0. The gradient at each example ($\Delta(x_i)$) is now simply the adjustment required for the probabilities to match the observed value for that example. If x_i is a negative example, the gradient for x_i is negative, thereby pushing the ψ value closer to $-\infty$ and the predicted probability of the example closer to 0. On the other hand if x_i is a positive example, the gradients push the probabilities closer to 1^3.

Recall that, in our current work, we are employing probabilistic labels that result from the weak supervision phase. Hence, we resort to optimizing a different function, namely, KL-divergence. So instead of optimizing PLL, we optimize the KL-divergence between the observed probabilities of the relations ($P_{obs}(y = \hat{y})$) and the corresponding predicted probabilities ($P_{pred}(y = \hat{y})$). We now derive the gradient for this objective function.

$$\Delta_m(x) = \frac{\partial}{\partial \psi_{m-1}} \sum_{\hat{y}} P_{obs}(y = \hat{y}) \log \left(\frac{P_{obs}(y = \hat{y})}{P_{pred}(y = \hat{y} | \psi_{m-1})} \right)$$

$$= P_{obs}(y = 1) - P_{pred}(y = 1 | \psi_{m-1})$$

A careful reader will see the similarity between the use of PLL and KL-divergence gradients. In the former case, the gradient is simply the difference between the observed label and the predicted probability of that label. In the latter case, the gradient is simply the difference between the observed probability of a label and its current predicted probability. Hence, similar to the original case, RFGB simply tries to push the examples in the direction of the observation.

Hence the key idea in our work is to use probabilistic examples that we obtain from the weakly supervised phase as input to our structure learning phase along with gold standard examples and their associated documents. Then an RDN is induced by learning to predict the target relations jointly, using features created by the Stanford NLP toolkit. Since we are learning a RDN, we do not have to explicitly check for acyclicity. We chose to employ RDNs as they have been demonstrated to have the state-of-the-art performance in many relational and noisy domains [12]. We use modified ordered Gibbs sampler [14] for inference.

4   Empirical Evaluation

In this section, we present the results of empirically validating our proposed approach on relation extraction from natural language text. Since the goal of

---

[3] With probabilistic training examples, it can be shown that minimizing the KL-divergence between the examples and the current model gives *true probability−predicted probability* as the gradient. This has the similar effect of pushing the predicted probabilities closer to the true probabilities.

this chapter is to analyze the use of world knowledge, we compare the use of our learning algorithm with weakly supervised examples. We did not consider other learning methods because most of them do not consider learning the model structure and secondly, the aim is to evaluate the use of weakly supervised examples and not the learning algorithm. Finally, we do not compare against other distant supervision methods as they have not yet been used on this task. Instead, in our paper [13], we also included another domain, that of New York Times articles in which we perform relation extraction and showed the state-of-the-art results using distant supervision techniques.

4.1 Experimental Domain: Knowledge Base Population Task

In this work, we apply our weak supervision framework to the TAC KBP 2015 Slot-Filling task. For results on other data sets, see our previous work [13] where we evaluate on joint prediction of game winners and losers from NFL news articles as well as extracting facts from the NYT articles data set.

KBP slot filling is a relation extraction task to identify slot-filler values (i.e., the arguments) and corresponding provenance information for a given relation. Together, these allow the population of a knowledge base from *raw, unannotated* texts. For example, consider the target relation *parents(a,b)*. Slot-filling aims to find pairs of entities, a and b, mentioned in the text corpus for which a has a parent b. In addition, the system must produce provenance to justify the prediction of a parent relationship (i.e., the supporting context in an article). The challenge posed by this task are two fold – the scale of the data (millions of documents of text) as well as the inherent complexity of extracting information from natural language sources.

Here, we analyze five specific slot-filling targets:

- $spouse(a, b)$ – person b is a spouse (e.g., husband or wife) of person a
- $parents(a, b)$ – person b is a parent of person a
- $siblings(a, b)$ – person b is a sibling (e.g., brother or sister) of a
- $age(a, b)$ – b is the reported age of person a
- $alternateName(a, b)$ – b is an alternate name/alias for person a

For each target, a human manually annotated a small subset of the KBP corpus of news articles for positive examples to create a "gold-standard" set. This same human generalized the linguistic patterns they observed in several "commonsense" rules that can be represented in the form of first-order logic rules. For each relation, the rules were converted to MLN format as in Table 1 and assigned weights based on their ranking. These rules were used to create a weakly-supervised set of examples for each target relation. In the next section, we compare the effectiveness of models constructed using weakly-supervised labels.

4.2 Experimental Setup

To evaluate our weak supervision approach, we consider three experimental conditions. First, we train our RDN models using weakly supervised examples only.

Second, we test our full approach by training our models on a small set of goal standard examples augmented with our weakly supervised labels. Last, as a baseline, we consider training our RDN models using gold-standard examples.

In all conditions, we utilize five-fold cross-validation for evaluation. Whenever training is done on gold-standard examples, we limit our training to 20 positive examples (i.e., entity pairs) for each relation and 40 negative examples. For cases where weak supervision is considered, the training set consists of 100 (most probable) inferred positive examples from the CMLN as well as 100 (least probable) negative examples. Each test set included roughly 20 positive and 40 negative examples for each relation.

Training was done as presented in the previous section, with a CMLN run on the training corpus to generated weakly supervised examples. All labels (weak and/or gold standard) were then used to train RDN models. Refer to [13] for details.

4.3   Initial Results

To evaluate our methods we utilize the area under a precision-recall curve and ROC curve. Results in Tables 2 and 3 present the average area across all five folds for ROC and PR curves, respectively

These results show that weakly-supervised labels obtain comparable results to fully-informed models that include expensive gold-standard labels and superior results to using only gold standard labels. For *siblings* and *parents*, we see equivalent, if not better, performance for using weak labels alone across both metrics. Two other relations, *alternateName* and *spouse*, produce modest

Table 2. Area under the ROC curve results

Target slot	Weak labels only	Weak+Gold labels	Gold labels only
age	0.838	0.972	0.346
alternateName	0.620	0.683	0.844
parents	0.668	0.685	0.482
siblings	0.810	0.744	0.641
spouse	0.730	0.745	0.606

Table 3. Area under the precision-recall curve results

Target slot	Weak labels only	Weak+Gold labels	Gold labels only
age	0.638	0.942	0.111
alternateName	0.055	0.068	0.445
parents	0.221	0.210	0.154
siblings	0.320	0.299	0.254
spouse	0.216	0.239	0.214

improvement with gold-standard examples. Only *age* demonstrates substantial benefits from including gold-standard labels. The only exception is alternate name where gold standard examples are much more informative than the weak supervision ones. We hypothesize that the concept of alternate names is hard to encapsulate with a good first-order logic clause.

5   Conclusion

One of the key challenges for applying learning methods in many real-world problems is the paucity of good quality labeled examples. While semi-supervised learning methods have been developed, we explore an alternative method of weak supervision – where the goal is to create examples of a quality that can be relied upon. We considered the NLP tasks of relation extraction and document extraction to demonstrate the usefulness of the weak supervision. Our key insight is that weak supervision can be provided by a "domain" expert instead of a "NLP" expert and thus the knowledge is independent of the underlying problem but is close to the average human thought process. In general, we are exploiting knowledge that the authors of articles assume their readers already know and hence the authors do not state it. We used the weighted logic representation of Markov Logic networks to model the expert knowledge, infer the relations in the unannotated articles, and adapted functional gradient boosting for predicting the target relations. Our results demonstrate that with high quality weak supervision, we can reduce the need for gold standard examples. In previous, smaller scale datasets, we actually see models produced with weakly-supervised labels outperforming training with gold-standard examples.

Our proposed method is closely related to distant supervision methods. So it will be an interesting future direction to combine the distant and weak supervision examples for structure learning. Combining weak supervision with advice taking methods [3, 9, 24] is another interesting direction. This method can be seen as giving advice about the examples, but AI has a long history of using advice on the model, the search space and examples. Hence, combining them might lead to a strong knowledge based system where the knowledge can be provided by a domain expert and not an AI/NLP expert. We envision that we should be able to "infer" the world knowledge from knowledge bases such as Cyc or Concept-Net and employ them to generate the weak supervision examples. Finally, it is important to evaluate the proposed model in similar tasks.

Acknowledgements. Sriraam Natarajan, Anurag Wazalwar and Dileep Viswanathan gratefully acknowledge the support of the DARPA Machine Reading Program and DEFT Program under the Air Force Research Laboratory (AFRL) prime contract nos. FA8750-09-C-0181 and FA8750-13-2-0039 respectively. Any opinions, findings, and conclusion or recommendations expressed in this material are those of the authors and do not necessarily reflect the view of the DARPA, AFRL, or the US government. Kristian Kersting was supported by the Fraunhofer ATTRACT fellowship STREAM and by the European Commission under contract number FP7-248258-First-MM.

References

1. Bell, B., Koren, Y., Volinsky, C.: The bellkor solution to the netflix grand prize (2009)
2. Craven, M., Kumlien, J.: Constructing biological knowledge bases by extracting information from text sources. In: ISMB (1999)
3. Devlin, S., Kudenko, D., Grzes, M.: An empirical study of potential-based reward shaping and advice in complex, multi-agent systems. Adv. Complex Syst. **14**(2), 251–278 (2011)
4. Domingos, P., Lowd, D.: Markov Logic: An Interface Layer for AI. Morgan & Claypool, San Rafael (2009)
5. Hoffmann, R., Zhang, C., Ling, X., Zettlemoyer, L., Weld, D.S.: Knowledge-based weak supervision for information extraction of overlapping relations. In: ACL (2011)
6. Kersting, K., Driessens, K.: Non-parametric policy gradients: a unified treatment of propositional and relational domains. In: ICML (2008)
7. Khot, T., Natarajan, S., Kersting, K., Shavlik, J.: Learning Markov logic networks via functional gradient boosting. In: ICDM (2011)
8. Kim, J., Ohta, T., Pyysalo, S., Kano, Y., Tsujii, J.: Overview of BioNLP'09 shared task on event extraction. In: BioNLP Workshop Companion Volume for Shared Task (2009)
9. Kuhlmann, G., Stone, P., Mooney, R.J., Shavlik, J.W.: Guiding a reinforcement learner with natural language advice: initial results in robocup soccer. In: AAAI Workshop on Supervisory Control of Learning and Adaptive Systems (2004)
10. Mintz, M., Bills, S., Snow, R., Jurafsky, D.: Distant supervision for relation extraction without labeled data. In: ACL and AFNLP (2009)
11. Natarajan, S., Kersting, K., Khot, T., Shavlik, J.: Boosted Statistical Relational Learners: From Benchmarks to Data-Driven Medicine. SpringerBriefs in Computer Science. Springer, Heidelberg (2015)
12. Natarajan, S., Khot, T., Kersting, K., Guttmann, B., Shavlik, J.: Gradient-based boosting for statistical relational learning: the relational dependency network case. Mach. Learn. **86**(1), 25–56 (2012)
13. Natarajan, S., Picado, J., Khot, T., Kersting, K., Re, C., Shavlik, J.: Effectively creating weakly labeled training examples via approximate domain knowledge. In: Davis, J., Ramon, J. (eds.) ILP 2014. LNCS, vol. 9046, pp. 92–107. Springer, Heidelberg (2015). doi:10.1007/978-3-319-23708-4_7
14. Neville, J., Jensen, D.: Relational dependency networks. In: Getoor, L., Taskar, B. (eds.) Introduction to Statistical Relational Learning, pp. 653–692. MIT Press, Cambridge (2007)
15. Niu, F., Ré, C., Doan, A., Shavlik, J.W.: Tuffy: scaling up statistical inference in Markov logic networks using an RDBMS. PVLDB **4**(6), 373–384 (2011)
16. Poon, H., Vanderwende, L.: Joint inference for knowledge extraction from biomedical literature. In: NAACL (2010)
17. Raghavan, S., Mooney, R.: Online inference-rule learning from natural-language extractions. In: International Workshop on Statistical Relational AI (2013)
18. Riedel, S., Chun, H., Takagi, T., Tsujii, J.: A Markov logic approach to biomolecular event extraction. In: BioNLP (2009)
19. Riedel, S., Yao, L., McCallum, A.: Modeling relations and their mentions without labeled text. In: Balcázar, J.L., Bonchi, F., Gionis, A., Sebag, M. (eds.) ECML PKDD 2010, Part III. LNCS, vol. 6323, pp. 148–163. Springer, Heidelberg (2010)

20. Sorower, S., Dietterich, T., Doppa, J., Orr, W., Tadepalli, P., Fern, X.: Inverting Grice's maxims to learn rules from natural language extractions. In: NIPS, pp. 1053–1061 (2011)
21. Surdeanu, M., Ciaramita, M.: Robust information extraction with perceptrons. In: NIST ACE (2007)
22. Surdeanu, M., Tibshirani, J., Nallapati, R., Manning, C.: Multi-instance multi-label learning for relation extraction. In: EMNLP-CoNLL (2012)
23. Takamatsu, S., Sato, I., Nakagawa, H.: Reducing wrong labels in distant supervision for relation extraction. In: ACL (2012)
24. Torrey, L., Shavlik, J., Walker, T., Maclin, R.: Transfer learning via advice taking. In: Koronacki, J., Raś, Z.W., Wierzchoń, S.T., Kacprzyk, J. (eds.) Advances in Machine Learning I. SCI, vol. 262, pp. 147–170. Springer, Heidelberg (2010)
25. Verhagen, M., Gaizauskas, R., Schilder, F., Hepple, M., Katz, G., Pustejovsky, J.: SemEval-2007 task 15: TempEval temporal relation identification. In: SemEval (2007)
26. Viola, P., Jones, M.: Rapid object detection using a boosted cascade of simple features. In: CVPR (2001)
27. Yoshikawa, K., Riedel, S., Asahara, M., Matsumoto, Y.: Jointly identifying temporal relations with Markov logic. In: ACL and AFNLP (2009)
28. Zhou, G., Su, J., Zhang, J., Zhang, M.: Exploring various knowledge in relation extraction. In: ACL (2005)

Supervised Extraction of Usage Patterns in Different Document Representations

Christian Pölitz[✉]

TU Dortmund University, Otto Hahn Str. 12, 44227 Dortmund, Germany
christian.poelitz@tu-dortmund.de

Abstract. Finding usage patterns of words in documents is an important task in language processing. Latent topic or latent factor models can be used to find hidden connections between words in documents based on correlations among them. Depending on the representation of the documents, correlations between different elements can be found. Given additional labels (either numeric or nominal) for the documents, we can further infer the usage patterns that reflect this given information. We present an empirical comparison of topic and factor models for different documents representations to find usage patterns of words in a large document collections that explain given label information.

1 Introduction

The investigation of contextual usages of words or meanings of documents is an important task in computer linguistic. For example the word "bank" is commonly used in two contexts. The word either appears in the context of a financial institute or as river bank. In each context the word "bank" co-occurs with specific other works like "money" or "manager" for one meaning, "sea" or "river" for the other meaning. Statistics on such co-occurrences lead to statistical relations among the words. Large text corpora like the German text archive (DTA[1]) provide large amounts of reference documents to extract statistical relation among words in the documents. A corpus means a large collection of documents in a certain language. The documents in a corpus are usually structured to enable text analysis. To avoid confusion, we use the term document also for texts and text snippets that might be only a small part of a whole document. This is important since our methods to extract usage patterns usually use only small snippets extracted from larger documents as input.

Factor models like Latent Semantic Analyse [4] or topic models like Latent Dirichlet Allocation [2] have been successfully used to extract such semantic usages. Depending on how we represent the documents, different methods can be used to find latent factors. Using the Bag-of-Words representation, topic models and factorization of the term-document matrix can be used. In case we want to include further structure from the documents, kernel methods are a possible choice. Polynomial kernels for instance include combinations of words

---

[1] www.DWDS.de.

© Springer International Publishing Switzerland 2016
S. Michaelis et al. (Eds.): Morik Festschrift, LNAI 9580, pp. 346–361, 2016.
DOI: 10.1007/978-3-319-41706-6_19

in the documents as representation of the documents. Further, Gaussian kernels even enable us to integrate all possible combinations of words in the documents.

Despite of the pure co-occurrences of the words in the documents or the intrinsic structures, additional information about the documents in the corpus provide additional inside. We might have additional information about the author and the time when the document was written. High level annotation might also be available. Such annotations could be the sentiment of some text snippets or words from the document.

In this paper, we investigate how to efficiently integrate such additional information in order to extract the semantic relation among words that also explain the additional information best. If we have information about the sentiments of certain words, we want to be able to extract from the usage patterns of the words the semantics that reflect positivity or negativity. Or, from given temporal information, we want to be able to extract patterns among words that reflect certain time episodes.

2   Related Work

There are several related approaches that incorporate supervision into topic models respectively latent factor models. A large number of recent work concentrates on integrating labels of documents into LDA topic models. Blei et al. introduce in [1] supervised Latent Dirichlet Allocation (sLDA). They model given labels for documents as Normally distributed random variables depending on the latent variables of the topic model. The parameters are estimated via Variational Inference. Using Gibbs sampling to train the topic model, EM-style algorithms can also be used for parameter estimation as done by Nguyeń et al. in [13]. Zhu et al. propose in [22] not to use a fully generative model for the labels of the documents but a max margin classification model similar to Support Vector Machines. Again, this supervised topic model can be either estimated via Variational Variance or as proposed by Zhu et al. in [23] via Gibbs sampling. Besides directly modelling the label information about the document, indirect methods are also investigated. In [12] Mimno and McCallum propose to make the prior on the document-topic distributions in an LDA topic model depending on document features (or labels). By this, documents with similar features or labels are more likely to be in the same topic. In [14] Rampage et al. integrate multiple nominal labels as binary vector. These vectors constrain an LDA topic model such that each document can only be assigned to topics that correspond to document labels.

For factor models, not much work has been done to integrate supervision. The most prominent factor models in text domains factorize the term-document matrices. A term-document matrix contains as rows word vectors for each document in the collection. The word vectors contain at each component frequency information about a certain word in the document. Now, factor models like Latent Semantic Analysis [4] extract subspaces with most of the variance in the space spanned by all word vectors. The basis of this space contains k (linear independent) "word vectors" and represent the latent factors. The components of these basis vectors respectively factors can be interpreted as term loadings telling how much variance the corresponding terms have in these dimensions. Previous approach integrating label information into such factor models mainly concentrated on Partial Least Squares (PLS) methods. Zeng et al. use in [21] PLS to find a low dimensional representation of documents for text classification. Yu et al. on the other hand proposed in [20] a supervised version of Latent Semantic Indexing. They find a projection that captures the correlation between input vectors and labels while retaining the information in the inputs.

Another factor model is Non-Negative Matrix Factorization (NNMF), see Lee and Seung in [8]. NNMF factorizes the term-document matrix such that the resulting factors contain only non-negative entries. This helps interpreting these factors. There are several previous approaches that integrate supervision into NNMF. Most of these approaches add constraints in the calculation of the non-negative factors. In [9] Liu and Wu add constraints forcing the representations of the documents in the space spanned by the factors to be the same for documents with the same label.

Latent factor models in Hilbert Spaces using kernels is rather unusual in text domains. This is due to the fact that the latent factors cannot be simply interpreted. Nonetheless, factorizations of Hilbert space using kernel methods have been successfully used. In [18] Schoelkopf et al. proposed to perform Principle Component Analysis (PCA) [6] in a kernel defined RKHS based on the eigenfunctions and eigenvalues of the covariance operator C which can be approximated by kernel matrices. Further, the kernelized version of PLS (kernel PLS) can be used to integrate labels in the extraction of latent factors in a Hilbert space. In [16], Rosipal and Trejo give a detailed description of this method.

3   Methods

Depending on the representation of the documents, different methods to extract contextual usages are possible. If we represent the documents simply as Bag-of-Words (see [10]), topic models like Latent Dirichlet Allocation (LDA) or linear factorization methods like Latent Semantic Analysis (LSA) or Partial Least Squares (PLS) can be used to extract latent topics or factors that correspond to usage patterns in these texts. For the linear models LSA and PLS, the words can be additionally weighted by frequency values like tf-idf values (see [15]). When we represent the documents as elements of a non-linear or high (possibly infinite) dimensional space, kernel methods can be used. Such representations can be all n-grams in the documents, all substrings or the parse trees of the contained sentences. Factorization methods in Reproducing Kernel Hilbert spaces (RKHS) can be used to extract bases corresponding to latent factors in the corresponding representation. Kernel Principle Component Analysis (kPCA) or kernel Partial Least Squares (kPLS) extract such bases as linear combinations of the elements of the Hilbert space by using positive definite kernels. In the next subsections, we introduce the methods to extract the latent topics and the latent factors with respect to the representation of the texts and given labels. First, we describe topic models with LDA as well as the supervised version of LDA. Next, we explain LSA as a factor model of the term-document matrix and PLS as a factor model that integrates labels into the extraction of the factors. Finally, we explain how we can extract latent factors in (possibly) infinite dimensional Hilbert spaces by kPCA and kPLS.

3.1   Topic Models

Topic Models are statistical models that group documents and words from a document collection into so called topics. The words and documents that are highly associated with a topic are statistically related based on co-occurrences of words. Latent Dirichlet Allocation (LDA) as introduced by Blei et al. [2] has been successfully used for the estimation of such topics. In LDA, it is assumed that the words in a document are drawn from a Multinomial distribution that depends on latent factors, later interpreted as topics. We briefly summaries the generative process of documents as the following:

1. For each topic t:
   (a) Draw $\theta_t \sim Dir(\beta)$
2. For each document d:
   (a) Draw $\phi_d \sim Dir(\alpha)$
   (b) For each word i:
      i. Draw $t_i \sim Mult(\phi_d)$
      ii. Draw $w_i \sim Mult(\theta_{t_i})$

Assuming a number of topics, we draw for each a Multinomial distribution of the words for this topic from a Dirichlet distribution $Dir(\beta)$ with metaparameter β. For each document we draw a Multinomial distribution of the topics in this document from a Dirichlet distribution $Dir(\alpha)$ with metaparameter α. Finally, for each word in the document we draw a topic with respect to the topic distribution in the document and a word based on the word distribution for the drawn topic. The metaparameter α and β are prior probabilities of the Multinomial distributions drawn from the Dirichlet distributions. These priors are the expected word probabilities in a topic before we have seen any data.

The generation of the LDA Topic Model is usually done by Variational Inference, as in the original work by Blei et al. [2], or via Gibbs samplers, as proposed by Griffiths et al. [5]. We use Gibbs sampler to sample topics directly from the topic distribution. Integrating θ and ϕ out, we get for the probability of a topic z_i, given a word w in a document d and all other topic assignments:

$$p(z_i|w, d, z_1, \cdots z_{i-1}, z_{i+1}, \cdots z_T)$$
$$\propto \frac{N_{w,z_i} - 1 + \beta}{N_{z_i} - 1 + W \cdot \beta} \cdot (N_{d,z_i} + \alpha)$$

We denote $N_{w,z}$ the number of times topic z has been assigned to word w, $N_{d,z}$ the number of times topic z has been assigned to any word in document d,

N_z the number of times topic z has been assigned to any word, W the number of words in the document collection and T the number of topics.

After a sufficient number of samples from the Gibbs sampler we get estimates of the word distributions for the topics and the topic distributions for the documents:

$$\theta_{w|t} = \frac{N_{w,t} + \beta}{N_t + W \cdot \beta}$$

$$\phi_{d|t} = \frac{N_{d,t} + \alpha}{N_d + T \cdot \alpha}$$

3.2 Supervised Topic Models

A simple extension of LDA to handle labels for the documents was proposed by Blei and McAuliffe in [1]. We briefly summarize the generative process of documents with labels as the following:

1. For each topic t:
   (a) Draw $\theta_t \sim Dir(\beta)$
2. For each document d:
   (a) Draw $\phi_d \sim Dir(\alpha)$
   (b) For each word i:
       i. Draw $t_i \sim Mult(\phi_d)$
       ii. Draw $w_i \sim Mult(\theta_{t_i})$
3. Draw $y_d \sim p(y_d)$

The difference to standard LDA is that we additionally draw labels from a distribution p for each document. Integrating θ and ϕ out, we get for the probability of a topic t_i, given a word w in a document d and all other topic assignments in a similar way as for standard LDA:

$$p(t_i|w, d, t_1, \cdots t_{i-1}, t_{i+1}, \cdots, t_T, y_d)$$
$$\propto \frac{N_{w,t_i} - 1 + \beta}{N_{t_i} - 1 + W\beta}(N_{d,t_i} + \alpha)p(y_d)$$

The last term comes from density of the label. Note that the label can be a class label for sentiments or a numeric value for time stamps.

For sentiments given for the documents, we assume that the density of the labels is $p(y_d) \propto exp(-\frac{(y_d - \mu_{w,d})^2}{2\rho})$, the Normal density and $\mu_{w,d} = \eta' \hat{t_d}' + \eta_{t_i}$ with $\hat{t_d}'$ the empirical topic frequencies removing the topic assignment of the current token. After each Gibbs iteration we estimate η by minimizing the likelihood:

$$L(\eta) = -\frac{1}{2\rho} \sum_d (y_d - \eta' \hat{t_d})^2 - -\frac{1}{2\sigma} \sum_k \eta_k^2$$

For time stamps given for the documents, we assume that the density of the labels is $p(y_d) \propto \frac{(1-t_d)^{a-1} \cdot t_d^{b-1}}{Beta(a,b)}$, the Beta density with $a = \hat{m} \cdot (\frac{\hat{m} \cdot (1-\hat{m})}{s^2} - 1)$ and $b = (1-\hat{m}) \cdot (\frac{\hat{m} \cdot (1-\hat{m})}{s^2} - 1)$ for each topic.

In several situations, we might not have labels for all documents in our corpus. In such a case, we would like to predict the label on the unlabeled texts. In order to do so we fold in the texts first. This means we estimate the empirical topic distributions $t_{\hat{d}_n}$ of the new text d_n by performing a few Gibbs sampling steps on these texts without changing the words and topic distributions any more. Finally, we assign the label $sign(\eta' z_{\hat{d}_n})$ to the new document for sentiments or $\frac{a}{a+b}$ for time steps. These predictions are simply derived by the expected values of the corresponding distributions of the labels.

The interpretation of the topic model at the end is quite straight forward. For each topic t, we extract the most likeliest words from the word distribution $\theta_{w|t}$ and use them as summarization of the corresponding topics.

3.3 Latent Semantic Analysis

Latent Semantic Analysis (LSA) as described by Landauer et al. in [4] extracts usage patterns in documents by grouping words into latent dimensions in the vector space spanned by the word vectors. The term-document matrix is factorized by a Singular Value Decomposition (SVD) to extract a low dimensional subspace in the space spanned by the documents and in the space spanned by the terms. Formally, we factorize the term-document matrix $D = UE_kV$, for U the left singular vectors, V the right singular vectors and E_k the diagonal matrix of the k largest singular values. The k left singular vectors that correspond to the largest singular values span the k dimensional subspace in the document space and the right singular values in the term space that contain most of the variance of the word vectors. The value of the components of the right singular vectors multiplied by the corresponding singular values indicate the variance of the terms in a certain direction of the subspace. The highest values can be interpreted as the terms most important of a certain usage pattern.

3.4 Partial Least Squares

Partial Least Squares (PLS) is a method that finds low dimensional subspaces that maximally align with given labels. Given texts as Word Vectors and labels of the texts as possible sentiments or temporal information, PLS finds low dimensional word vector representations that are the optimal covariates for a linear regressor to predict the labels. Algorithm 1 describes the steps of PLS for a given term document matrix X of word vectors and a label vector y as described by Rosipal and Trejo in [16]. The algorithm successively extracts latent variables or components as linear combinations of the input word vectors. These components are removed from the word vectors by deflating the term document matrix by $X - tt'X$. This process is repeated until we have found all k components. The results of the algorithm are so called loadings vectors. Each loadings vector is a

---

Algorithm 1. Partial Least Squares to extract the latent factors.

---

function GETCOMPONENT(X, y, k)
    for $i = 1 : k$ **do**
        $u = rand$
        repeat
            $w = X'u$
            $t_i = Xw,\ t_i = t_i/\|t_i\|$
            $c = y't$
            $u = yc,\ u = u/\|u\|$
        until convergence
        $X = X - tt'X$
        $y = y - tt'y$
    end for
    return $XL = [X't_i]_{i=1\cdots k}$
end function

---

low dimensional representation of a corresponding word vector. These loadings can be used to estimate the amount of rotation the words in the vector space experience when mapped onto the loadings.

Similar as before, we want to be able to also predict new unlabeled documents. PLS is a linear regression model. The label is simple modelled as: $y = X\beta + r$ for the term document matrix X, the regression coefficients β mapping onto the latent factors and a residual vector r. For $T = [t_1 \cdots t_k]$ from PLS we can simple estimate the coefficients as shown in Eq. 1. Now, a new document represented as Bag-of-Words vector x_n gets assigned label $sign(x_n'\beta)$ for nominal labels like sentiments, respectively $x_n'\beta$ for numeric labels like time stamps. The regression coefficient (including the mapping onto the latent subspace) as defined as:

$$\beta = X'U(T'XX'U)^{-1}T'y \tag{1}$$

The interpretation of the latent factors respectively loading vectors is not as simple as for topic models. One important factor for the importance of a word for one latent factor is the value of the corresponding component in the loadings vector $l_i = XL_{i,:}$ for the i_{th} loadings vector. The amount of the j component of l_i tells how much weight the corresponding word has to predict the label when we project it onto the latent factor t_i. Figure 1 illustrates this idea for a two dimensional example. The loadings vector t spans a latent dimension $\alpha \cdot t$ that keeps enough variance in the data and separates the classes well. The i_{th} component of t is the length of the projection of the i_{th} basis vector in the vector space spanned by the words. This can be interpreted as the importance of word i in this latent dimension.

In order to better interpret the importance of some words for the latent factors we can further rotate the loading vectors such that the variance in these vectors is maximized. Intuitively, we want loadings that have a few large components and near zeros components elsewhere. The method called Varimax

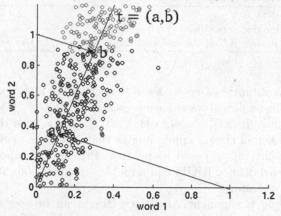

Fig. 1. Illustration of a latent dimension spanned by a loadings vector t in two dimensions of the vector space spanned by the words. Blue and red indicate different class labels of points. (Color figure online)

Rotation by [7] can be used for such a transformation. This method rotates the coordinate system spanned by the latent factors such that the loadings in the new coordinate system have maximum variance in their components.

3.5 Kernel Principle Component Analysis

Kernel methods accomplish to apply linear methods on non-linear representations of data. Any kernel method uses a map $X \rightarrow \phi(X)$ from a compact input space X, for example $\Re^n$, into a so called Reproducing Kernel Hilbert Space (RKHS). In this space, linear methods are applied to the mapped elements like Linear Regressions or Support Vector Machines. The RKHS is a space of functions $f(y) = \phi(x)(y) \ \forall x \in X$ that allows point evaluations by an inner product, hence $f(y) = \phi(x)(y) = < \phi(x), \phi(y) >$. $\phi(x)$ is a function and $\phi(x)(y)$ mean the function value at y.

For the mapping ϕ from above, K_ϕ is the integral operator as defined in Eq. 2 for a probability distribution P on the input space X.

$$K_\phi(f)(t) = \int f(x) \cdot < \phi(x), \phi(t) > \cdot dP(x) \tag{2}$$

For this integral operator, we denote $< \phi(x), \phi(y) >= k(x, y)$ with kernel k. By Mercer Theorem [11] there is a one to one correspondence of the above defined RKHS and the integral operator via the kernel k. This correspondence is given by the expansion $k(x, y) = \sum_{i=1}^{\infty} \phi_i(x) \cdot \phi_j(y)$ for $\{\phi_i\}$ an orthonormal basis in the RKHS.

Now, the covariance operator C on a Hilbert space H is is defined as $E[Z \times Z^*]$ the outer product of a random elements $Z \in H$ with its adjoint Z^*. This is

Algorithm 2. Kernel Principle Component Analysis.

Center kernel Matrix K

Perform Eigenvalue decomposition: $[V, \Lambda] = eig(K)$

Calculate kernel matrix K^P of the mapped data samples into the subspace: K^P

---

analogue to the covariance of centred random elements in $\Re^n$ where we have $C = E[X \cdot X^T]$. The empirical covariance is estimated via $\hat{C} = \frac{1}{m} \sum \phi(x_i) \cdot \phi(x_j)$ for a centred sample $\{\phi(x_1), \cdots, \phi(x_m)\}$ with x_i drawn from distribution P. Consequently the kernel matrix approximates the covariance operator: $K \sim C$.

Schoelkopf et al. [18] proposed to perform Principle Component Analysis (PCA) [6] in a kernel defined RKHS based on the eigenfunctions and eigenvalues of the covariance operator C.

Kernel Principle Component Analysis extracts an orthogonal basis, also called principle components, in a kernel induced RKHS. Projecting the data onto the subspace spanned by the first k components captures most of the variance among the data compared to all other possible subspaces where the data lies in.

The k components are exactly the eigenfunctions corresponding to the largest k eigenvalues of the covariance operator of the kernel.

The covariance operator is approximated by the empirical covariance matrix $C = \frac{1}{b} \sum_i \phi(x_i) \cdot \phi(x_i)^T$. An eigenvalue decomposition on C results in a set of eigenvalues $\{\lambda_i\}$ and eigenvectors $\{v_i\}$ such that $\lambda_i \cdot v_i = C \cdot v_i$.

A projection of a sample x in the RKHS onto $U = \{v_i\}$ is done by $P_U(\phi(x)) = (< v_i, \phi(x) >, \cdots, < v_k, \phi(x) >) \in U$. Since the v_i lie in the span of the $\{\phi(x_i)\}$, each component is given by $v_i = \sum_j \alpha_{j,i} \cdot \phi(x_j)$. This results in the projection $P_U(\phi(x)) = (\sum_j \alpha_{j,1} < \phi(x_i), \phi(x) >, \cdots, \sum_j \alpha_{j,k} < \phi(x_i), \phi(x) >) \in U$. From the eigenvalue decomposition we have $\alpha_{i,j} = (\frac{1}{\sqrt{\lambda_i}} \cdot v_i)_j$.

The steps of kernel PCA are summarized in Algorithm 2 as described by Shawe-Taylor and Cristianini in [19].

3.6 Kernel Partial Least Squares

Kernel Partial Least Squares (kPLS) performs PLS in a kernel defined Reproducing Kernel Hilbert Space (RKHS). From PLS we see that computing a component t is done by $t = XX'u$. The matrix XX' is the empirical covariance matrix between the word vectors in X. This matrix is an approximation of the true covariance matrix for random word vectors drawn from the same distributions as the word vectors. The idea now is to apply kernel methods for the extractions of the latent factors.

The algorithm of kPLS is analogue to PLS. In Algorithm 3 we shortly state the difference compared to the standard PLS. The only differences are that we directly calculate t as Ku and that the projection onto the orthogonal complement respectively the deflation of t is done by $(I - tt')K(I - tt')$.

Algorithm 3. Kernel Partial Least Squares to extract the latent factors.

---

function GETCOMPONENT(K, Y)

    ...

    $u = rand$

    repeat

        $t = Ku, t = t/\|t\|$

        $c = y't$

        $u = yc, u = u/\|u\|$

    until convergence

    $K = (I - tt')K(I - tt')$

    $Y = (I - tt')Y(I - tt')$

    ...

    return t_i

end function

Like PLS, kPLS can be used as regression to prediction the labels for unlabeled documents. The regression is simply: $y = \Phi\beta + r$ and the coefficients β can be estimated as described in Eq. 3.

$$\beta = \Phi'U(T'KU)^{-1}T'y \tag{3}$$

Now, the interpretation of the latent factors t is even more difficult than before. The factors t are linear combinations of possible infinite dimensional Hilbert space elements. In order to interpret them we investigate the documents that are mapped the closest to the one dimensional subspace spanned by each t. The idea is that these documents contain the (possible not countable) structures that are important for the corresponding factors. In Fig. 2 we illustrate this idea.

Assuming we have a two class classification problem and a mapping ϕ of the data points, a latent factor t should span a dimension in the RKHS that keeps the structure of the data (hence most of the variance) and best separates the classes. A mapped data point can be written as $p = (p_1, p_2, \cdots, p_n, \cdots)'$ a (possible infinite) vectors. This is always possible for Hilbert spaces since they are isomorph to l_2 (cf. [17]). The components p_i can be interpreted as the length of the projection onto a basis $\{\phi_i\}$ which spans the RKHS. Unlike the Vector-Space spanned by the words, we can have infinitely many ϕ_i's. Instead of investigating each basis vector, we investigate only the data points that are closest to a latent factor t. A mapped data point p with large distance to t will most likely not have large differences in its components compared to t. For points with very small distances, we expect many very similar components.

This is indeed the pre-image problem in an RKHS where I look at a list of k closest points. For a latent factor t we find the k closest data points $x_1, \cdots x_k$ such that $\|t - \phi(x_i)\|_H \leq \|t - \phi(x_j)\|_H$ for $i < j$ and $\nexists k' > k : \|t - \phi(x_k)\|_H \leq \|t - \phi(x_{k'})\|_H$.

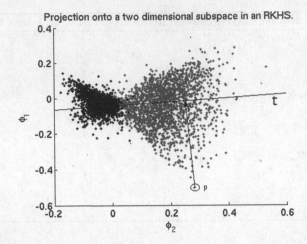

Fig. 2. Illustration of a latent dimension t in a two dimensional RKHS after projecting the data onto a two dimensional subspace spanned by ϕ_1 and ϕ_2. Blue and red indicate different class labels of points. (Color figure online)

4   Experiments

We show the alignments of latent variable models to given supervisions on two data sets. As the first data set, we use the Amazon reviews [3] about products from the categories books (B), DVDs (D), electronics (E) and kitchen (K). The classification task is to predict a given document as being written in a positive or negative context. We use stop word removal and keep only the words that appear less than 95 % and more often than 5 % of the time on all documents.

For the second data set, we retrieve from the DWDS corpus snippet lists for the word Platte (with meanings board / disc / hard disc / plate / conductor) of small document parts containing this word, including time stamps of the publication date of the corresponding document.

In the first experiment, we investigate how good the topics respectively latent factors extracted on the reviews data set align with given sentiments for the Amazon reviews data set. We performed supervised LDA, PLS, and kPLS on reviews from books and kitchen and used the given sentiments as labels. For each methods we set the number of topics respectively the number of latent factors to 2. By this, we want to reflect the bipolarity of the sentiments given for each review.

Tables 1 and 2 show the top 10 words for the topics respectively the latent factors on the book and kitchen reviews from the Amazon data set. For supervised LDA the top words are simply the most likeliest words for the topics. For PLS the top words are those words that have the largest absolute value in the corresponding loadings. For kPLS the top words are those words that appear most often in those document snippets that are closest to the extracted components in the RKHS.

Table 1. Results of the supervised LDA, PLS and kPLS on the reviews data about books with sentiment information. We extracted two topics by sLDA, respectively two latent factors by (kernel) PLS.

sLDA		PLS		kPLS	
topic 1	topic 2	factor 1	factor 2	factor 1	factor 2
read	read	quot	quot	quot	book
num	num	waste	recipe	great	edition
story	great	great	character	love	expected
great	time	boring	story	highly	black
good	story	bad	novel	recommend	wow
life	get	money	num	excellent	reference
people	people	excellent	read	easy	purchase
reading	reading	love	series	favorit	wait
world	author	num	plot	recipe	date
work	life	pages	time	read	modern

Table 2. Results of the supervised LDA, PLS and kPLS on the reviews data about kitchens with sentiment information. We extracted two topics by sLDA, respectively two latent factors by (kernel) PLS.

sLDA		PLS		kPLS	
topic 1	topic 2	factor 1	factor 2	factor 1	factor 2
num	num	easy	pan	num	loves
great	use	great	coffee	month	wine
use	get	love	knife	waste	comfortable
time	coffee	clean	knives	broke	glasses
good	time	num	pans	worked	filters
pan	great	perfect	stick	told	hope
easy	good	waste	set	service	red
get	product	broke	machine	costumer	kettle
coffee	buy	works	num	produce	ice
love	pan	price	non	item	died

The extracted topics respectively factors, all align quite well with judging adjectives like "great", "good", "excellent" or "waste". Comparing the three methods, PLS and kPLS seem to favour the adjectives more than sLDA. Further, the overlap of the top 10 words for sLDA is larger than on the other methods. Such cases are more likely to happen on (supervised) LDA since the topics a not orthogonal as the latent factors from (kernel) PLS.

Table 3. Accuracies for predicting the sentiment of the book reviews.

	sLDA	PLS	kPLS
books	75.20 ± 2.39	78.02 ± 4.98	77.05 ± 5.20
DVDs	78.35 ± 2.08	80.00 ± 7.00	80.42 ± 5.92
electronics	77.18 ± 1.96	81.82 ± 4.82	80.16 ± 5.16
kitchen	79.34 ± 1.93	82.55 ± 4.05	81.00 ± 5.00

Besides the descriptive power of the topics and latent factor, we are also interested in how good we can generalize the sentiments to given unlabeled reviews.
For the Amazon reviews data set we might receive gradually new reviews. In
Table 3, we show the accuracies on held out test sets for the reviews from the
Amazon reviews data set. PLS and kPLS outperform supervised LDS on all four
data sets. The generative model from sLDA obviously cannot discriminate the
sentiments as good as the pure discriminative models PLS and kPLS.

In the next experiment, we investigate how good the topic respectively latent
factor models align to temporal information about the texts. We use snippets
about the word "Platte" with temporal information about the publication of
the corresponding document from the DWDS core corpus. The dates range from
1900 to 1999. Since we have no idea of how many topics respectively factors
might be appropriate for the models, we simply use a larger number of 20 topics
respectively factors. First, we apply sLDA with 20 topics and visualize the results
to validate that we find interesting topics at all when we use 20 topics.

In Fig. 3 we show the distribution of the topics extracted by supervised LDA
over the time. We see that using supervised LDA, we get a clear distinction
of the topics over time. We can directly read off the topics and the temporal
period when this topic was prominent. The results indicate three possible main
meanings that clearly separate over time. Among the 20 topics respectively latent
factors we identify these three main meanings to validate how good the method
finds word usage patterns over time.

The topics respectively the latent factors extracted by sLDA, PLS and kPLS
are summarize in Table 4. There we show again the most likeliest words for the
topic respectively the latent factors. First, in topic 1 from sLDA, in the latent
factor 1 from PLS and in the latent factor 1 from kLDA, we find computer
related words as most likely. This corresponds "Platte" in the meaning of "hard
drive". The distribution of the time stamps from sLDA shows a peak between
1990 and 2000. Before this period, this topic has not appeared. For topic 7 from
sLDA, factor 14 from PLS or factor 10 from kPLS, the most probable words
indicate the meaning of a photographic plate for the word "Platte". The two
most likeliest words are "Abb" which is short for "Abbildung" (engl. picture)
and "zeigt" (engl. to show). The distribution of the time stamps shows a major
usage of this meaning till the 50. Topic 10, respectively factors 16 from PLS
or 3 from kPLS associated corresponds to the meaning conductor for the word

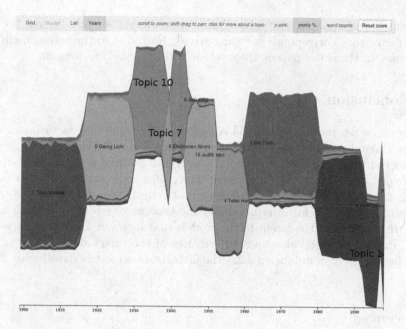

Fig. 3. Supervised LDA topics extracted from text snippets about the word "Platte" over a time period from 1900 to 1999. (Color figure online)

"Platte" that is most used in 1920 and 1930. The two most likeliest words are "Elektronen" (engl. electrons) and "Strom" (engl. current).

Comparing the three methods, sLDA results in a much clearer separation of topics over time. The alignment of the term-document distributions to the temporal information really reflects usage patterns in certain temporal periods. The meaning of "hard drive" is well covered by all methods, but the meanings "photographic plate" and "conductor" are not very pure represented in the factors from PLS and kernel PLS. Again, this is likely to be due to the orthogonality of

Table 4. Three topics respectively latent factors extracted from text snippets about the word "Platte" over a time period from 1900 to 1999.

sLDA			PLS			kPLS		
topic 1	topic 7	topic 10	factor 1	factor 14	factor 16	factor 1	factor 2	factor 3
BIOS	Abb	Elektronen	BIOS	Licht	Bild	BIOS	Elektronen	Abb
EIDE	zeigt	Strom	SCSI	Bild	Elektronen	SCSI	Herr	Herr
DOS	Negativ	Abb	DOS	läßt	Tisch	IDE	Art	Judith
MByte	Hintergrund	Achse	MByte	Hintergrund	Stromquelle	MByte	Paper	Achse
SCSI	Hand	Stromquelle	Zyliner	Vater	Kellner	Sektoren	Platz	Elektronen
IDE	Aufnahme	Hand	Sektoren	Person	Röhre	Treiber	Form	Gast
Partition	läßt	Strahlen	IDE	Art	Strom	Adapter	Bild	Strom
Sektoren	Stelle	parallel	Partition	Strahlen	Strahlen	Controller	Bilder	Licht
Windows	Person	senkrecht	Adapter	Aufnahme	Herr	DOS	Farben	Richtung
Daten	Licht	Röhre	EIDI	negativ	Stück	IDE	Abb	Stromquelle

the factors. For example electricity related words might already be well present in the factor that corresponds to "hard drive". Now, due to the orthogonality of all factors, in the other factors these words will likely be less present.

5   Conclusion

In this paper we perform empirical comparisons of topic and factor models to extract textual usage patterns with respect to given supervisions. We give an introduction in topic and factor models with and without additional labels for documents. We perform a comparative study of a topic model and two factor models to extract textual usage patterns that reflect given labels for the texts. The results show that for interpretability the generative latent topic models like LDA are better than the latent factor models that factorize the data in a vector space. In terms of generalization performance of the topics respectively factors for a classification on unlabeled data, the factorization models show higher accuracies.

References

1. Blei, D.M., McAuliffe, J.D.: Supervised topic models. In: Advances in Neural Information Processing Systems 20, Proceedings of the Twenty-First Annual Conference on Neural Information Processing Systems, Vancouver, 3–6 December 2007, pp. 121–128 (2007). http://papers.nips.cc/paper/3328-supervised-topic-models
2. Blei, D.M., Ng, A.Y., Jordan, M.I.: Latent dirichlet allocation. J. Mach. Learn. Res. **3**, 993–1022 (2003)
3. Blitzer, J., Dredze, M., Pereira, F.: Biographies, bollywood, boom-boxes and blenders: domain adaptation for sentiment classification. In: Proceedings of the 45th Annual Meeting of the Association of Computational Linguistics, pp. 440–447. Association for Computational Linguistics, Prague (June 2007)
4. Deerwester, S., Dumais, S.T., Furnas, G.W., Landauer, T.K., Harshman, R.: Indexing by latent semantic analysis. J. Am. Soc. Inform. Sci. Technol. **41**(6), 391–407 (1990)
5. Griffiths, T.L., Steyvers, M.: Finding scientific topics. Proc. Nat. Acad. Sci. **101**(Suppl. 1), 5228–5235 (2004)
6. Hotelling, H.: Analysis of a complex of statistical variables into principal components (1933)
7. Kaiser, H.: The varimax criterion for analytic rotation in factor analysis. Psychometrika **23**(3), 187–200 (1958). http://dx.doi.org/10.1007/BF02289233
8. Lee, D.D., Seung, H.S.: Algorithms for non-negative matrix factorization. In: NIPS, pp. 556–562 (2000). citeseer.ist.psu.edu/lee01algorithms.html
9. Liu, H., Wu, Z.: Non-negative matrix factorization with constraints (2010). https://www.aaai.org/ocs/index.php/AAAI/AAAI10/paper/view/1820/2027
10. Manning, C.D., Raghavan, P., Schütze, H.: Introduction to Information Retrieval. Cambridge University Press, New York (2008)
11. Mercer, J.: Functions of positive and negative type, and their connection with the theory of integral equations. Philos. Trans. R. Soc. Lond. **209**, 415–446 (1909)

12. Mimno, D.M., McCallum, A.: Topic models conditioned on arbitrary features with dirichlet-multinomial regression. CoRR abs/1206.3278 (2012). http://arxiv.org/abs/1206.3278

13. Nguyen, V.A., Boyd-Graber, J., Resnik, P.: Sometimes average is best: the importance of averaging for prediction using mcmc inference in topic modeling. In: Proceedings of the 2014 Conference on Empirical Methods in Natural Language Processing (EMNLP), pp. 1752–1757. Association for Computational Linguistics (2014). http://aclweb.org/anthology/D14-1182

14. Ramage, D., Hall, D., Nallapati, R., Manning, C.D.: Labeled lda: a supervised topic model for credit attribution in multi-labeled corpora. In: Proceedings of the 2009 Conference on Empirical Methods in Natural Language Processing, EMNLP 2009, vol. 1, pp. 248–256. Association for Computational Linguistics, Stroudsburg (2009). http://dl.acm.org/citation.cfm?id=1699510.1699543

15. Robertson, S.: Understanding inverse document frequency: on theoretical arguments for idf. J. Documentation **60**(5), 503–520 (2004). http://dx.doi.org/10.1108/00220410410560582

16. Rosipal, R., Trejo, L.J.: Kernel partial least squares regression in reproducing kernel hilbert space. J. Mach. Learn. Res. **2**, 97–123 (2002). http://dl.acm.org/citation.cfm?id=944790.944806

17. Rudin, W.: Real and Complex Analysis, 3rd edn. McGraw-Hill Inc., New York (1987)

18. Schölkopf, B., Smola, A.J., Müller, K.R.: Advances in kernel methods. In: Kernel Principal Component Analysis, pp. 327–352. MIT Press, Cambridge (1999). http://dl.acm.org/citation.cfm?id=299094.299113

19. Shawe-Taylor, J., Cristianini, N.: Kernel Methods for Pattern Analysis. Cambridge University Press, New York (2004)

20. Yu, K., Yu, S., Tresp, V.: Multi-label informed latent semantic indexing. In: Proceedings of the 28th Annual International ACM SIGIR Conference on Research and Development in Information Retrieval, SIGIR 2005, pp. 258–265. ACM, New York (2005). http://doi.acm.org/10.1145/1076034.1076080

21. Zeng, X.Q., Wang, M.W., Nie, J.Y.: Text classification based on partial least square analysis. In: Proceedings of the 2007 ACM Symposium on Applied Computing, SAC 2007, pp. 834–838. ACM, New York (2007). http://doi.acm.org/10.1145/1244002.1244187

22. Zhu, J., Ahmed, A., Xing, E.P.: Medlda: maximum margin supervised topic models for regression and classification. In: Proceedings of the 26th Annual International Conference on Machine Learning, pp. 1257–1264, ICML 2009. ACM, New York (2009). http://doi.acm.org/10.1145/1553374.1553535

23. Zhu, J., Chen, N., Perkins, H., Zhang, B.: Gibbs max-margin topic models with fast sampling algorithms. In: ICML (1). JMLR Proceedings, vol. 28, pp. 124–132. JMLR.org (2013). http://dblp.uni-trier.de/db/conf/icml/icml2013.htmlZhuCPZ13

Data-Driven Analyses of Electronic Text Books

Ahcène Boubekki[1], Ulf Kröhne[1], Frank Goldhammer[1],
Waltraud Schreiber[2], and Ulf Brefeld[1(✉)]

[1] Leuphana University of Lüneburg, Lüneburg, Germany
{boubekki,kroehne,goldhammer}@dipf.de, brefeld@cs.tu-darmstadt.de
[2] Faculty of History and Social Science, KU Eichstätt, Eichstätt, Germany
waltraud.schreiber@ku.de

Abstract. We present data-driven log file analyses of an electronic text book for history called the mBook to support teachers in preparing lessons for their students. We represent user sessions as contextualised Markov processes of user sessions and propose a probabilistic clustering using expectation maximisation to detect groups of similar (i) sessions and (ii) users. We compare our approach to a standard K-means clustering and report on findings that may have a direct impact on preparing and revising lessons.

1 Introduction

Electronic text books may offer a multitude of benefits to both teachers and students. They allow to combine text, images, interactive maps and audiovisual content in an appealing way, and the usage is supported by hyperlinks, search functions, and integrated glossaries. By representing learning content in various ways and enabling alternative trajectories of accessing learning objects, electronic text books offer great potentials for individualised teaching and learning. Although technological progress passed by schools for a long time, inexpensive electronic devices and handhelds have found their way into schools and are now deployed to complement traditional (paper-based) learning materials.

Particularly text books may benefit from cheap electronic devices. Electronic versions of text books may revolutionise rigour presentations of learning content by linking maps, animations, movies, and other multimedia content. However, these new degrees of freedom in presenting and combining learning materials may bring about also new challenges for teachers and learners. For instance, learners need to regulate and direct their learning process to a greater extent if there are many more options they can choose from. Thus, the ultimate goal is not only an enriched and more flexible presentation of the content but to effectively support teachers in preparing lessons and children in learning. To this end, not only the linkage encourages users to quickly jump through different chapters but intelligent components such as recommender systems [35] may highlight alternative pages of interest to the user. Unfortunately, little is known on the impact of these methods on learning as such and even little is known on how such electronic text books are used by students.

© Springer International Publishing Switzerland 2016
S. Michaelis et al. (Eds.): Morik Festschrift, LNAI 9580, pp. 362–376, 2016.
DOI: 10.1007/978-3-319-41706-6_20

In this article, we present insights on the usage of an electronic text book for history called the mBook [36]. Among others, the book has been successfully deployed in the German-speaking Community of Belgium. We show how data-driven analyses may support history teachers in preparing their lessons and showcase possibilities for recommending resources to children. Our approach is twofold: Firstly, we analyse user sessions to find common behavioural patterns across children and their sessions. Secondly, we aggregate sessions belonging to the same user to identify similar types of users. This step could help to detect deviating learners requiring additional attention and instructional support.

In this paper, we argue that conclusions on an individual session or user basis can only be drawn by taking the respective population into account and propose a contextualised clustering of user sessions. We represent user sessions as fully observed Markov processes that are enriched by context variables such as timestamps and types of resources. We derive an expectation maximisation algorithm to group (user-aggregated) sessions according to the learners' behaviour when using the text book. To showcase the expressivity of our approach, we compare the results to a standard K-means-based [30] solution. While the latter leads to trivial and insignificant groups, our methodology allows to project similar sessions (users) onto arbitrary subsets of variables that can easily be visualised and interpreted. We report on observations that can be used to support teacher instructions and students learning.

The remainder is organised as follows. Section 2 reviews related work. We introduce the *mBook* in Sect. 3 and present our probabilistic model in Sect. 4. We report on empirical results in Sects. 5 and 6 provides a discussion of the results and Sect. 7 concludes.

2   Related Work

The analysis of log files is common in computer science and widely used to understand user navigation on the web [1,24]. Often, sequential approaches, such as Markov models and/or clustering techniques, are used to detect browsing patterns that are predictive for future events [18,34,38] or interests of the user [3]. However, previous approaches to modelling user interaction on the Web mainly focus on the pure sequence of page views or categories thereof, without taking contextual information into account. Patterns in page view sequences have been analysed using all sorts of techniques, including relational models [2], association rule mining [14,15], higher-order Markov models [18], and k-nearest neighbours [6].

A useful step toward interpretable patterns is to partition navigation behaviour into several clusters, each with its own characteristics. Hobble and Zicari [23] use a hierarchical clustering to group website-visitors and Chevalier et al. [11] correlate navigation patterns with demographic information about users. Other heuristic approaches to identify clusterings of user interactions include sequence alignments [21], graph-mining [19]. The advantage of model-based clusterings is that the cluster parameters itself serve as a starting point for interpreting the results. Prior work in this direction focuses on modelling navigational sequences using Markov processes [10,32] and hidden Markov models [8,18,39]. Haider et al. [20] cluster user sessions with Markov processes for Yahoo! News. Their approach is similar to ours as they also propose a nested EM algorithm, however, we model timestamps with periodic distributions while Haider et al. resort to simulate periodicity by external filtering processes. They focus on presenting clustering results and do not compare their methods to alternative ones.

In recent years, logfile analyses attract more and more researchers from other disciplines such as educational research [5]. Although the analysis of log and process related data is still a new and emerging field in educational research, two methodologies can be described [4]: *Educational data mining* (EDM) and *learning analytics*. Their common goal is to discover knowledge in educational data, however, the former is purely data-driven while the latter keeps the user/expert in the loop to guide the (semi-automatic) analysis. Many approaches are related to web-based learning environments such as MOOCs [9] and other learning management systems [31].

Köck and Paramythis [27] cluster learners in a web-based learning environment according to their performance in exercises and the usage of an interactive help. Lemieux et al. [29] develop an online exerciser for first year students and visualise identified patterns of usage and different behaviours. Sheard et al. [37] and Merceron and Yacef [33] analyse logfile data not only to describe user behaviour but also to provide information and feedback to learners and tutors. Generally, much work is put into visualising and organising the discovered knowledge as relationships and correlations are often complex and difficult to communicate [16].

Another line of research deals with the assessment of the level of motivation [22,26]; the time spent on a page and filling in exercises turn out to be predictive indicators. Cocea [12] and Cocea and Weibelzahl [13] studied disengagement criteria as a counterpart of motivation. One of their findings showed that a user's history needs to be taken into account for predicting engagement/disengagement as exploratory phases always precede learning phases and vice versa.

3   The mBook

History as a subject is especially promising for a prototype of a multimedia textbook. The re-construction of past events and the de-construction of *historical narrations* is ubiquitously present in our historical consciousness; different narrations about the past contain and provoke a great deal of different intentions and interpretations. It is therefore crucial to deal with history in various and different perspectives as a single narration is always a retrospective, subjective, selective, and thus only a partial re-construction of the past.

The *mBook* is guided on a constructivist and instructional-driven design. Predominantly, the procedural model of historical thinking is implemented by a structural competence model that consists of four competence areas that are deduced from processes of historical thinking: (i) the competency of posing and answering historical questions, (ii) the competency of working with historical methodologies, and (iii) the competency of capturing history's potential for human orientation and identity. The fourth competency includes to acquire and apply historical terminologies, categories, and scripts and is best summarised as (iv) declarative, conceptual and procedural knowledge. It is often referred to as the foundation of structural historical thinking in terms of premises and results and systematised by principles, conceptions of identity and alterity as well as categories and methodological scripts [28].

Imparting knowledge in this understanding is therefore not about swotting historic facts but aims at fostering a reflected and (self-)reflexive way of dealing with our past. The underlying concept of the multimedia history schoolbook implements well-known postulations about self-directed learning process in practice. The use of the mBook allows an open-minded approach to history and fosters contextualised and detached views of our past (cf. [25]). To this end, it is crucial that a purely text-based narration is augmented with multimedia elements such as historic maps, pictures, audio and video tracks, etc. Additionally, the elements of the main narration are transparent to the learners. Learners quickly realise that the narration of the author of the mBook is also constructed, as the author reveals his or her construction principle.

The mBook consists of 5+1 chapters, *Antiquity, Middle Age, Renaissance, 19th Century, 20th and 21st Century* and a chapter on methods. In the German-speaking Community in Belgium, the mBook has about 1300 regular users. In our analysis, we focus on about 330.000 sessions collected in Belgium between March and November 2014 containing approximately 5 million events (clicks, scrolls, key press, etc.). The book encompasses 648 pages including 462 galleries and 531 exercises among others.

4   Methodology

4.1   Preliminaries

Let X denote the set of N user sessions given by $X = \{X^i\}_{i=1}^N$. A session is assembled by user events such as page views, clicks, scrolls, text edit, etc. In this work, we focus on the *connection time* t, the *sequence of the visited page* in terms of the chapter they belong to $\mathbf{x} = \langle x_1^i, \ldots, x_{T^i}^i \rangle$, and the *sequence of categories* realised by the viewed pages $\mathbf{c} = \langle c_1^i, \ldots, c_{T^i}^i \rangle$. The six chapters of the book together with the *homepage* and a *termination* page that encodes the end of a session form 8 possible realisations for every visited page, i.e., the values for the variable x_t^i. There are five different categories, *summary, text, gallery* and the auxiliary variables representing the categories for the *homepage* and the *termination* page.

4.2 Representation

We deploy a parameterised mixture model with K components to compute the probability of a session.

$$p(X^i|\Theta) = \sum_{k=1}^{K} \kappa_k p(X^i|\Theta_k).$$

The variables κ_k represent the probability that a *random* session is generated by the k-th component and also known as the *prior probability* for cluster k. The term $p(X^i|\Theta_k)$ is the *likelihood* of the session given that it belongs to cluster k with parameters Θ_k. Defining sessions in terms of time, chapters, and categories allows to assemble the likelihood of a session as

$$p(X^i|\Theta_k) = p(t^i|\beta_k)p(\mathbf{x}^i|\alpha_k)p(\mathbf{c}^i|\gamma_k).$$

The browsing process through chapters is modelled by a first-order Markov chain so that pages are addressed only by their chapter. We have,

$$p(\mathbf{x}|\alpha_k) = p(x_1|\alpha_k^{init}) \prod_{l=2}^{L} p(x_l|x_{l-1}, \alpha_k^{tr})$$

where $\alpha_k = (\alpha^{init}, \alpha^{tr})$ is split up into parameters α^{init} for the first page view and the transition parameters α^{tr} for the process.

The category model depends on the chapters as we aim to observe correlations between different types of pages. This may show for example whether galleries of some of the chapters are more often visited (and thus more attractive) than others and thus generate feedback for the teachers (e.g., to draw students attention to some neglected resources) and developers (e.g., to re-think the accessibility or even usefulness of resources). Categories are modelled by

$$p(\mathbf{c}|\gamma_k) = p(c_1|x_1, \gamma_k^{init}) \prod_{l=2}^{L} p(c_l|c_{l-1}, x_{l-1}, \gamma_k^{tr})$$

where again γ^{init} is used for the prior category and γ^{tr} for the subsequent transitions.

4.3 Modeling Time

The model for the connection times is inspired by the approach described in [20]. The goal is to project the continuous time space into a multinomial space to ease the estimation process. For this purpose, we introduce fixed unique and periodic components that serve as a new basis for generating continuous time events. To capture periodic behaviours, 90 time components are defined: 48 daily and 42 weekly components. The connection model is a multinomial law over each component with parameters $\beta_{k,j}^d$ and $\beta_{k,j}^w$ where j encodes the component and

d and w refer to a daily or weekly setting, respectively. The probability for a session to start at a certain time t is therefore given by

$$p(t|\beta_k) = \sum_{j=1}^{48} \beta_{k,j}^d d_j(t_d) + \sum_{j=1}^{42} \beta_{k,j}^w w_j(t_w).$$

The components are derived from the normal distribution. Periodic constraints are embedded in the probabilities, so that the density is composed with the tangent function which, besides of being periodic, conserves the symmetry. The generic form can be written as

$$p_{\mu,\sigma,T}(t) = \frac{1}{\text{erfc}(\frac{1}{\sigma})T} \exp\left(-\frac{1 + \tan^2(\frac{\pi}{T}(t - \mu))}{\sigma^2}\right).$$

The period is governed by the parameter T, and erfc is the complementary error function. The parameter μ represents the expectation similarly to the normal law. A component is said to recover a time slot if its density in this interval is higher than half of its maximum. This condition is parametrised by standard deviation σ. For a component covering time unit Δ we obtain $\sigma = \tan\left(\frac{\pi}{T}\Delta\right)/\sqrt{\log(2)}$.

The daily components are centred every 30 min, have a duration of 30 min, and the first component is centred at 12:00am. The weekly components are centred every four hours with a duration of four hours, and the first one is centred on Monday at 2:00am. This shift allows a synchronisation with the schools working hours, as we have a slice for the morning between 8:00am and 12:00pm, one for the afternoon between 12:00pm and 4:00pm and another one for the evening between 4:00pm and 8:00pm.

To capture the daily and weekly behaviors, connection times are considered modulo the 48 slices of a day (t_d) or the 336 slices of a week (t_w). The daily d_j and weekly w_j distributions are described as follows

$$d_j(t) = \frac{1}{48 \, \text{erfc}(\frac{1}{\sigma_d})} \exp\left(-\frac{1 + \tan^2\left(\frac{\pi}{48}(x - j)\right)}{\sigma_d^2}\right)$$

$$w_j(t) = \frac{1}{336 \, \text{erfc}(\frac{1}{\sigma_w})} \exp\left(-\frac{1 + \tan^2\left(\frac{\pi}{336}(x - 4 - 8j)\right)}{\sigma_w^2}\right)$$

using the variances $\sigma_d = \tan\left(\frac{\pi}{48}\frac{1}{2}\right)/\sqrt{\log(2)} \simeq 0.039$ and $\sigma_w = \tan\left(\frac{4\pi}{336}\right)/\sqrt{\log(2)} \simeq 0.045$.

4.4 Optimisation

Given our mixture model and assuming independence of the user sessions, the likelihood of the sessions is given by

$$p(X|\Theta) = \prod_{i=1}^{N} \sum_{k=1}^{K} \kappa_k p(t^i|\beta_k) p(\mathbf{x}^i|\alpha_k) p(\mathbf{c}^i|\mathbf{x}^i, \gamma_k).$$

The joint likelihood needs to be maximised with respect to the parameters $\Theta = (\pi_k, \beta_k, \alpha_k, \gamma_k)$. For computational reasons, we address the equivalent problem of maximising the log of the likelihood. The optimisation problem becomes

$$\Theta^* = \underset{\Theta}{\operatorname{argmax}} \, \log p(X|\Theta).$$

We develop an expectation maximisation (EM)-like algorithm [10,17]. Expectation maximisation is an iterative approach to approximate a local maximum from a given set of parameters. The procedure works in two steps: the expectation step (E-Step) computes the expectation of the objective function related to the problem from the actual set of parameters and deduces two temporary distributions of the Markov sequences over the cluster and the time components called *class-conditional probability distribution* and *time component-conditional probability distribution* denoted by $P_{i,k}$ and $Z_{i,k,j}^\bullet$, respectively,

$$P_{i,k} = \frac{\kappa_k p(X^i|\Theta_k)}{\sum_{k'=1}^{K} \kappa_{k'} p(X^i|\Theta_{k'})}$$

$$Z_{i,k,j}^d = \frac{\beta_{k,j}^d d_j(t_d^i)}{\sum_{j'=1}^{48} \beta_{k,j'}^d d_{j'}(t_d^i)}$$

$$Z_{i,k,j}^w = \frac{\beta_{k,j}^w w_j(t_w^i)}{\sum_{j'=1}^{42} \beta_{k,j'}^w w_{j'}(t_w^i)}$$

The maximisation step (M-Step) re-estimates the parameters from these proposal distributions to increase the value of the objective function and thus also the likelihood. We give the update formulas of four of the parameters, as they can be easily translated to the other parameters:

$$\kappa_k = \frac{\sum_{i=1}^{N} P_{i,k}}{\sum_{k'=1}^{K} \sum_{i=1}^{N} P_{i,k'}}$$

$$\beta_{k,j}^d = \frac{\sum_{i=1}^{N} Z_{i,k,j}^d P_{i,k}}{\sum_{j'=1}^{42} \sum_{i=1}^{N} Z_{i,k,j'}^d P_{i,k}}$$

$$\gamma_{k,g}^{init} = \frac{\sum_{i=1}^{N} P_{i,k} \delta(x_1^i, g)}{\sum_{g'=1}^{5} \sum_{i=1}^{N} P_{i,k} \delta(x_1^i, g')}$$

$$\gamma_{k,g,h}^{tr} = \frac{\sum_{i=1}^{N} P_{i,k} \eta_{g,h}(x^i)}{\sum_{h'=1}^{5} \sum_{i=1}^{N} P_{i,k} \eta_{g,h'}(x^i)},$$

where g and h take their values in the fives possible types of pages. The function $\delta(x_1^i, g)$ is the Kronecker delta that equals 1 if the two arguments are equal and 0 otherwise. The function $\eta_{g,h}(x^i)$ returns the number of transitions in session x^i from a page of type g to a page of type h.

5   Empirical Results

In our empirical analysis, we focus on about 330.000 sessions collected in Belgium between March and November 2014 containing approximately 5 million events including clicks, scrolls, key presses, etc. In the remainder, we show results for $K = 8$ clusters to trade-off expressivity and interpretability, however, other choices are possible.[1]

5.1   Comparison with K-Means

The first experiment demonstrates the expressivity of our approach. We compare our probabilistic solution with a winner-takes-all clustering by K-means [30]. Since K-means acts in vector spaces, user session are represented as vectors in a 354 dimensional space, so that both algorithms have access to identical information.

Figure 1 shows the results from clustering sessions. For lack of space, we focus on a projection of the final clusterings on the daily components capturing repeti-

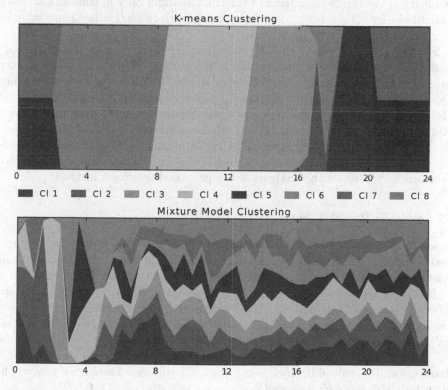

Fig. 1. Results for K-means and the proposed model. (Color figure online)

---

[1] Note that we obtain similar results for all $1 \leq K \leq 30$; this holds in particular for the comparison with K-means.

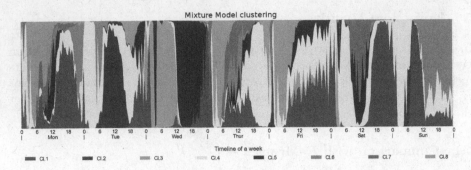

Fig. 2. Cluster distribution across a week. (Color figure online)

tive behaviour across days. The geometric nature of K-means is clearly visible in Fig. 1 (top): the clusters separate the day into six time slots of about four hours. The more complex colouring between 4pm and 8pm indicates that more variables than the connection time are active during that period. Nevertheless, the simplicity of the result (e.g., most of the clusters differ only in connection time), particularly for school hours, is clearly inappropriate for further processing or interpretation.

Figure 1 (bottom) shows the corresponding results for our probabilistic approach. The distribution of the clusters is fairly more interesting and balanced across the day. Clusters clearly specialise on dependencies across week days which is also shown in Fig. 2. Cluster $C1$ and $C4$ capture recurrent behaviour and cover a large part of the user activity. Cluster $C6$ focuses on the activity on Sunday afternoon and similarly, cluster $C5$ specialises on Wednesday afternoon. Clusters $C2$ and $C3$ have a similar shape as they occur mainly on Tuesday and Wednesday morning, respectively, during school hours. In the remainder, we discard K-means and focus on the analysis of the proposed approach instead.

5.2   Session-Based View

Figure 3 shows the results of a session-based clustering. User sessions are distributed across the whole clustering according to the expressed behaviour. Clusters can therefore be interpreted as similar user behaviours at similar times.

Before we go into details, recall Fig. 2, where the Sunday afternoon is shared between cluster $C1$, $C4$ and $C6$. The latter aggregates most of the activity and also most of the text page views. Figure 3 allows for a clearer view on the clustering. According to Fig. 3, a similar observation can be made for clusters $C7$ and $C8$. Both of them have *Antics* as the main chapter, and have a similar weekly distribution, the only difference being that $C8$ contains more text views. The latter indicates more experienced users as we will discuss in the following.

The visualisation shows that all categories are clearly visible for all clusters, indicating a frequent usage of all possible types of resources by the users. Cluster $C6$ possesses half of the mass on the weekend of category *text*. This indicates

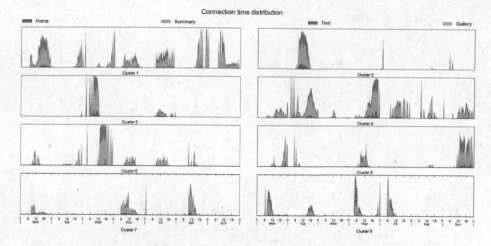

Fig. 3. Resulting clusters for the session-based clustering (Color figure online)

more experienced users who like to form their opinion themselves instead of going to summary pages. The same holds for cluster $C8$ that possesses in addition only a vanishing proportion of the *home* category. Small probabilities of category *home* as well as large quantities of category *text* indicate that users continuously read pages and do not rely on the top-level menu for navigation.

5.3   User-Based View

Our approach can also be used to group similar users. To this end, we change the expectation step of the algorithm so that sessions by the same user are processed together. That is, there is only a single expectation for the sessions being in one of the clusters. Clusters therefore encode similar users rather than similar behaviour as in the previous section.

Figure 4 shows the results. Apparently, the main difference of the clusters is the intensity of usage during working days and weekends. Cluster $C2$ for instance clearly focuses on working day users who hardly work on weekends compared to Cluster $C1$ whose users place a high emphasise on Saturdays and Sundays. Cluster $C3$ contains low frequency users who rarely use the mBook and exhibit the smallest amount of sessions and page views per session (see also Fig. 5). Cluster $C8$ contains heavy (at night) users with high proportions of category *text*. In general, we note that transition matrices are consistent between chapters in contrast to the session-based clustering, that is, test takers interact with most of the chapters.

Figure 5 confirms our interpretations with descriptive statistics. Cluster $C8$ containing the power users possess the highest number of sessions and also the highest number of sessions per user. Clusters $C3$ and $C7$ are the smallest. As cluster $C3$ has been identified as encoding low frequency users, it may be comforting to know that these users constitute a clear minority. Nevertheless teachers

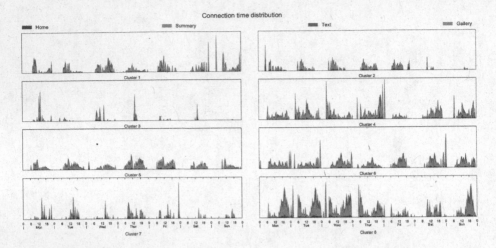

Fig. 4. Resulting clusters for the user-based clustering. (Color figure online)

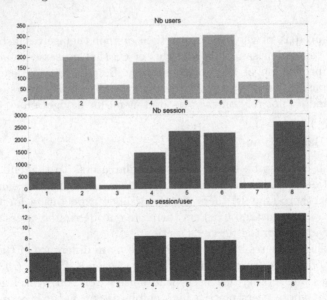

Fig. 5. Number of users (top), sessions (center), and sessions per user (bottom) for the user-view.

may be well advised to keep an eye on these children and individually support them by all means.

Figure 6 finally shows differences in clicking behaviour of users. For two clusters, transition matrices for types of resources are visualised, darker colours indicate more probable transitions. The two bars on top visualise the distribution of chapters. Both distributions are quite different. While users in cluster *C1* exhibit a broad interest and visit chapters uniformly, their peers in cluster

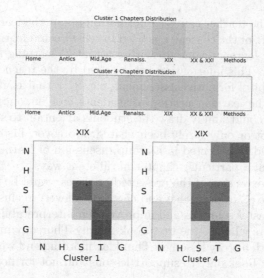

Fig. 6. Transitions (row → column) realised by two clusters.

$C4$ focus clearly on the three chapters *Renaissance* and the *XIX* and *XX & XXI* centuries. However, the observation also shows that linkage is exploited by the users who seem to like browsing and learning about the book and thus also about history.

Users in clusters $C1$ and $C4$ exhibit very different click behaviours. While users in both clusters prefer viewing galleries, users in cluster $C1$ move deterministically on to a text page, while users in cluster $C4$ visit summary pages or terminate the session. This reflects the two possibilities of browsing inside the mBook: a hierarchical and a flat one. The former is realised by performing transitions between text pages always through a summary page. By contrast, a flat navigation makes use of the page to page navigation possibility and hence reflects the way a regular paper book is read. Knowing these relationships is an important means to personalise electronic books like the mBook. For instance, identifying an active session as a member of cluster $C1$ allows to replace links to summary pages by other content as these users will almost always go back to a text page. On the other hand, knowing that a user of cluster $C4$ is viewing a gallery may be utilised to actively recommend other resources to prevent her from churning.

6 Discussion

Our results illustrate potential benefits from clustering learners for instructional purposes. In the first place, the probabilistic clustering approach shows a way how to condense a huge amount of logfile information to meaningful patterns of learner interaction. Classifying a student into one of several clusters reveals whether, when, and how the learner used the materials offered by the electronic text book. Thus, the teacher can get information about the learners' navigation speed, whether part of the content was used in self-directed learning processes as expected, whether learners came up with alternative learning trajectories, and so on and so forth. This information can be used by the teacher in a formative way (cf. the concept of formative assessment, e.g., [7]), that is, it is directly used to further shape the learning process of students. For instance, in a follow-up lesson the teacher could simply draw the students attention to some parts of the book that have not or only rarely been visited. Moreover, history and learning about history could be reflected in a group discussion of learners who used the mBook resources of a particular chapter in different ways.

An important extension of the presented analyses would be to relate contextual information (e.g., from teacher and class room level) to clusters. This would help to validate cluster solutions and improve their interpretability. For instance, a cluster of learners who use the text book mainly Thursday morning may consist of students with history lesson on Thursday morning and with teachers using the electronic text book only to support lessons and not for homework.

7   Conclusion

We presented contextualised Markov models to represent user sessions and proposed an Expectation Maximisation algorithm for optimisation. We applied our approach to clustering user sessions of the mBook, an electronic text book for history. Our results may have a direct impact on teachers and learners and can be used together with outlier analyses to find students who need individual support.

References

1. Agosti, M., Crivellari, F., Di Nunzio, G.: Web log analysis: a review of a decade of studies about information acquisition, inspection and interpretation of user interaction. Data Mining and Knowledge Discovery pp. 1–34 (2011)
2. Anderson, C.R., Domingos, P., Weld, D.S.: Relational markov models and their application to adaptive web navigation. In: Proceedings of the ACM SIGKDD International Conference on Knowledge Discovery and Data Mining (2002)
3. Armentano, M., Amandi, A.: Modeling sequences of user actions for statistical goal recognition. User Model. User-Adap. Inter. 22(3), 281–311 (2012)
4. Baker, R., Siemens, G.: Cambridge Handbook of the Learning Sciences. In: Sawyer, R.K. (ed.) Educational data mining and learning analytics, 2nd edn., pp. 253–274. Cambridge University Pres, New York (2014)
5. Baker, R., Yacef, K.: The state of educational data mining in 2009: a review and future visions. J. Educ. Data Min. 1(1), 3–17 (2009)
6. Billsus, D., Pazzani, M.: User modeling for adaptive news access. User Model. User-Adap. Inter. 10(2), 147–180 (2000)
7. Black, P., Wiliam, D.: Assessment and classroom learning. Assess. Educ. 5(1), 7–74 (1998)
8. Borges, J., Levene, M.: Evaluating variable-length markov chain models for analysis of user web navigation sessions. IEEE Trans. Knowl. Data Eng. 19(4), 441–452 (2007)

9. Brinton, C.G., Chiang, M., Jain, S., Lam, H., Liu, Z., Wong, F.M.F.: Learning about social learning in moocs: From statistical analysis to generative model. Technical Report (2013). arXiv:1312.2159

10. Cadez, I., Heckerman, D., Meek, C., Smyth, P., White, S.: Visualization of navigation patterns on a web site using model-based clustering. In: Proceedings of the ACM SIGKDD International Conference on Knowledge Discovery and Data Mining (2000)

11. Chevalier, K., Bothorel, C., Corruble, V.: Discovering rich navigation patterns on a web site. In: Grieser, G., Tanaka, Y., Yamamoto, A. (eds.) DS 2003. LNCS (LNAI), vol. 2843, pp. 62–75. Springer, Heidelberg (2003)

12. Cocea, M.: Can log files analysis estimate learner's level of motivation? In: Proceedings of the Workshop on Lernen - Wissensentdeckung - Adaptivität (2006)

13. Cocea, M., Weibelzahl, S.: Log file analysis for disengagement detection in e-learning environments. User Model. User-Adap. Inter. **19**(4), 341–385 (2009)

14. Daş, R., Türkoğlu, İ.: Creating meaningful data from web logs for improving the impressiveness of a website by using path analysis method. Expert Syst. Appl. **36**(3), 6635–6644 (2009)

15. Daş, R., Türkoğlu, İ.: Extraction of interesting patterns through association rule mining for improvement of website usability. Istanbul Univ. J. Electr. Electron. Eng. **9**(18), 1037–1046 (2010)

16. Delestre, N., Malandain, N.: Analyse et représentation en deux dimensions de traces pour le suivi de l'apprenant. Revue des Sciences et Technologies de l'Information et de la Communication pour l'Education et la Formation (STICEF) 14 (2007)

17. Dempster, A., Laird, N., Rubin, D.: Maximum likelihood from incomplete data via the em algorithm. J. R. Stat. Soc. Ser. B (Methodological) **39**, 1–38 (1977)

18. Deshpande, M., Karypis, G.: Selective markov models for predicting web page accesses. ACM Trans. Internet Technol. (TOIT) **4**(2), 163–184 (2004)

19. Gündüz, Ş., Özsu, M.T.: A web page prediction model based on click-stream tree representation of user behavior. In: Proceedings of the Ninth ACM SIGKDD International Conference on Knowledge Discovery and Data Mining (2003)

20. Haider, P., Chiarandini, L., Brefeld, U., Jaimes, A.: Contextual models for user interaction on the web. In: ECML/PKDD Workshop on Mining and Exploiting Interpretable Local Patterns (I-PAT) (2012)

21. Hay, B., Wets, G., Vanhoof, K.: Mining navigation patterns using a sequence alignment method. Knowl. Inf. Syst. **6**, 150–163 (2004)

22. Hershkovitz, A., Nachmias, R.: Developing a log-based motivation measuring tool. In: Proceedings of the International Conference on Educational Data Mining (2008)

23. Wegrzyn-Wolska, K.M., Szczepaniak, P.S.: On clustering visitors of a web site by behavior and interests. In: Hoebel, N., Zicari, R.V. (eds.) Advances in Intelligent Web Mastering. Advances in Soft Computing, vol. 43, pp. 160–167. Springer, Heidelberg (2007)

24. Jansen, B.J.: Understanding user-web interactions via web analytics. Synth. Lect. Inf. Concepts Retrieval Serv. **1**(1), 1–102 (2009)

25. Karagiorgi, Y., Symeou, L.: Translating constructivism into instructional design: potential and limitations. Educ. Technol. Soc. **8**(1), 17–27 (2005)

26. Kay, J., Maisonneuve, N., Yacef, K., Zaïane, O.: Mining patterns of events in students' teamwork data. In: Proceedings of the ITS Workshop on Educational Data Mining (2006)

27. Köck, M., Paramythis, A.: Activity sequence modelling and dynamic clustering for personalized e-learning. User Model. User-Adap. Inter. **21**(1–2), 51–97 (2011)

28. Körber, A., Schreiber, W., Schöner, A. (eds.): Kompetenzen historischen Denkens: Ein Strukturmodell als Beitrag zur Kompetenzorientierung in der Geschichtsdidaktik. Neuried, Ars una (2007)

29. Lemieux, F., Desmarais, M.C., Robillard, P.N.: Analyse chronologique des traces journalisées d'un guide d'étude pour apprentissage autonome. Revue des Sciences et Technologies de l'Information et de la Communication pour l'Education et la Formation (STICEF) 20 (2014)

30. Lloyd, S.P.: Least squares quantization in PCM. IEEE Trans. Inf. Theory **28**(2), 129–137 (1982)

31. Macfadyen, L.P., Dawson, S.: Mining LMS data to develop an "early warning system" for educators: a proof of concept. Comput. Educ. **54**(2), 588–599 (2010)

32. Manavoglu, E., Pavlov, D., Giles, C.L.: Probabilistic user behavior models. In: Proceedings of the Third IEEE International Conference on Data Mining (2003)

33. Merceron, A., Yacef, K.: A web-based tutoring tool with mining facilities to improve learning and teaching. In: 11th International Conference on Artificial Intelligence in Education (AIED03), pp. 201–208. IOS Press (2003)

34. Qiqi, J., Chuan-Hoo, T., Chee Wei, P., Wei, K.K.: Using sequence analysis to classify web usage patterns across websites. In: Proceedings of the 45th Hawaii International Conference on System Science (HICSS), pp. 3600–3609 (2012)

35. Ricci, F., Rokach, L., Shapira, B., Kantor, P.B. (eds.): Recommender Systems Handbook. Springer, US (2015)

36. Schreiber, W., Sochatzy, F., Ventzke, M.: Das multimediale schulbuch - kompetenzorientiert, individualisierbar und konstruktionstransparent. In: Schreiber, W., Schöner, A., Sochatzy, F. (eds.) Analyse von Schulbüchern als Grundlage empirischer Geschichtsdidaktik, pp. 212–232. Kohlhammer (2013)

37. Sheard, J., Ceddia, J., Hurst, J., Tuovinen, J.: Inferring student learning behaviour from website interactions: a usage analysis. Educ. Inf. Technol. **8**(3), 245–266 (2003)

38. Srivastava, J., Cooley, R., Deshpande, M., Tan, P.N.: Web usage mining: discovery and applications of usage patterns from web data. ACM SIGKDD Explor. Newsl. **1**(2), 12–23 (2000)

39. Ypma, A., Heskes, T.: Automatic categorization of web pages and user clustering with mixtures of hidden markov models. In: Zaïane, O.R., Srivastava, J., Spiliopoulou, M., Masand, B. (eds.) WebKDD 2003. LNCS (LNAI), vol. 2703, pp. 35–49. Springer, Heidelberg (2003)

k-Morik: Mining Patterns to Classify Cartified Images of Katharina

Elisa Fromont[1] and Bart Goethals[1,2](✉)

[1] UMR CNRS 5516, Laboratoire Hubert-Curien,
Université de Lyon, Université de St-Etienne, 42000 St-Etienne, France
elisa.fromont@univ-st-etienne.fr
[2] Department of Math and Computer Science,
University of Antwerp, Antwerp, Belgium
goethals@gmail.com

Abstract. When building traditional Bag of Visual Words (BOW) for image classification, the k-Means algorithm is usually used on a large set of high dimensional local descriptors to build a visual dictionary. However, it is very likely that, to find a good visual vocabulary, only a sub-part of the descriptor space of each visual word is truly relevant for a given classification problem. In this paper, we explore a novel framework for creating a visual dictionary based on Cartification and Pattern Mining instead of the traditional k-Means algorithm. Preliminary experimental results on face images show that our method is able to successfully differentiate photos of Elisa Fromont, and Bart Goethals from Katharina Morik.

1 Introduction

Classification of images is of considerable interest in many image processing and computer vision applications. A common approach to represent the image content is to use histograms of color, texture and edge direction features [8,29]. Although they are computationally efficient, such histograms only use global information and thus only provide a crude representation of the image content. One trend in image classification is towards the use of *bag-of-visual-words* (BOW) features [11] that come from the *bag-of-words* representation of text documents [27]. The creation of these features requires four basic steps: (i) keypoints detection (ii) keypoints description, (iii) codebook creation and (iv) image representation. Keypoints refer to small regions of interest in the image. They can be sampled densely [16], randomly [31] or extracted with various detectors [21] commonly used in computer vision. Once extracted, the keypoints are characterized using a local descriptor which encodes a small region of the image in a D-dimensional vector. The most widely used keypoint descriptor is the 128-dimensional SIFT descriptor [20]. Once the keypoints are described, the collection of descriptors of all images of a training set are clustered, often using the k-Means algorithm, to obtain a visual codebook. Each cluster representative (typically the centroid) is considered as a visual word in a visual dictionary

© Springer International Publishing Switzerland 2016
S. Michaelis et al. (Eds.): Morik Festschrift, LNAI 9580, pp. 377–385, 2016.
DOI: 10.1007/978-3-319-41706-6_21

and each image can be mapped into this new space of visual words leading to a *bag-of-visual-words* (or a histogram of visual words) representation. The k-Means algorithm considers all the feature dimensions (128 for SIFT descriptors) for computing the distances to estimate clusters. As a consequence, the nearest neighbor estimation can get affected by noisy information from irrelevant feature dimensions [6,18]. Furthermore, the k-Means algorithm forces every keypoint to be assigned to a single visual word, while in practice, multiple visual words could be relevant for a particular keypoint.

Our goal is to use the recently developed Cartification methodology [4,5] to obtain better subspace clusters that could be used in the clustering phase of the BOW creation process. Then, standard machine learning classification algorithms can be used over the set of image descriptions using our BOW representation to tackle the image classification problem. As a result, we present the novel algorithm k-Morik (**M**ining patterns t**O** classify ca**R**tified **I**mages of **K**atharina), which takes only one parameter, k, to classify images of Katharina Morik using the proposed methodology. This paper presents the main ideas of this methodology and provides initial experimental results.

2   Related Work

2.1   Pattern Mining in Computer Vision

Frequent pattern mining techniques have been used to tackle a variety of computer vision problems, including image classification [13,17,23,33,37,38], action recognition [14,26], scene understanding [36], object recognition and object-part detection [25]. All these methods use pattern mining to build a set of mid-level features from low-level image descriptors. Apart from the application, these methods mostly differ in the image representation used, the way they convert the original image representation into a transactional description suitable for pattern mining techniques and the way they select relevant or discriminative patterns. In this work, we are not interested in building a new set of mid-level features but to improve the low level BOW-based representation using pattern mining.

2.2   Subspace Clustering

Subspace clustering methods have only seldom been explored in the context of visual dictionary creation (for BOW models). In recent work, Chandra et al. [7] constructed nonlinear subspaces from the raw image space using a Restricted Boltzmann Machine (RBM) [15]. In the experimental part, they incorporate additional structural information [19] which makes their method difficult to compare to a standard clustering method when aiming to classify images after a BOW creation step. $\mathcal{S}$pectral clustering methods [32] have shown to give very good results in comparison to other existing unsupervised methods for object discovery [28]. The main difference between existing spectral clustering methods comes from the used *affinity matrix*. Most spectral clustering methods start by constructing a similarity matrix $W \in \mathbb{R}^{N \times N}$ between the N data points that we want to cluster. Let $G(V; E)$ be an undirected graph where V is the set of N vertices and E is the set of weighted edges. W might be the adjacency matrix of G where $W_{i,j} = w, w > 0$, when the vertices i and j are in the neighborhood of each other and $W_{i,j} = 0$ otherwise. A degree matrix $D \in \mathbb{R}^{N \times N}$ is constructed from W using the degree d_i of each vertex, $v_i \in V$ and defined as $d_i = \sum_{j=1}^{N} w_{ij}$. Then, a Laplacian matrix $L \in \mathbb{R}^{N \times N}$ is constructed from D and W (for example by taking $L = D - W$). Finally, in most spectral subspace clustering algorithms [9,12,30,39], the k clusters are found by applying the k-means algorithm to the k first Eigen vectors of L.

Subspace clustering methods have also been developed in the context of pattern mining. **Clique** [2] is one of the first subspace clustering algorithm introduced. **Clique** partitions each dimension into the same number of equal length intervals i.e. it partitions an m-dimensional data space into non-overlapping rectangular units. A unit is considered *dense* if the number of data points which fall in this unit exceeds a given threshold. A cluster is a maximal set of connected dense units within a subspace. In the algorithm, the size of the intervals and the density threshold are the input parameters. In Clique, clusters may be found overlapping or disjoint subspaces which is suitable for our image classification task as one image patch can be described by different overlapping descriptors in different relevant subspaces. However, the choice of the parameters (difficult to tune) can have a dramatic effect on the resulting clusters, and the algorithm (based on Apriori [3]) does not scale very well. **Enclus** [10] follows the same procedure as Clique but uses an *entropy* and a *correlation* measures (among dimensions) rather than the coverage to define dense units. **MAFIA** [22] is another Apriori-based algorithm which uses an adaptive grid instead of an equal length grid for finding dense units based on the distribution of data. *MAFIA* is faster and produces more subspace clusters than *Clique* but still has strong scaling problems. Other algorithms such as **Proclus** [1], **Findit** [34], δ-**Cluster** [35] follow a top-down approach (contrary to the previous bottom-up algorithms). They all assign each instance into only one cluster and allow outliers.

Each of the methods presented above requires the user to set the number and the size of the subspaces in advance and they all tend to find equal-sized redundant clusters. Moreover, these methods are very sensitive to parameter tunings which again might lead to overlook the actual interesting subspaces. In the rest of the paper, we explore the use of the recently developed *Cartification* technique to obtain a more robust visual dictionary.

3   Cartification for Images

In a nutshell, the Cartification approach to subspace clustering developed by Aksehirli et al. [5] transforms the data into a collection of local neighborhoods, in which it detects clusters by finding the re-occurring object sets. Cartification allows to exploit a property of most subspace clusters: if a set of objects forms a cluster structure in a combination of dimensions, then this set of objects is likely to also form a cluster structure in subsets of these dimensions. Therefore, the **CLON** algorithm, proposed by Aksehirli et al. [4], efficiently discovers the cluster structures in one-dimensional projections using Cartification, and then iteratively refines these clusters to find higher-dimensional subspace clusters. More precisely, for each dimension, Cartification operates as follows:

1. For each data point, create a set consisting of its k nearest data points, resulting in a neighborhood database in which the i-th row contains a set of objects representing the k points that are closest to the i-th point. This neighborhood database is a transaction database and can be represented as a binary matrix in which each row has k columns that have the value 1 and the other columns have value 0.
2. Search for large sets of objects that are frequently repeated in the rows of the resulting binary matrix; each such set represents a set of data points that are close to each other, and hence represents a cluster for this attribute.

For one dimension, the above algorithm returns a set of (possibly overlapping) clusters. Then, in a second stage, CLON combines clusters over multiple dimensions, resulting in sets of points that are frequent in multiple dimensions and as such, subspace clusters are found [4]. An advantage of the Cartification approach is that in step 2, it does not use the distance measure itself, which makes the approach less scale dependent; for instance, whether a logarithmic scale or a linear scale is used for an attribute has no impact on the results. Another advantage of Cartification is the interpretability of its parameters: neighborhood size k and minimum frequency. Setting k and a lower bound on the frequency, both of which are functions of the expected cluster size, is much easier than determining settings for parameters such as density or distribution. Aksehirli et al. have shown that after parameter tuning of all methods, Cartification performs typically better than other subspace clustering approaches [4,5]. Another advantage of Cartification is that it can very efficiently find cluster centers and outliers immediately after transforming the data, that is, already after the first step of the above procedure, as the cluster centers are represented by the very frequent singleton objects in the data, and the outliers by the infrequent objects [5]. It is mainly this aspect that we will explore in this paper. Here, our goal is to study whether Cartification can give better results for creating a dictionary of so called visual words. As already explained in the Introduction, such words are typically discovered using k-Means clustering after which only the centroids of the clusters are retained. As Cartification already immediately provides the cluster centers, i.e. the frequent items in the cartified data, without the need to perform actual clustering, we will use only these items as visual words. Furthermore, for every item, instead of considering all 128 dimensions (of the SIFT descriptors), we will only consider those (cartified) dimensions in which the item has a high support (i.e., it is a frequent item in that cartified dimension). In this way, the relevant subspaces for each cluster center are automatically obtained. The remaining procedure remains the same as is standard in image classification (see Sect. 1). When computing the distances of each descriptor to the cluster centers, or visual words, only the relevant subspaces are being considered.

4 Experiments

4.1 Experimental Setting

Our dataset consists of 20 pictures of 3 different persons: 8 photos of Katharina Morik (Fig. 1), 6 photos of Elisa Fromont (Fig. 2), and 6 photos of Bart Goethals (Fig. 3).

Our goal is to build a classifier that can automatically detect the photos of Katharina Morik. In order to represent the images in terms of feature descriptors, SIFT descriptors and detectors are used [20]. The parameters of the SIFT descriptors are tuned to obtain around 400 descriptors per image. The size of the SIFT patch is kept to 16 × 16 pixels. For computing the SIFT descriptors the implementation by Lazebnik [19] is used. Our descriptor space contains 8 546 128-dimensional descriptors.

These descriptors are transformed (cartified) to more than a million transactions ($8\,546 * 128$). More specifically, for each descriptor and each dimension a transaction is created consisting of its k-nearest neighbors among the complete set of descriptors in that dimension. Then, we select the top-k most frequent descriptors in this transaction database, representing the descriptors that occur in the most nearest neighborhoods, and therefore must be central to clusters of descriptors. Next, for each of these frequent descriptors, we only retain the dimensions (of the original SIFT vector) in which the descriptor has a support larger than a given minimum threshold, as these must be the sub-spaces of the cluster in which the descriptor is central. Then, for every picture, we compute the histograms of all visual words built from those clusters. That is, for each original descriptor, we determine to which cluster center, or visual word its distance,

Fig. 1. Photos of Katharina Morik

Fig. 2. Photos of Elisa Fromont

Fig. 3. Photos of Bart Goethals

computed over the relevant subspaces only, is the smallest. These histograms are then used as feature vectors for the classifiers. As such, we obtain a k-dimensional histogram for each photo which serves as input to the classifiers.

For our experiments, we implemented k-Morik in Python, and used two classifiers implemented in the SciKit-learn Python modules [24], namely Decision-TreeClassifier, a basic Decision Tree Classifier implementation using Gini impurity as splitting criterion, and GaussianNB, a Gaussian Naive Bayes implementation. For both algorithms, we used all default parameter settings. To evaluate the performance, we used a leave-one-out cross validation.

4.2   Results

Due to time constraints, we have not been able to run extensive experiments, or collect more data. We did, however, manipulate the data and ran enough experiments to find the good parameter settings. Entirely as expected, k-Morik, is showing very promising results. More specifically, the neighborhood size used for cartification is set to 50, which is a relatively small number, but large enough to obtain enough and good cluster centers [5]. The parameter k which determines the number of visual words, i.e. the top most frequent descriptors occurring in the nearest neighborhoods, is set to 10. Note, that this is a very small number as compared to the typical image classification setting where the number of visual words is typically a few hundreds. The resulting 10 descriptors covered photos of all three persons in the data. Then, for each of these visual words, only the dimensions for which the descriptor has a support larger than 100 were considered when computing the distances to each of them.

Finally, the Decision Tree Classifier classified all 20 photos correctly, resulting in an accuracy of 100 %. The Naive Bayes Classifier classified 18 out of 20 photos correct.

5   Conclusion

We have presented K-MORIK, a new algorithm based on Cartification and dedicated to learn better visual vocabularies for image classification. Based on our preliminary experiments, we are convinced that subspace clustering methods can be interesting when building visual vocabularies for image classification especially when the size of the original descriptor is large and when all dimensions might not be relevant. k-Morik is a a very promising method in this context as it provides at the same time the relevant subspace clusters, their underlying dimensions and their centroïds.

References

1. Aggarwal, C.C., Han, J., Wang, J., Yu, P.S.: A framework for projected clustering of high dimensional data streams. In: Proceedings of the Thirtieth International Conference on Very Large Data Bases-Volume 30, VLDB Endowment, pp. 852–863 (2004)
2. Agrawal, R., Gehrke, J., Gunopulos, D., Raghavan, P.: Automatic subspace clustering of high dimensional data for data mining applications, vol. 27. ACM (1998)
3. Agrawal, R., Srikant, R., et al.: Fast algorithms for mining association rules. In: Proceedings of 20th International Conference on Very Large Data Bases, VLDB, vol. 1215, pp. 487–499 (1994)
4. Aksehirli, E., Goethals, B., Müller, E.: Efficient cluster detection by ordered neighborhoods. In: Madria, S., Hara, T. (eds.) DaWaK 2015. LNCS, vol. 9263, pp. 15–27. Springer, Heidelberg (2015)

5. Aksehirli, E., Goethals, B., Müller, E., Vreeken, J.: Cartification: a neighborhood preserving transformation for mining high dimensional data. In: 2013 IEEE 13th International Conference on Data Mining, Dallas, TX, USA, 7–10 December 2013, pp. 937–942 (2013)

6. Beyer, K., Goldstein, J., Ramakrishnan, R., Shaft, U.: When is nearest neighbor meaningful? In: Beeri, C., Bruneman, P. (eds.) ICDT 1999. LNCS, vol. 1540, pp. 217–235. Springer, Heidelberg (1998)

7. Chandra, S., Kumar, S., Jawahar, C.V.: Learning multiple non-linear sub-spaces using k-rbms. In: 2013 IEEE Conference on Computer Vision and Pattern Recognition (CVPR), pp. 2778–2785. IEEE (2013)

8. Chapelle, O., Haffner, P., Vapnik, V.: SVMs for histogram-based image classification. IEEE Trans. Neural Networks 10(5), 1055 (1999)

9. Chen, G., Lerman, G.: Spectral curvature clustering (scc). Int. J. Comput. Vis. 81(3), 317–330 (2009)

10. Cheng, C.-H., Ada Waichee, F., Zhang, Y.: Entropy-based subspace clustering for mining numerical data. In: Proceedings of the Fifth ACM SIGKDD International Conference on Knowledge Discovery and Data Mining, pp. 84–93. ACM (1999)

11. Csurka, G., Dance, C.R., Fan, L., Willamowski, J., Bray, C.: Visual categorization with bags of keypoints. In: Workshop on Statistical Learning in Computer Vision, ECCV, pp. 1–22 (2004)

12. Elhamifar, E., Vidal, R.: Sparse subspace clustering: Algorithm, theory, and applications (2012)

13. Fernando, B., Fromont, É., Tuytelaars, T.: Mining mid-level features for image classification. Int. J. Comput. Vis. 108(3), 186–203 (2014)

14. Gilbert, A., Illingworth, J., Bowden, R.: Fast realistic multi-action recognition using mined dense spatio-temporal features. In: ICCV, pp. 925–931 (2009)

15. Hinton, G.E.: A practical guide to training restricted boltzmann machines. In: Montavon, G., Orr, G.B., Müller, K.-R. (eds.) Neural Networks: Tricks of the Trade, 2nd edn. LNCS, vol. 7700, pp. 599–619. Springer, Heidelberg (2012)

16. Jurie, F., Triggs, B.: Creating efficient codebooks for visual recognition. In: Proceedings of the Tenth IEEE International Conference on Computer Vision (ICCV'05), vol. 1, pp. 604–610. IEEE Computer Society, Washington (2005)

17. Kim, S., Jin, X., Han, J.: Disiclass: discriminative frequent pattern-based image classification. In: Tenth International Workshop on Multimedia Data Mining (2010)

18. Kriegel, H.-P., Kröger, P., Zimek, A.: Lustering high-dimensional data: a survey on subspace clustering, pattern-based clustering, and correlation clustering. ACM Trans. Knowl. Disc. Data (TKDD) 3(1), 1 (2009)

19. Lazebnik, S., Schmid, C., Ponce, J.: Beyond bags of features: spatial pyramid matching for recognizing natural scene categories. In: IEEE Computer Society Conference on Computer Vision and Pattern Recognition (CVPR), vol. 2, pp. 2169–2178. IEEE (2006)

20. David, G.: Lowe: distinctive image features from scale-invariant keypoints. Int. J. Comput. Vis. 60(2), 91–110 (2004)

21. Mikolajczyk, K., Schmid, C.: Scale and affine invariant interest point detectors. Int. J. Comput. Vis. 60(1), 63–86 (2004)

22. Nagesh, H., Goil, S., Choudhary, A.: Adaptive grids for clustering massive data sets. In: Proceedings of the 1st SIAM ICDM, Chicago, IL, vol. 477 (2001)

23. Nowozin, S., Tsuda, K., Uno, T., Kudo, T., Bakir, G.: Weighted substructure mining for image analysis. In: CVPR (2007)

24. Pedregosa, F., Varoquaux, G., Gramfort, A., Michel, V., Thirion, B., Grisel, O., Blondel, M., Prettenhofer, P., Weiss, R., Dubourg, V., Vanderplas, J., Passos, A., Cournapeau, D., Brucher, M., Perrot, M., Duchesnay, E.: Scikit-learn: machine learning in Python. J. Mach. Learn. Res. **12**, 2825–2830 (2011)
25. Quack, T., Ferrari, V., Leibe, B., Van Gool, L.: Efficient mining of frequent and distinctive feature configurations. In: ICCV (2007)
26. Quack, T., Ferrari, V., Van Gool, L.: Video mining with frequent itemset configurations. In: Sundaram, H., Naphade, M., Smith, J.R., Rui, Y. (eds.) CIVR 2006. LNCS, vol. 4071, pp. 360–369. Springer, Heidelberg (2006)
27. Salton, G., Wong, A., Yang, C.S.: A vector space model for automatic indexing. Commun. ACM **18**(11), 613–620 (1975)
28. Tuytelaars, T., Lampert, C.H., Blaschko, M.B., Buntine, W.L.: Unsupervised object discovery: a comparison. Int. J. Comput. Vis. **88**(2), 284–302 (2010)
29. Vailaya, A., Figueiredo, M.A.T., Jain, A.K., Zhang, H.J.: Image classification for content-based indexing. IEEE Trans. Image Process. **10**(1), 117–130 (2001)
30. Vidal, R., Favaro, P.: Low rank subspace clustering (lrsc). Pattern Recogn. Lett. **43**, 47–61 (2014). ICPR2012 Awarded Papers
31. Vidal-Naquet, M., Ullman, S.: Object recognition with informative features and linear classification. In: Proceedings of the Ninth IEEE International Conference on Computer Vision, ICCV 2003, p. 281. IEEE Computer Society, Washington (2003)
32. Von Luxburg, U.: A tutorial on spectral clustering. Stat. comput. **17**(4), 395–416 (2007)
33. Voravuthikunchai, W., Crémilleux, B., Jurie, F.: Histograms of pattern sets for image classification and object recognition. In: IEEE Conference on Computer Vision and Pattern Recognition, Columbus, Ohio, United States, pp. 1–8 (2014)
34. Woo, K.-G., Lee, J.-H., Kim, M.-H., Lee, Y.-J.: Findit: a fast and intelligent subspace clustering algorithm using dimension voting. Inf. Softw. Technol. **46**(4), 255–271 (2004)
35. Yang, J., Wang, W., Wang, H., Philip, Y.: δ-clusters: capturing subspace correlation in a large data set. In: Proceedings of 18th International Conference on Data Engineering, pp. 517–528. IEEE (2002)
36. Yao, B., Fei-Fei, L.: Grouplet: a structured image representation for recognizing human and object interactions. In: CVPR (2010)
37. Yuan, J., Ying, W., Yang, M.: Discovery of collocation patterns: from visual words to visual phrases. In: CVPR (2007)
38. Yuan, J., Yang, M., Ying, W.: Mining discriminative co-occurrence patterns for visual recognition. In: CVPR, pp. 2777–2784, June 2011
39. Zhang, T., Szlam, A., Wang, Y., Lerman, G.: Hybrid linear modeling via local best-fit flats. Int. J. Comput. Vis. **100**(3), 217–240 (2012)

Author Index

Printed in the United States
By Bookmasters